UNDERSTANDING CIRCUITS AND OP-AMPS

PRENTICE HALL UNDERSTANDING ELECTRONICS TECHNOLOGY SERIES

UNDERSTANDING CIRCUITS AND OP-AMPS

Concepts, Experiments, and Troubleshooting

Dale R. Patrick
Stephen W. Fardo
Eastern Kentucky University

Prentice Hall
Englewood Cliffs, New Jersey 07632

Library of Congress Cataloging-in-Publication Data

Patrick, Dale R.
  Understanding circuits and op-amps : concepts, experiments, and
troubleshooting / Dale R. Patrick, Stephen W. Fardo.
    p.   cm. — (Prentice Hall understanding electronics technology
series ; bk. 5)
  Includes index.
  ISBN 0-13-943515-8
  1. Electronic circuits. 2. Operational amplifiers. I. Fardo,
Stephen W. II. Title. III. Series
TK7867.P36   1989
621.3815′3—dc19   88-31902
                    CIP

Editorial/production supervision
  and interior design: *Denise Gannon*
Cover design: *Lundgren Graphics, Ltd.*
Manufacturing buyer: *Robert Anderson*

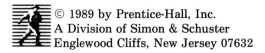 © 1989 by Prentice-Hall, Inc.
A Division of Simon & Schuster
Englewood Cliffs, New Jersey 07632

Printed in the United States of America

10 9 8 7 6 5 4 3 2 1

ISBN 0-13-943515-8

PRENTICE-HALL INTERNATIONAL (UK) LIMITED, *London*
PRENTICE-HALL OF AUSTRALIA PTY. LIMITED, *Sydney*
PRENTICE-HALL CANADA INC., *Toronto*
PRENTICE-HALL HISPANOAMERICANA, S.A., *Mexico*
PRENTICE-HALL OF INDIA PRIVATE LIMITED, *New Delhi*
PRENTICE-HALL OF JAPAN, INC., *Tokyo*
SIMON & SCHUSTER ASIA PTE. LTD., *Singapore*
EDITORA PRENTICE-HALL DO BRASIL, LTDA., *Rio de Janeiro*

Contents

UNIT FOUR   POWER SUPPLY CIRCUITS   **203**

UNIT FIVE   OSCILLATOR CIRCUITS   **271**

Preface

Understanding Circuits and Op-Amps: Concepts, Experiments, and Troubleshooting is an introductory text that provides coverage of the various topics in the study of electronic circuits and operational amplifiers (op-amps). The key concepts presented in this book are discussed using a simplified approach that greatly enhances learning. The use of mathematics is kept to the very minimum and is discussed clearly through applications and illustrations.

Every unit is organized in a step-by-step progression of concepts and theory. Each unit begins with a *unit introduction* and *unit objectives*. A discussion of important *concepts* and *theories* follows. Numerous *self-examinations*, with answers provided, are integrated into each chapter to reinforce learning. In addition, *experimental activities*, with components and equipment listed, are included with each unit to help students learn electronics through practical experimental applications. The final learning activity for each unit is a *unit examination*, which includes at least twenty multiple choice objective questions.

Definitions of *important terms* are presented at the beginning of each unit. Several *appendices* are used to aid students in performing experimental activities. It should be noted that the expense of the equipment required for the experiments is kept to a minimum. A comprehensive parts list is provided, as well as information on electronics distributors. The experiments suggested are low-cost activities that can be performed in the home or a school laboratory. They are very simple and easy to understand and stress troubleshooting concepts. Electronics can be learned experimentally at a low cost by completing these labs. Appendices dealing with electronic symbols, safety, tools, and soldering are provided for easy reference.

This book is the fifth in Prentice Hall's "Understanding Electronics Technology" series. This book covers electronic circuits and operational amplifiers, which are important *applications* in the study of electronics. Prentice Hall's "Understanding Electronics Technology" series includes the following books:

> Book 1—*Understanding Electricity and Electronics: Concepts, Experiments, and Troubleshooting*
>
> Book 2—*Understanding DC Circuits: Concepts, Experiments, and Troubleshooting*
>
> Book 3—*Understanding AC Circuits: Concepts, Experiments, and Troubleshooting*
>
> Book 4—*Understanding Semiconductor Devices: Concepts, Experiments, and Troubleshooting*
>
> Book 5—*Understanding Circuits and Op-Amps: Concepts, Experiments, and Troubleshooting*
>
> Book 6—*Understanding Digital Electronics: Concepts, Experiments, and Troubleshooting*

Each of these books is organized in the same easy-to-understand format. They can be used to acquire an understanding of electronics in the home, school, or work place. The sequence of the books allows the student to progress at a desired pace in the study of electronics and to perform low-cost experiments with equipment and supplies (depending on availability). As students progress, they may wish to purchase various types of test equipment with varying degrees of expense. The experiments allow students to further develop an understanding of the topics discussed in each unit. They are intended as an important supplement to student learning.

Several *supplemental materials* are available to provide an aid to effective learning. These include the following:

1. *Instructor's Resource Manual*—provides the instructor with answers to all unit examinations and suggested data for experimental activities, including a comprehensive analysis of each experiment.

2. *Instructor's Transparency Masters*—enlarged reproductions of selected illustrations used in the textbook, which are suitable for use for transparency preparation for class presentations.

3. *Instructor's Test Item File*—provides the instructor with many suggested multiple choice objective questions that may be used with each unit of instruction.

These supplements are extremely valuable for instructors in organizing electronics classes. The complete instructional cycle, from objectives to evaluation, is included in this series of books. The authors hope you will find the Understanding Electronics Technology Series easy to understand and that you are successful in your pursuit of knowledge in an exciting technical area. Electronics is an extremely vast and interesting field of study. This book provides *applications* for *understanding electronics technology.*

Dale R. Patrick
Stephen W. Fardo

Richmond, Kentucky

Course
Objectives

Upon completion of this course on *Understanding Circuits and Op-Amps,* you should be able to:

1. Identify and describe the operational characteristics of common-base, common-emitter, and common-collector transistor amplifier circuits.

2. Construct experimental electronic circuits using schematic diagrams and perform tests and measurements with a multimeter and oscilloscope.

3. Demonstrate a knowledge of FET amplifiers and operational amplifier circuits.

4. Understand amplifier characteristic curves and be able to perform load-line analysis of amplifier circuits.

5. Explain the operation of power amplifiers, push-pull amplifiers, integrated circuit (IC) amplifiers, and methods of coupling amplifier circuits.

6. Explain the operation of rectifier, filter, and regulator circuits in power supplies.

7. Discuss the operation of oscillator circuits and be able to identify types of oscillators.

8. Explain amplitude modulation (AM), frequency modulation (FM), and the basic principles of communications circuits.

Parts List
for Experiments

Various components and equipment are needed to perform the experimental activities in this course. These parts may be obtained from electronic suppliers, mail-order warehouses, or educational supply vendors. A list of several of these is included in Appendix C. A specific parts package for this course is available through Heath/Zenith Company of Benton Harbor, MI 49022. The component hardware package number is EB 6104-31.

For the more venturesome learner, these parts may be obtained through a variety of electronic suppliers. As a rule, there is a standard part number that is used to obtain parts. In many cases, however, there is an equivalent part made by many other manufacturers.

The following list of equipment and components is necessary for the successful completion of the activities included in this course.

1. Transistors: NPN (2N3397), PNP
2. Circuit board and connecting wires
3. Resistors (in Ohms): 4.7, 10, 100, 150, 220, 330, 560, 620, 1k(3), 1.5k, 1.8k, 2.2k, 3.3k, 4.7k, 8.2k, 10k, 15k, 22k, 33k, 47k, 56k, 100k, 470k, 1M
4. Inductors: 305 μH, 47 μH, 25 mH
5. Oscillator coil

6. Potentiometer (in Ohms): 1.5M, 100k, 5k, 1k, 200
7. JFET (GE FET1)
8. Microphone
9. Capacitors: 0.0022 μF, 0.01 μF, 0.02 μF, 0.039 μF, 4.7 pF, 24 pF, 47 pF, 100 pF, 470 pF, .001 μF, 0.1 μF, 0.68 μF, 10 μF, 33 μF, 100 μF(2), 250 μF
10. Batteries: 1.5 V(2)
11. Diodes: 4 silicon; germanium, varicap
12. Zener diodes: 2.7 V, 6.2 V
13. Switches: SPST (2)
14. Lamp: 10 V (with socket)
15. Crystal: 3579.545 kHz
16. Integrated circuits (ICs): 741 op-amp, 555 timer
17. Oscilloscope
18. Multimeter
19. Transformers: Audio, IF
20. Speaker
21. Soldering iron
22. Darlington transistor
23. AM radio

UNIT ONE

Amplifier Circuits

UNIT INTRODUCTION

Electronic systems must perform a variety of basic functions in order to accomplish a particular operation. An understanding of these functions is essential. In this unit we investigate the amplification function. Amplification is achieved by devices that produce a change in signal amplitude. Transistors and integrated circuits are typical amplifying devices.

In the first part of this unit, we look at *amplification* in general. No particular amplifying device is considered. Second, we discuss how a particular type of device achieves amplification. Circuit operation, biasing methods, circuit configurations, classes of amplification, and operational conditions are considered. As a general rule, this unit is directed toward *small-signal amplification.* The following unit deals with large-signal, or *power amplification.*

UNIT OBJECTIVES

Upon completion of this unit you will be able to:

1. List basic types of amplifiers and explain the function of each.
2. Identify the three basic amplifier configurations and describe their characteristics.
3. Explain the basic biasing arrangements used with transistor circuits.
4. Design basic amplifier circuits.
5. Determine if a transistor amplifier is operating in the class A, AB, B, or C mode.
6. Understand thermal stability of basic amplifier circuits.
7. Discuss the operation and characteristics of a common-emitter amplifier circuit.
8. Explain how the input resistance of an amplifier is affected by the beta of the transistor and the value of the emitter resistor.

IMPORTANT TERMS

Amplification. The ability to control a large force by a smaller force. There is voltage, current, and power amplification.

Audio amplifiers. Amplifiers that respond to the human range of hearing or frequencies from approximately 20 Hz to 20 kHz.

Beta-dependent biasing. A method of biasing that responds in some way to the amplifying capabilities of the device.

Bypass capacitor. A capacitor that provides an alternative path around a component or to ground.

Circuit configuration. A method of connecting a circuit, such as common-emitter, common-source, and so on.

Digital integrated circuit. An integrated circuit (IC) that responds to two-state (on-off) data.

Dual-in-line package (DIP). A packaging method for ICs.

Dynamic characteristics of operation. Electronic device operational characteristics that show how a device responds to ac or changing voltage values.

Feedback network. A resistor combination that returns some of the output signal to the input of an IC.

Impedance. A form of opposition to alternating current measured in ohms. Electronic devices have input and output impedance values.

Linear amplifiers. Amplifying circuits that increase the amplitude of a signal that is a duplicate of the input. These devices operate in the straight-line part of the characteristic curves.

Load line. A line drawn on a family of characteristic curves that shows how the device will respond in a circuit with a specific load resistor value.

Microphone. A device that changes sound energy into electrical energy.

Noninverting input. An input lead for an integrated circuit that does not shift the phase of a signal.

Open loop. An operation amplifier connection method that causes full or maximum amplification.

Power dissipation. An electronic device characteristic that indicates the ability of the device to give off heat.

Q point. An operating point for an electronic device that indicates its dc or static operation with no ac signal applied.

Radio frequency (RF). An alternating current (ac) frequency in excess of the human range of hearing which starts at 30,00 Hz.

Static state. A direct current (dc) operating condition of an electronic device with operating energy but no signal applied.

Thermal stability. The condition of an electronic device that indicates its ability to remain at an operating point without variation due to temperature.

Voltage follower. An amplifying condition where the input and output voltages are of approximately the same value across different impedance.

AMPLIFICATION PRINCIPLES

Amplification is achieved by an electronic device and its associated components. As a general rule, the amplifying device is placed in a circuit. The components of the circuit usually have about as much influence on amplification as the device itself. The circuit and the device must be supplied electrical energy for it to function. Typically, *amplifiers* are energized by dc. This may be derived from a battery or a rectified ac power supply. The amplifier then processes a signal of some type when it is placed in operation. The signal may be either ac or dc, depending on the application.

Reproduction and Amplification

The signal to be processed by an amplifier is first applied to the *input* part of the circuit. After being processed the signal appears in the *output* circuitry. The output signal may be reproduced in its exact form, amplified, or both amplified and reproduced. The value and type of input signal, operating source energy, device characteristics, and circuit components all have some influence on the output signal.

 Figure 1-1 illustrates the process of reproduction, amplification, and the combined amplification-reproduction functions of an amplifier. In part (a), the amplifier performs only the reproduction function. Note that the input and output signals have the same size and shape. Part (b) shows only the amplification function. In this case, the input signal is increased in amplitude. The output signal is amplified but does not necessarily resemble the input sig-

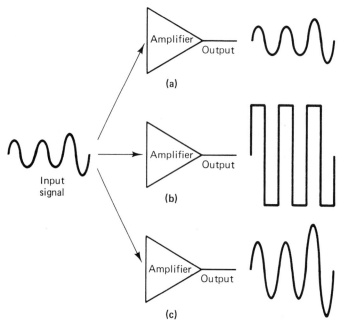

FIGURE 1-1 Amplification and reproduction: (a) reproduction; (b) amplification; (c) combined amplification and reproduction.

nal. In many applications, amplification and reproduction must both be achieved at the same time. Part (c) shows this function of an amplifier. Depending on the application, an amplifier must be capable of developing any of these three output signals.

Voltage Amplification

A system designed to develop an output voltage that is greater than its input voltage is called a *voltage amplifier.* A voltage amplifier has voltage gain. This function is defined as a ratio of the output signal voltage to the input signal voltage. In equation form, voltage amplification is expressed as

$$\text{Amplification voltage} = \frac{\text{output voltage}}{\text{input voltage}}$$

or

$$A_V = \frac{V_{\text{out}}}{V_{\text{in}}} \quad \text{or} \quad \frac{\Delta V_{\text{out}}}{\Delta V_{\text{in}}}$$

The uppercase letters in V_{out} and V_{in} denote dc voltage values. Ac voltages are usually expressed by lowercase letters. The capital Greek letter delta (Δ) preceding a voltage indicates a changing value. Voltage amplifiers are capable of a wide range of amplification values.

Current Amplification

An amplifying system designed to develop output current that is greater than the input current is called a *current amplifier.* A current amplifier has current *gain.* This type of amplification is defined as a ratio of output current to input current. In equation form, it is expressed as:

$$\text{Current amplification} = \frac{\text{output current}}{\text{input current}}$$

or

$$A_i = \frac{I_{\text{out}}}{I_{\text{in}}} \quad \text{or} \quad \frac{\Delta i_{\text{out}}}{\Delta i_{\text{in}}}$$

Current gain takes into account the *beta* of a transistor and all the associated components that make the device operational.

Power Amplification

Power amplification is a ratio of the developed output signal power to the input signal power. Note that this refers only to the *power gain* of the signal. All amplifiers consume a certain amount of power from the energy source during operation. This is usually not included in an expression of power gain. An equation of power gain shows that

$$\text{Power amplification} = \frac{\text{output signal power}}{\text{input signal power}}$$

or

$$A_P = \frac{P_{\text{out}}}{P_{\text{in}}}$$

Power signal gain can also be expressed as the product of voltage gain and current gain. In this regard, the equation is

Power amplification = voltage amplification

× current amplification

or

$$A_p = A_V \times A_i$$

It should be apparent from the A_p equation that A_v and A_i do not both need to be large to have power gain. Typical power amplifiers may have a large current gain and a voltage gain of less than 1. Power gain must take into account both signal voltage and current gain in order to be meaningful.

Bipolar Transistor Amplifiers

The primary function of an amplifier is to provide *voltage gain, current gain,* or *power gain.* The resulting gain depends a great deal on the device being used and its circuit components. Normally, the circuit is energized by dc voltage. An ac signal is then applied to the input of the amplifier. After passing through the transistor, an amplified version of the signal appears in the output. Operating conditions of the device and the circuit determine the level of amplification to be achieved.

In order for a bipolar transistor to respond as an amplifier, the emitter-base junction must be *forward biased,* and the collector-base junction must be *reverse biased.* Specific operating voltage values must then be selected that will permit amplification. If both reproduction and amplification are to be achieved, the device must operate in the center of its active region. Remember that this is between the saturation and cut off regions of the collector family of characteristic curves. Proper circuit component selection permits a transistor to operate in this characteristic region.

Basic Amplifiers

Suppose now that we look at the circuitry of a basic amplifier. Figure 1-2 shows a schematic diagram of this type of circuit. Notice that a schematic symbol of the NPN transistor is used in this diagram. Schematic symbols are normally used in all circuit diagrams. Keep in mind the crystal structure of the device represented by the symbol.

The basic amplifier has a number of parts that are needed to

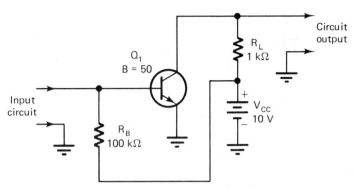

FIGURE 1-2   Basic bipolar amplifier.

make it operational. V_{CC} is the dc energy source. The negative side of V_{CC} is connected to ground. The emitter is also connected to ground. This type of circuit configuration is called a grounded-emitter, or *common-emitter, amplifier.* The emitter, one side of the input, and one side of the output are all commonly connected together. In an actual circuit these points are usually connected to ground.

Resistor R_B of the basic amplifier is connected to the positive side of V_{CC}. This connection makes the base positive with respect to the emitter. The emitter-base junction is forward biased by R_B. Resistor R_L connects the positive side of V_{CC} to the collector. This connection reverse-biases the collector. Through the connection of R_B and R_L the transistor is properly biased for operation. A transistor connected in this manner is considered to be in its *static* or *dc operating state.* It has the necessary dc energy applied for it to be operational. No signal is applied to the input for amplification.

Let us now consider the operation of the basic amplifier in its static state. With the given values of Fig. 1-2 base current (I_B) can be calculated. I_B, in this case, is limited by the value R_B and the base-emitter junction resistance. When the emitter-base junction is *forward biased,* its resistance becomes very small. The value of R_B is, therefore, the primary limiting factor of I_B. I_B can be determined by an application of Ohm's law. The equation is

$$\text{Base current} = \frac{\text{source voltage}}{\text{base resistance}}$$

or

$$I_B = \frac{V_{CC}}{R_B}$$

For our basic transistor amplifier, this is determined to be

$$I_B = \frac{V_{CC}}{R_B} = \frac{10 \text{ V}}{100,000 \ \Omega} \quad \text{or} \quad \frac{10 \text{ V}}{100 \times 10^3 \ \Omega}$$

$$= 100 \times 10^{-6}, \text{ or } 100 \ \mu A$$

The value of base current, in this case, is extremely small. Remember that only a small amount of I_B is needed to produce I_C.

The *beta* of the transistor used in Fig. 1-2 has a given value. With beta and the calculated value of I_B, it is possible to determine the collector current of the circuit. Beta is the current gain of a common-emitter amplifier. Beta is described by the formula

$$\text{Beta} = \frac{\text{collector current}}{\text{base current}}$$

or

$$\beta = \frac{I_C}{I_B}$$

In our transistor amplifier I_B has been determined and beta has a given value of 50. By transposing the beta formula, I_C can be determined by the equation

$$\text{Collector current} = \text{beta} \times \text{base current}$$

or

$$I_C = \beta \times I_B$$

For the amplifier circuit, collector current is determined to be

$$I_C = \beta \times I_B = 50 \times 100 \times 10^{-6}$$
$$= 5 \times 10^{-3}, \text{ or 5 mA}$$

This means that 5 mA of I_C will flow through R_L when an I_B of 100 μA flows.

In any electric circuit, we know that current flow through a resistor will cause a corresponding voltage drop. In a basic transistor amplifier, I_C will cause a voltage drop across R_C. In the preceding step I_C was calculated to be 5 mA. By using Ohm's law again, the voltage drop across R_L can be determined. This is figured by the equation

$$R_L \text{ voltage} = \text{collector current} \times \text{collector resistance}$$

or

$$V_{RL} = I_C \times R_L$$

For the amplifier circuit, the voltage drop across R_L is

$$V_{RL} = I_C \times R_L$$
$$= 5 \times 10^{-3} \times 1 \times 10^3$$
$$= 5 \text{ V}$$

This means that half, or 5 V, of V_{CC} will appear across R_L. With a V_{CC} value of 10 V, the other 5 V will appear across the collector-emitter of the transistor. A V_{CE} of 5 V from a V_{CC} of 10 V means that the transistor is operating near the center of its active region. Ideally, the transistor should respond as a *linear amplifier*. When a suitable signal is applied to the input, it should amplify and re-produce the signal in the output.

The operational voltage and current values of the basic transistor amplifier in its static state are summarized in Fig. 1-3. Note that an I_B of 100 μA will cause an I_C of 5 mA. With a beta of 50, the amplifier has the capability of a rather substantial amount of output current. The collector current passing through R_L causes V_{CC} to be divided. This establishes operation in the approximate center of the collector family of characteristic curves. Ideally, this static condition should cause the amplifier to respond as a linear device when a signal is applied.

Signal Amplification

In order for the basic transistor circuit of Fig. 1-3 to respond as an amplifier, it must have a *signal* applied. The signal may be either voltage or current. An applied signal causes the transistor to change from its *static* state to a *dynamic* condition. Dynamic conditions involve changing values. All ac amplifiers respond in the dynamic state. The output of this should develop an ac signal.

Figure 1-4 shows a basic transistor amplifier with an ac signal applied. A capacitor is used, in this case, to *couple* the ac signal source to the amplifier. Remember that ac passes easily through a capacitor while dc is blocked. As a result of this, the ac signal is injected into the base-emitter junction. Dc does not flow back into the signal source. The ac signal is therefore added to the dc operating voltage. The emitter-base voltage is a dc value that changes at an ac rate.

Consider now how the applied ac signal alters the emitter-base junction voltage. Figure 1-5 shows some representative voltage and current values that appear at the emitter-base junction of the transistor. They can be measured with a voltmeter or observed with an oscilloscope. Part (a) shows the dc operating voltage and base current. This occurs when the amplifier is in its steady or

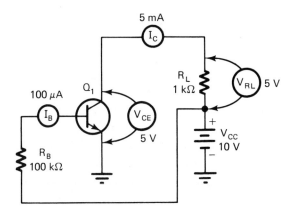

FIGURE 1-3   Static operating condition of a basic amplifier.

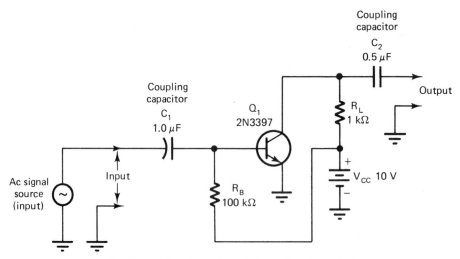

FIGURE 1-4   Amplifier with ac signal applied.

static state. Part (b) shows the ac signal that is applied to the emitter-base junction. Note that the amplitude change of this signal is a very small value. This also shows how the resulting base current and voltage change with ac applied. The ac signal is essentially riding on the dc voltage and current.

Refer now to the schematic diagram of the basic amplifier in Fig. 1-6. Note in particular the waveform inserts that appear in the diagram. These show how the current and voltage values respond when an ac signal is applied.

The ac signal applied to *input* of our basic amplifier rises in value during the positive alternation and falls during the negative alternation. Initially, this causes an increase and decrease in the value of V_{BE}. This voltage has a dc level with an ac signal riding on it. The indicated I_B is developed as a result of this voltage. The changing value of I_B causes a corresponding change in I_C. Note that I_B and I_C both appear to be the same. There is, however, a very noticeable difference in values. I_B is in microamperes, whereas I_C is in milliamperes. The resulting I_C passing through R_L causes a corresponding voltage drop (V_{RL}) across R_L. V_{CE} appearing across the transistor is the reverse of V_{RL}. These signals are both ac values riding upon a dc level. The output, V_{out}, is changed to an ac value. Capacitor C_2 blocks the dc component and passes only the ac signal.

It is interesting to note in Fig. 1-6 that input and output signals of the amplifier are reversed. When the ac input signal rises in value it causes the output to fall in value. The negative alternation of the input causes the *output* to rise in value. This condition of operation is called *phase inversion*. Phase inversion is a distinguishing characteristic of the common-emitter amplifier.

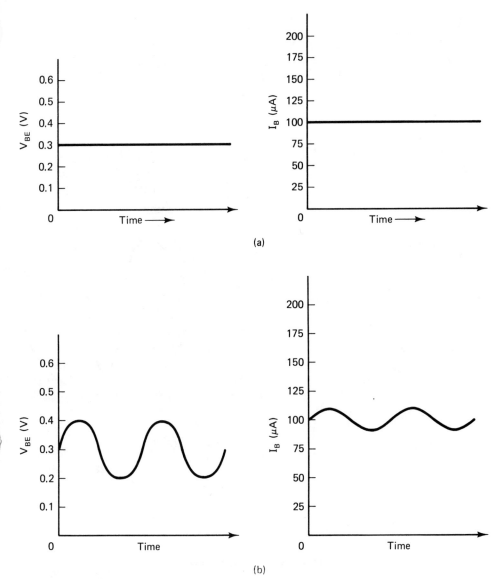

FIGURE 1-5 I_B and V_{BE} conditions: (a) static; (b) dynamic.

Amplifier Bias

If a transistor is to operate as a *linear amplifier*, it must be properly *biased* and have a suitable *operating point*. Its steady state of operation depends a great deal on base current, collector voltage, and collector current. Establishing a desired operating point requires proper selection of bias resistors and a load resistor to provide proper input current and collector voltage.

Stability of operation is a very important consideration. If the operating point of a transistor is permitted to shift with temperature changes, unwanted distortion may be introduced. In addition to distortion, operating point changes may also cause dam-

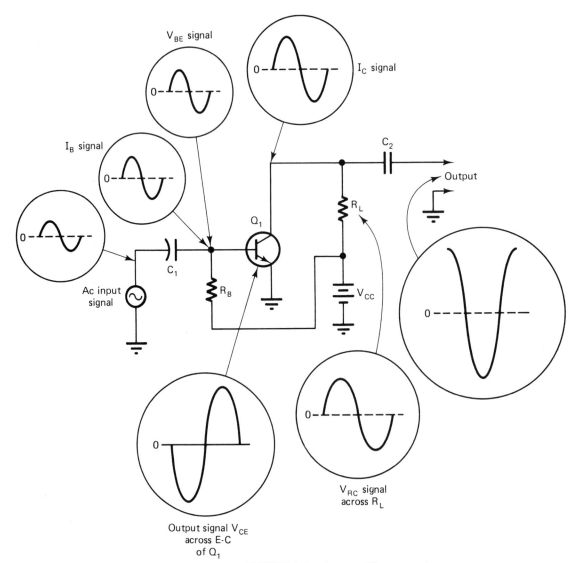

FIGURE 1-6   Ac amplifier operation.

age to a transistor. Excessive collector current, for example, may cause the device to develop too much heat.

The method of *biasing* a transistor has a great deal to do with its thermal stability. Several different methods of achieving bias are in common use today. One method of biasing is considered to be *beta dependent*. Bias voltages are largely dependent on transistor beta. A problem with this method of biasing is transistor response. Transistor beta is rarely ever the same for a specific device. Biasing set up for one transistor will not necessarily be the same for another transistor.

Biasing that is *independent* of beta is very important. This type of biasing responds to fixed voltage values. Beta does not alter these voltage values. As a general rule, this form of biasing is more reliable. The input and output of a transistor is very stable and the results are very predictable.

Beta-dependent Biasing. Four common methods of beta-dependent biasing are shown in Fig. 1-7. The steady-state, or static, conditions of operation are a function of the transistor's beta. These methods are rather easy to achieve with a minimum of parts. They are not very widely used today.

Circuit (*a*) is a very simple form of biasing for a common-emitter amplifier. This method of biasing was used for our basic transistor amplifier. The base is simply connected to V_{CC} through R_B. This causes the emitter-base junction to be forward biased. The resulting collector current is beta times the value of I_B. The value of I_B is V_{CC}/R_B. As a general rule, biasing of this type is very sensitive to changes in temperature. The resulting output of the circuit is rather difficult to predict. This method of biasing is often called *fixed biasing*. It is used primarily because of its simplicity.

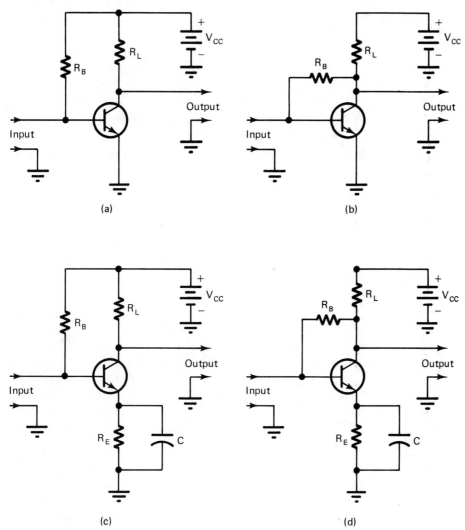

FIGURE 1-7 Methods of beta-dependent biasing: (a) fixed biasing; (b) self-biasing; (c) fixed-emitter biasing; (d) self-emitter biasing.

Circuit (b) of Fig. 1-7 was developed to compensate for the temperature sensitivity of circuit (a). Bias current is used to counteract changes in temperature. In a sense, this method of biasing has negative feedback. R_B is connected to the collector rather than V_{CC}. The voltage available for base biasing is the remaining voltage after a voltage drop across the load resistor. If the temperature rises, it causes an increase in I_B and I_C. With more I_C, there is more voltage drop across R_L. Reduced V_C voltage causes a corresponding drop in I_B. This in turn brings I_C back to normal. The opposite reaction occurs when transistor temperature becomes less. This method of biasing is called *self-biasing*.

Circuit (c) is an example of *emitter biasing*. Thermal stability is improved with this type of construction. I_B is again determined by the value of R_B and V_{CC}. An additional resistor is placed in series with the emitter of this circuit. Emitter current passing through R_E produces emitter voltage (V_E). This voltage opposes the base voltage developed by R_B. Proper values R_B and R_E are selected so that I_B and I_E will flow under ordinary operating conditions. If a change in temperature should occur, V_E will increase in value. This action will oppose the base bias. As a result, collector current will drop to its normal value. The capacitor connected across R_E is called an emitter bypass capacitor. It provides an ac path for signal voltages around R_E. With C in the circuit, an average dc level is maintained at the emitter. Without C in the circuit, amplifier gain would be reduced. The value of C is dependent on the frequencies being amplified. At the lowest possible frequency being amplified, the capacitive reactance (X_C) must be 10 times smaller than the resistance of R_E.

Circuit (d) of Fig. 1-7 is a combination of circuits (b) and (c). It is often called *self-emitter bias*. In the same regard, circuit (c) could be called *fixed-emitter bias*. As a general rule, emitter biasing is not very effective when used independently. Circuit (d) has good thermal stability. The output has reduced gain because of the base resistor connection.

Independent Beta Biasing. Two methods of biasing a transistor that is independent of beta are shown in Fig. 1-8. These circuits are extremely important because they do not change operation with beta. As a general rule, these circuits have very reliable operating characteristics. The *output* is very predictable and *stability* is excellent.

Circuit (a) is described as the *voltage divider method* of biasing. It is widely used today. The base voltage (V_B) is developed by a voltage-divider network made up of R_B and R_1. This network makes the circuit independent of beta changes. Voltage at the base, emitter, and collector all depend on external circuit values. By proper selection of components, the emitter-base junction is forward biased, with the collector being reverse biased. Normally, these bias voltages are all referenced to ground. The base, in this case, is made slightly positive with respect to ground. This voltage

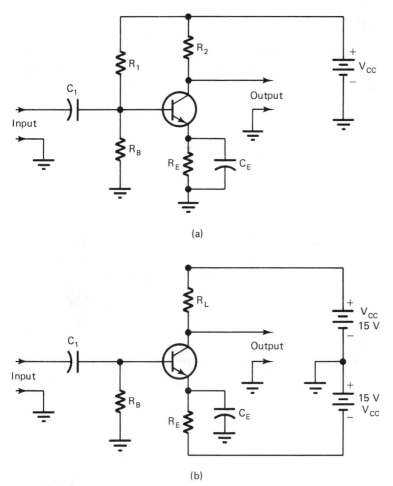

(a)

(b)

FIGURE 1-8 Independent beta biasing: (a) divider bias; (b) divider bias with split supply.

is somewhat critical. A voltage that is too positive, for example, will drive the transistor into saturation. With proper selection of bias voltage, however, the transistor can be made to operate in any part of the active region. The temperature stability of the circuit is excellent. With proper R_E bypass capacitor selection, this method of biasing produces very high gain. The divider method of biasing is often a *universal biasing circuit*. It can be used to bias all transistor amplifier circuit configurations.

Circuit (b) is very similar in construction and operation to circuit (a). One less resistor is used. The power supply requires two voltage values with reference to ground. A *split power supply* is used as an energy source for this circuit. Note the indication of $+V_{CC}$ and $-V_{CC}$. R_B is connected to ground. In this case, the value of R_B would determine the value of I_B with only half of total supply voltage. The values of R_L and R_E are usually larger to accommodate the increased supply voltage. If R_E is properly bypassed, the gain of this circuit is very high. Thermal stability is excellent.

Load-Line Analysis

Earlier in the unit we looked at the operation of an amplifier with respect to its beta. Current and voltage were calculated and operation was related to these values. This method of analysis is very important. Amplifier operation can also be determined graphically. This method employs a *collector family of characteristic curves*. A *load line* is developed for the graph. With the load-line method of analysis, it is possible to predict how the circuit will respond graphically. A great deal about the operation of an amplifier can be observed through this method.

The *load-line method* of circuit analysis is widely used today. Engineers use this method in designing new circuits. The operation of a specific circuit can also be visualized by this method. In circuit design, a specific transistor is selected for an amplifier. Source voltage, load resistance, and input signal levels may be given values in the design of the circuit. The transistor is made to fit the limitation of the circuit.

For our application of the load line, assume that a circuit is to be analyzed. Figure 1-9(a) shows the circuit being analyzed. A collector family of characteristic curves for the transistor is shown in Fig. 1-9(b). Note that the *power dissipation rating* of the transistor is included in the diagram.

Power Dissipation Curve

A common practice in load-line analysis is to first develop a power dissipation curve. This gives some indication of the maximum operating limits of the transistor. *Power-dissipation (Pd)* refers to maximum heat that can be given off by the base-collector junction. Usually, this value is rated at 25°C. Pd is the product of I_C and V_{CE}. In our circuit the Pd rating for the transistor is 300 mW.

To develop a Pd curve each value of V_{CE} is used with the Pd rating to determine an I_C value. The formula is

$$\text{Collector current} = \frac{\text{power dissipation}}{\text{collector-emitter voltage}}$$

or

$$I_C = \frac{\text{Pd}}{V_{CE}}$$

Using this formula, calculate the I_C value for each of the V_{CE} values of the family of curves. Using the V_{CE} value and the corresponding calculated I_C value, note the location on the curve. These points connected together are representative of the 300-mW Pd curve. In practice, the load line must be located to the left of the established Pd curve. Satisfactory operation without excessive heat generation can be assured in that area of the curve.

Static Load-line Analysis. The load line of a transistor amplifier represents two extreme conditions of operation. One of these

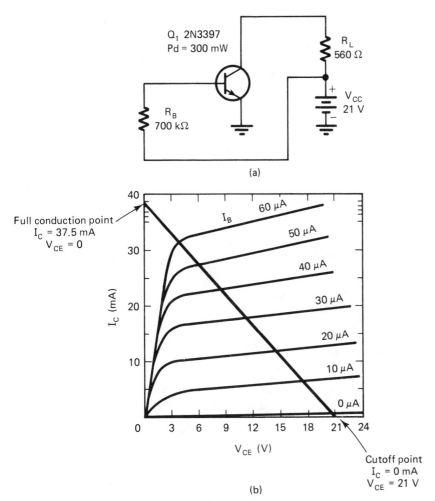

FIGURE 1-9 (a) Circuit and (b) characteristic curves of a circuit to be analyzed.

is in the cutoff region. When the transistor is cut off, there is no I_C flowing through the device. V_{CE} equals the source voltage with zero I_C. The second load-line point is in the saturation region. This point assumes full conduction of I_C. Ideally, when a transistor is fully conductive, $V_{CE} = 0$ and $I_C = V_{CC}/R_L$.

The two load-line construction points for the analysis circuit are shown on the curves of Fig. 1-9. The cutoff point is located at the zero I_C, $21 V_{CE}$ point. The value of V_{CC} determines this point. At cutoff, $V_{CE} = V_{CC}$. The saturation point is located at the 37.5 mA I_C and zero V_{CE}. The I_C value is calculated using V_{CC} and the value of R_L. The formula is

$$I_C = \frac{V_{CC}}{R_L} = \frac{21 \text{ V}}{560 \text{ }\Omega}$$

$$= 0.0375 \text{ A or } 37.5 \text{ mA}$$

These two points are connected with a straight line.

Figure 1-10 shows a *family of characteristic curves* with a load line for the circuit of Fig. 1-9. Development of the *load line* makes

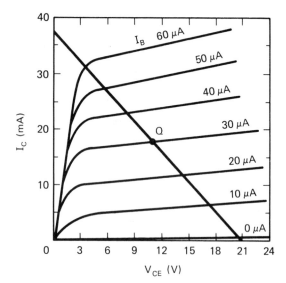

FIGURE 1-10 Family of characteristic curves for Fig. 1-9.

it possible to determine the operating conditions of the amplifier. For linear amplification, the operating point should be located near the center of the load line. In our circuit, an operating point of 30 μA is used. In the circuit diagram the value of R_B determines I_B. It is calculated by the equation $I_B = V_{CC}/R_B$. The value is

$$I_B = \frac{V_{CC}}{R_B} = \frac{21 \text{ V}}{700 \text{ k}\Omega}$$

$$= 0.00003 \text{ A, or } 30 \text{ }\mu\text{A}$$

The *operating point* for this value is located at the intersection of the load line and the 30-μA I_B curve. It is indicated at point Q. Knowing this much about an amplifier shows how it will respond in its steady or static state. The Q point shows how the amplifier will respond without a signal applied.

Operation of our amplifier in its static state is displayed by the family of curves in Fig. 1-9. Projecting a line from the Q point to the I_C scale shows the resulting collector current. In this case the I_C is 17.5 mA. The *dc beta* for the transistor at this point is determined by the formula $\beta = I_C/I_B$. This value is

$$\beta = \frac{I_C = 17.5 \text{ mA}}{I_B = 30 \text{ }\mu\text{A}} = \frac{0.0175 \text{ A}}{0.000030 \text{ A}}$$

$$= 583.3$$

The resulting V_{CE} that will occur for the amplifier can also be determined graphically. Projecting a line directly down from the Q point will show the value of V_{CE}. In our circuit the V_{CE} is approximately 11 V. This means that 10 V (21 V − 11 V) will appear across R_L when the transistor is in its static state.

Dynamic Load-line Analysis. Dynamic load-line analysis shows how an amplifier will respond to an ac signal. In this case the collector curves and circuit of Fig. 1-11 are used. A load line and Q point for the circuit have been developed on the curves. This establishes the static operation of the amplifier. Note the values of V_{CE} and I_C in the static state.

Assume now that a 0.1-V peak-to-peak ac signal is applied to the input of the amplifier. In this case the signal will cause a 20-μA p-p change in I_B. During the positive alternation of the input, I_B will change from 30 μA to 40 μA. This is shown as point P on the load line. During the negative alternation, I_B will drop from 30 μA to 20 μA. This is indicated as point N on the load line. In effect, this means that 0.0 V p-p causes I_B to change 20 μA p-p. The I_B signal extends to the right of the load line. Its value is shown as ΔI_B.

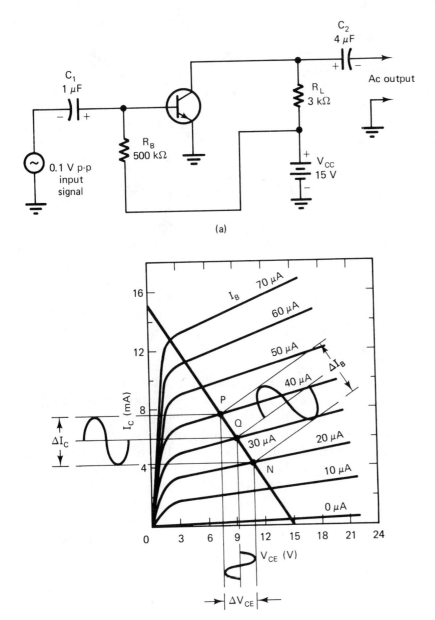

FIGURE 1-11   Dynamic load line: (a) circuit; (b) characteristic curves.

To show how a change in I_B influences I_C, lines are projected to the left of the load line. Note the projection of lines P, Q, and N toward the I_C values. The changing value of I_C is indicated as ΔI_C. An increase and decrease in I_B causes a corresponding increase and decrease in I_C. This shows that I_B and I_C are in phase.

The ac beta of the amplifier can be determined by ΔI_C and ΔI_B. First determine the peak-to-peak I_C and I_B values. Then divide ΔI_C by ΔI_B. Determine the *beta* of the amplifier circuit. Using the same procedure, determine the dc beta of the transistor at point Q. How do the ac and dc beta values of this circuit compare?

Projecting points P, Q, and N downward from the load line shows how V_{CE} changes with I_B. The value of V_{CE} is indicated as ΔV_{CE}. Note that an increase in I_B causes a decrease in the value of V_{CE}. A decrease in I_B causes V_{CE} to increase. This shows that I_B and V_{CE} are 180° out of phase. The difference in V_{CE} at any point appears across R_L.

The ac voltage gain of the amplifier can be determined from the dynamic load line. Remember that 0.1 V p-p input caused a change of 20 μA p-p in the I_B signal. The ac voltage gain can be determined by dividing ΔV_{CE} by ΔV_B. The ΔV_B value is 0.1 V p-p. Using the ΔV_{CE} value from the graph and ΔV_B, determine the *ac voltage gain* of the amplifier circuit.

LINEAR AND NONLINEAR OPERATION

The V_{CE} or *output* of an amplifier should be a duplicate of the input with some gain. When a sine wave is applied, the output should develop a sine wave. When an amplifier operates in this manner it is considered to be *linear*. For this to be achieved, the amplifier must operate in or near the center or linear area of the collector curves. As a rule, *nonlinearity* occurs near the saturation and cutoff regions. If the operating point is adjusted near these regions, it usually causes the output to be *distorted*. Normally, this is called *nonlinear distortion*.

Figure 1-12 shows how an amplifier will respond at three different operating points. Part (a) shows *linear* operation. The input and output are duplicates in this case. Part (b) shows operation in the *saturation* region. Note that the top of the input wave distorts the negative alternation of the output voltage. Part (c) shows operation near the *cutoff* region. The lower part of the input wave causes distortion of the positive alternation. As a general rule, nonlinear distortion can be tolerated in some applications. In other applications nonlinear distortion is very obvious.

An input signal that is too large can also produce distortion. Figure 1-13 shows how an amplifier will respond to a *large signal*. Note that the input signal swings into the saturation region during the positive alternation and into the cutoff region during the negative alternation. This condition causes distortion of both alternations of the output. Normally, this condition is described as

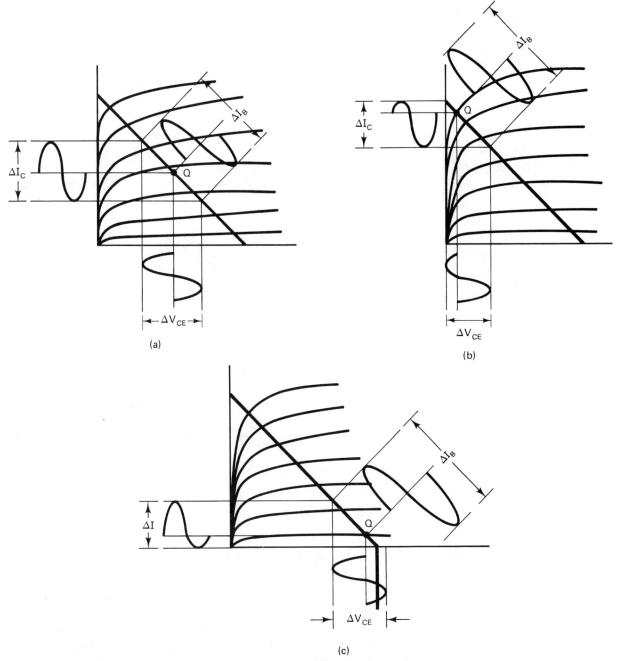

FIGURE 1-12 Amplifier operation: (a) linear operation; (b) operation near saturation; (c) operation near cutoff.

overdriving. In a radio receiver or stereo amplifier, overdriving causes speech or music to sound bad. This usually occurs when the volume control is turned too high. The audio or sound signal has its peaks *clipped*. The volume level may be higher, but the quality of reproduction is usually very poor when this occurs. It is interesting to note that the operating point is near the center of the active region. Overdriving can occur even with a properly selected operating point.

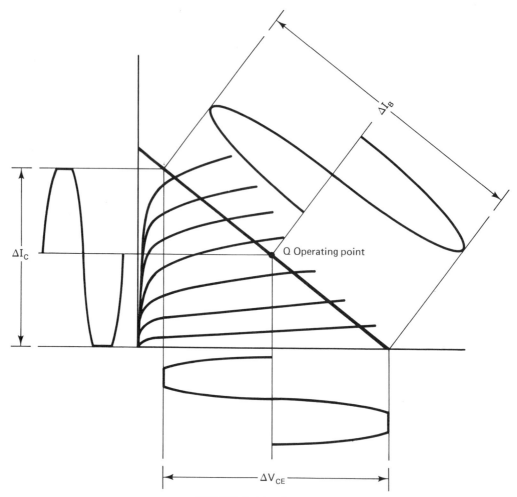

ΔI_B

ΔI_C

Q Operating point

ΔV_{CE}

FIGURE 1-13   Overdriven amplifier.

CLASSES OF AMPLIFICATION

Transistor amplifiers are frequently classified according to their bias operating point. This means of classification describes the shape of the output wave. Three general groups of amplifiers are class A, class B, and class C. Figure 1-14 shows a graphical display of these three amplifier classes.

Class A amplifiers generally have linear operation. The bias operating point is set near the center of the active region. With a sine wave applied to the input, the output is a complete sine wave. Figure 1-14(a) shows the input-output waveforms of a class A amplifier. This type of amplifier is used when a true reproduction of the input signal is required.

Figure 1-14(b) shows the input-output waveforms of a *class B amplifier.* The operating point of this amplifier is adjusted near the cutoff point. With a sine wave applied to the input, only one alternation of the signal is reproduced. When using class B amplifiers it is possible to get a large change in output for one alter-

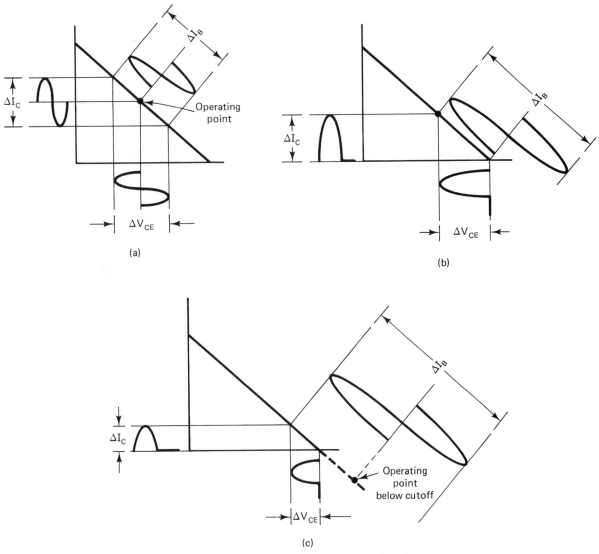

FIGURE 1-14 Classes of amplification: (a) class A; (b) class B; (c) class C.

nation. Two class B amplifiers working together can each amplify one alternation of a sine wave. When the wave is restored, a complete sine wave is developed. Class B amplifiers are commonly used in *push-pull audio output circuits*. These amplifiers are usually concerned with power amplification. A class AB amplifier has a bias operating point between the center of the active region and the cutoff point.

A class C amplifier is shown in Figure 1-14(c). In this amplifier the bias operating point is below cutoff. With a sine wave applied to the input, the output is less than half of one alternation. *Radio-frequency circuits* are the primary application of class C amplifiers today. The operational efficiency of this amplifier is quite high. It consumes energy only for a small portion of the applied sine wave.

TRANSISTOR CIRCUIT CONFIGURATIONS

The elements of a transistor can be connected together in one of three different circuit configurations. These are usually described as *common-emitter, common-base,* and *common-collector* circuits. Of the three transistor leads, one is connected to the input and one to the output. The third lead is commonly connected to both the input and output. The common lead is generally used as a reference point for the circuit. It is usually connected to the circuit *ground* or *common* point. This has brought about the terms *grounded emitter, grounded base,* and *grounded collector.* The terms common and ground mean the same thing. In some circuit configurations the emitter, base, or collector may be connected directly to ground. When this occurs, the lead is at both dc and ac ground potential. When the lead goes to ground through a battery or resistor that is bypassed by a capacitor, it is at an *ac ground* only.

Common-emitter Amplifiers

The *common-emitter amplifier* is a very important transistor circuit. A very high percentage of all amplifiers in use today are of the common-emitter type. The *input* signal of this amplifier is applied to the base and the *output* is taken from the collector. The emitter is the common or grounded element.

Figure 1-15 shows a circuit diagram of the common-emitter amplifier. This circuit is very similar to the basic amplifier used in the first part of the chapter. In effect, we have used this circuit

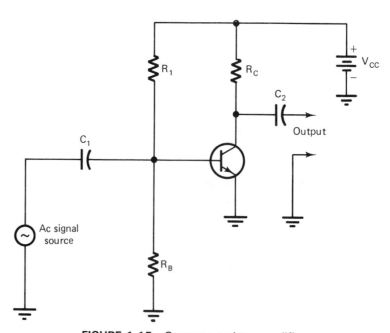

FIGURE 1-15   Common-emitter amplifier.

to acquaint you with amplifiers in general. It is presented here to make a comparison with the other circuit configurations.

The *signal* being amplified by the common-emitter amplifier is applied to the emitter-base junction. This signal is superimposed on the dc bias of the emitter-base. The base current then varies at an ac rate. This action causes a corresponding change in collector current. The *output* voltage developed across the collector-emitter is *inverted 180°*. The current gain of the circuit is determined by beta. Typical beta values are in the range of 50. Voltage gain of the circuit ranges from 250 to 500. Power gain is in the range of 20,000. *Input impedance* is moderately high, typical values being 100 Ω. *Output impedance* is moderate, typical values being 2000 Ω. In general, common-emitter amplifiers are used in small-signal applications or as voltage amplifiers.

Common-base Amplifiers

A *common-base amplifier* is shown in Figure 1-16. This type of amplifier has the emitter-base junction forward biased and the collector-base junction reverse biased. In this circuit configuration, the emitter is the *input*. An applied input signal changes the circuit value of I_E. The *output* signal is developed across R_L by changes in collector current. For each value change in I_E, there is a corresponding change in I_C.

In a common-base amplifier the current gain is called alpha. *Alpha* is determined by the formula I_C/I_E. In a common-base amplifier the gain is always less than 1. Remember that $I_E = I_B + I_C$. Therefore, I_C will always be slightly less than I_E by the value of I_B. Typical values of alpha are 0.98 to 0.99.

In Figure 1-16 V_{EE} forward-biases the emitter-base, whereas V_{CC} reverse-biases the collector-base. Resistor R_E is an emitter-

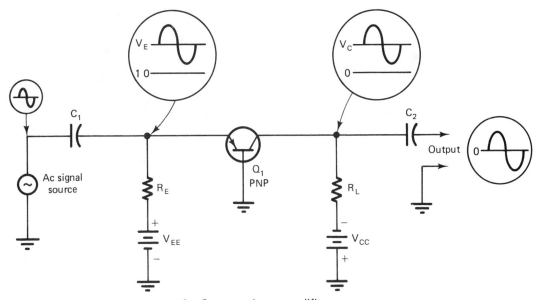

FIGURE 1-16 Common-base amplifier.

current-limiting resistor and R_L is the load resistor. Note that the transistor is a PNP type.

When an ac signal is applied to the *input*, it is added to the dc operating value of V_E. The positive alternation therefore adds to the forward-bias voltage of V_E. This condition causes an increase in I_E. A corresponding increase in I_C also takes place. With more I_C through R_L, there is an increase in its voltage drop. The collector therefore becomes less negative or swings positive. In effect, the positive alternation of the input produces a positive alternation in the *output*. The input and output of this amplifier in the *output*. The input and output of this amplifier are *inphase*. See the waveform inserts in the diagram. The negative alternation causes the same reaction, the only difference being a reverse in polarity.

A common-base amplifier has a number of rather unique characteristics that should be considered. Typical values of the current gain are 0.98 to 0.99. *Voltage gain* is usually very high. Typical values range from 100 to 2500, depending on the value of R_L. *Power gain* is the same as the voltage gain. The *input impedance* of the circuit is quite low. Values of 10 to 200 Ω are very common. The *output impedance* of the amplifier is somewhat moderate. Values range from 10 to 40 kΩ. This amplifier does not invert the applied signal.

Common-base amplifiers are used primarily to *match* a low-impedance input device to a circuit. This type of circuit configuration is also used in radio-frequency amplifier applications. As a general rule, the common-base amplifier is not used very commonly today.

Common-collector Amplifiers

A *common-collector amplifier* is shown in Figure 1-17. In this circuit the base serves as the signal input point. The *input* of this amplifier is primarily the same as the common-emitter circuit. The collector is connected to ground through V_{CC}. Note that the input, output, and collector are all commonly connected. The unique part of this circuit is the output. It is developed across the load resistor R_L in the emitter. There is no resistor in the collector circuit.

When an ac signal is applied to the input, it adds to the base current. The positive alternation increases the value of I_B above its static operating point. An increase in I_B causes a corresponding increase in I_E and I_C. With an increase in I_E, there is more voltage drop across R_L. The top side of R_L, the emitter, becomes more positive. A positive input alternation therefore causes a positive output alternation across R_L. Essentially, this means that the input and output are *in phase*. The negative alternation reduces I_B, I_E, and I_C. With less I_E through R_L, the output swings negative. The output is again in phase for this alternation.

The common-collector amplifier is capable of *current gain*. A

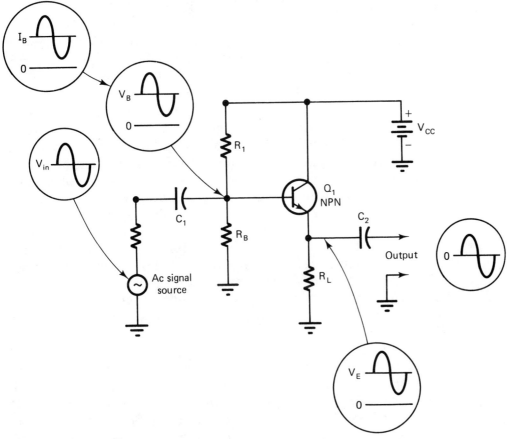

FIGURE 1-17 Common-collector amplifier.

small change in I_B causes a large change in I_E. In general, the current gain is assumed to be greater than 1. Values approximate those of the common-emitter amplifier.

Other factors of importance are voltage gain, power gain, input impedance, and output impedance. *Voltage gain* is less than 1. *Power gain* is moderately high. Typical values are in the range of 50. *Input impedance* is in the range of 50 kΩ. The *output impedance* is quite low because of the location of R_L. Typical values are 50 Ω.

Common-collector amplifiers are used primarily as *impedance-matching* devices. They can match a high-impedance device to a low-impedance load. Practical applications include preamplifiers operated from a high-impedance microphone or phonograph pickup. Common-collector amplifiers are also used as driver transistors for the last stage of amplification. In this application power transistors generally require large amounts of input current to deliver maximum power to the load device. Since the emitter output follows the base, this type of amplifier is often called an *emitter follower*.

FIELD-EFFECT TRANSISTOR AMPLIFIERS

The primary function of an amplifier is to reproduce the applied signal and provide some level of amplification. Unipolar transistors can be used to achieve this function. The junction field-effect transistor (JFET) and metal-oxide semiconductor field-effect transistor (MOSFET) are examples of unipolar amplifying devices. Amplification depends on the device being used and its circuit components. The composite circuit is normally energized by dc voltage. An ac signal is then applied to the input of the amplifier. After passing through the device, an amplified version of the signal appears in the output. Operating conditions of the device and the circuit combine to determine the level of amplification to be achieved.

For a FET to respond as an amplifier, it must be energized with dc voltage. Specific voltage values and polarities must be applied to each type of device. Forward and reverse biasing are not as meaningful in FET circuits. Current conduction takes place through a channel. The value of voltage and its polarity determines the amount of control being achieved.

An illustration of the supply voltages and polarities needed for different types of FETs is shown in Figure 1-18. JFETs must have the gate-source junction reverse biased to achieve control. The gate voltage of an *enhancement MOSFET* is opposite that of the source. *Depletion MOSFETs* have some flexibility in the polarity of the gate voltage. Some of these devices have a zero value of V_G and swing positive or negative during operation.

Basic FET Amplifier Operation

We will now take a look at the circuitry of a basic *FET amplifier*. In this example an *N-channel* JFET is used. A *P-channel* device could also have been selected. The primary difference in operation is the polarity of the source voltage and the type of current carrier in the channel.

Figure 1-19 shows an *N-channel JFET amplifier* connected in the *common-source* configuration. The source of this device is common to both the input signal and the output. This circuit is very similar to the common-emitter amplifier. V_{DD} serves as the dc energy source for the source and drain. V_{GS} reverse-biases the gate with respect to the source. The value of V_{GS} establishes the static operating point of the circuit. R_g is normally extremely high. No gate current flows through R_g. The input signal is applied to the gate through capacitor C. Selecting a large value of R_g keeps the input impedance of the amplifier high. If R_g is 1 MΩ, the signal source sees an impedance of 1 MΩ.

Let us now consider the operation of the JFET amplifier of Fig. 1-19 in its static state. The family of drain curves of Fig. 1-20 will be used in this explanation. A load line for the circuit must be established. Remember that the two extreme conditions

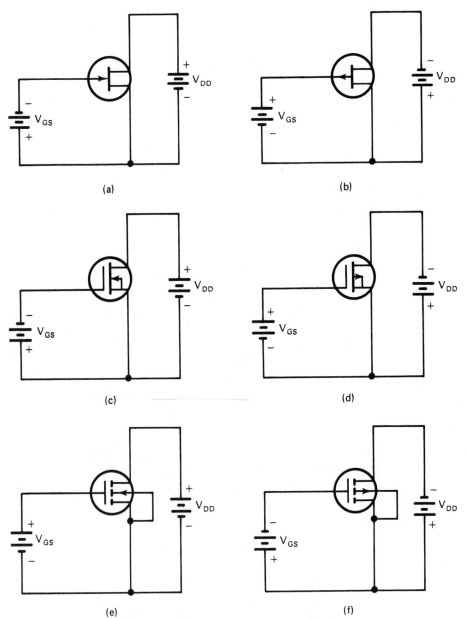

FIGURE 1-18 FET operational voltages: (a) N-channel JFET; (b) P-channel JFET; (c) N-channel D-MOSFET; (d) P-channel D-MOSFET; (e) N-channel E-MOSFET; (f) P-channel E-MOSFET.

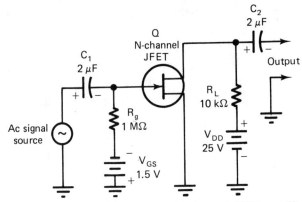

FIGURE 1-19 Common-source N-channel JFET amplifier.

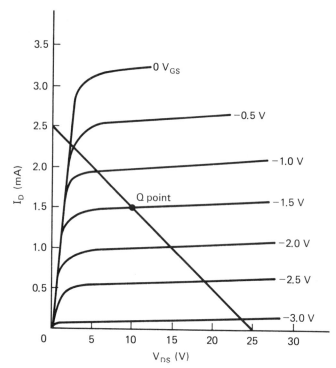

FIGURE 1-20 Family of drain curves.

of operation are needed for the load line. In a JFET, these extremes are *full conduction* and *cutoff.* At the cutoff point, no current passes through the channel. The V_{DD} voltage all appears across V_{DS}. Full conduction occurs when a maximum value of I_D flows through R_L. This value is determined by the formula

$$\text{Drain current} = \frac{\text{source voltage}}{\text{load resistance}} \quad \text{or} \quad I_D = \frac{V_{DD}}{R_L}$$

For the amplifier circuit, the maximum value of I_D is

$$I_D = \frac{25 \text{ V}}{10 \text{ k}\Omega} = \frac{25 \text{ V}}{10 \times 10^3}$$

$$= 0.0025 \text{ A or } 2.5 \text{ mA}$$

With this conduction through R_L, the V_{DS} will be zero. The two locating points of the load line are $V_{DS} = 25$ V with 0 mA of I_D and $V_{DS} = 0$ V at 2.5 mA I_D. Check these two location points on the load line. A straight line connects the points.

With the *load line* developed, it is possible to see how the JFET will respond in its static state. Static operation occurs without a signal applied. For linear operation, the amplifier should respond near the center of the active region. In our basic circuit, V_{GS} is -1.5 V. Point Q on the load line denotes the operating point. With an appropriate input signal, the amplifier should respond in the linear region.

To see how the JFET responds in its static state, lines are projected from the Q point. Projecting a line from the Q point to the I_D scale shows the drain-source voltage. V_{DS} is approximately 10 V. This means that 15 V appear across R_L at this point of operation. The dc voltage gain can be determined with this value. Voltage amplification (A_V) equals the drain-source voltage (V_{DS}) divided by the gate-source voltage. For our circuit the gain is

$$A_V = \frac{V_{DS}}{V_{GS}} = \frac{10\text{ V}}{1.5\text{ V}}$$

$$= 6.667$$

Dynamic JFET Amplifier Analysis

Dynamic JFET amplifier analysis shows how the device will respond when an ac signal is applied. For this explanation we will use the *drain curves* and circuit of Figure 1-21. A load line and Q point for the circuit has been developed on the curves. This shows how the JFET responds in its static state. Check the load-line location points and Q point for accuracy.

The circuit diagram shows a 1-V p-p ac signal applied to the input of the JFET amplifier. With this signal the V_{GS} operating point changes in value from -2.0 V to -3.0 V. For the positive alternation, V_{GS} swings from -2.5 V to -2.0 V. This change is shown as point P on the load line. During the negative alternation, V_{GS} drops from -2.5 V to -3 V. This value is shown as point N on the load line. In effect, this means that a 1-V p-p input signal causes V_{GS} to change from -2.0 V to -3.0 V. This is considered to be the ΔV_{GS} value.

To show how ΔV_{GS} alters I_D, lines are projected to the left of points P, Q, and N. Note the projection of these points toward the I_D values in Fig. 1-21. The changing value of I_D is shown as ΔI_D. An increase or decrease in V_{GS} causes a corresponding change in I_D. This shows that V_{GS} and I_D are *in phase*.

Projecting lines P, Q, and N downward from the load line shows how V_{DS} changes with V_{GS}. The value of V_{DS} is shown as ΔV_{DS}. Note that an increase in V_{GS} causes a decrease in V_{DS}. This tells us that V_{GS} and V_{DS} are 180° *out of phase*. The value difference in V_{DS} and V_{DD} appears across the load resistor as V_{RL}.

The *voltage gain* of a JFET amplifier can be readily observed by load-line data. A ΔV_{GS} of 1 V p-p causes a ΔV_{DS} of approximately 18V p-p. This shows an obvious gain of 18. Using the formula:

$$A_V = \frac{\Delta V_{DS}}{\Delta V_{GS}} = \frac{18\text{ V p-p}}{1\text{ V p-p}}$$

$$= 18$$

MOSFET Circuit Operation

D-type (depletion) MOSFET and *E-type (enhancement) MOSFET* operation is very similar to that of the JFET. There is an obvious

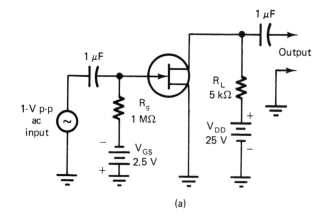

(a)

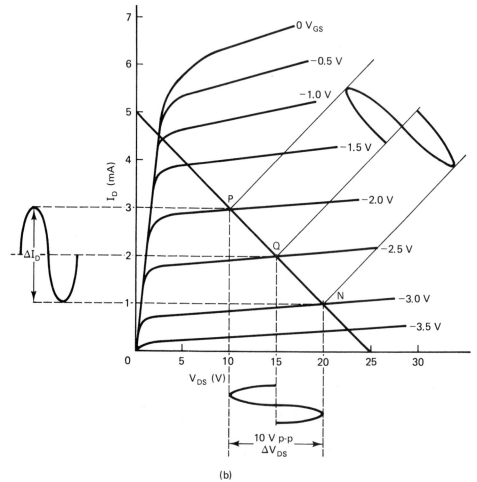

(b)

FIGURE 1-21 JFET amplifier; (a) N-channel amplifier; (b) drain family
of curves.

difference in the required operational voltage values and polarity.
A load-line analysis of either amplifier type shows the same basic
principle of operation. The gate is highly resistant or infinite for
all devices. Current flow in the channel is controlled by the voltage
value and polarity of the gate signal. The only difference of any
concern is with the D-type MOSFET. It is frequently biased at the

zero V_{GS} point. An ac input signal would cause V_{GS} to vary above and below the zero value. It is important to remember that all FETs are voltage-sensitive devices. They produce only voltage amplification. No current flows in the gate circuit.

FET Biasing Methods

For FET operation biasing refers to the necessary dc voltage needed by the gate with respect to the source lead. Essentially, *biasing* refers to the development of a suitable V_{GS} value. By selection of a proper bias voltage, it becomes possible to force a device to operate at a chosen Q point. The V_{DD} voltage applied to the source-drain is generally not considered to be bias voltage. The biasing method selected for a specific circuit has a great deal to do with the type of device involved. Enhancement MOSFETs and depletion MOSFETs require different biasing procedures.

Fixed Biasing. The simplest way to bias an FET is with *fixed biasing.* This method of biasing can be used equally well with E- or D-MOSFETs or JFETs. In fixed biasing, the correct voltage value and polarity are supplied by a small battery or cell. An electronic power supply could also be used to perform the same operation. Figure 1-22 shows fixed biasing applied to three different N-channel FET types. If P-channel devices were used, the polarity of V_{CC} and V_{DD} would be reversed. V_{CC} is used as the fixed bias source for the three FET types.

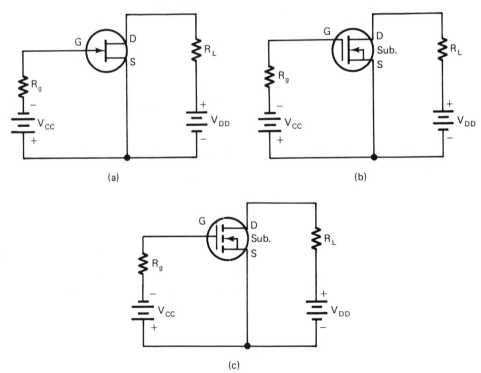

FIGURE 1-22 Methods of fixed biasing: (a) N-channel JFET; (b) N-channel D-type MOSFET; (c) N-channel E-type MOSFET.

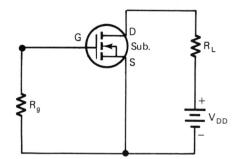

FIGURE 1-23 Fixed bias of an N-channel
E-D-type MOSFET.

If an enhancement-depletion type of MOSFET is to be fixed-biased, the circuit of Figure 1-23 would be used. In this type of device the operating point is usually $V_{GS} = 0$ V. No voltage source is needed to establish the operating point. R_g is normally of a high value. Typically, values of 1 to 10 MΩ are used. R_g must be included in the circuit. I serves as a return for the gate to the common connection point.

Voltage-divider Biasing. The voltage-divider method of biasing FETs is rather easy to achieve. It is applicable only to enhancement-type MOSFETs. With these devices, V_{GS} and V_{DD} are of the same polarity. As a result of this condition , a single voltage source can be used as a supply. The V_{GS} voltage is only a fraction of the value of V_{DD}. The divider resistors are used to develop the necessary voltage values. Figure 1-24 shows the divider method of biasing for N- and P-channel E-MOSFETs.

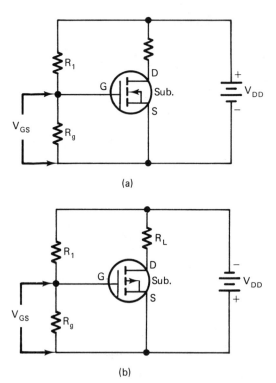

(a)

(b)

FIGURE 1-24 Voltage-divider method of biasing for E-MOSFETs: (a)
N-channel; (b) P-channel.

The values of R_l and R_g must be properly chosen to establish a desired operating point. By the voltage-divider equation, we find that the bias voltage is

$$\text{Gate-source voltage} = R_g \text{ voltage}$$

$$= \text{source voltage } \frac{R_g}{R_1 + R_g}$$

or

$$V_{GS} = V_{Rg} = V_{DD} \frac{R_g}{R_1 + R_g}$$

Self-biasing of FETs.

Self-biasing is commonly called source biasing. *Self-biasing* is rather easy to recognize. The source current of an FET is used to develop the bias voltage. The voltage developed is achieved by a resistor (R_s) connected in series with the source. Current flow through the source-drain causes a voltage drop across R_s. The polarity of the developed voltage is based on the direction of current flow through R_s. Figure 1-25 shows biasing for N- and P-channel JFETs. D-type MOSFETs can also be biased by the same method.

For the N-channel circuit, current flow through R_s causes the source to be slightly positive with respect to the gate. With R_s connected to the lower side of R_s, this makes R_g negative and the source positive. The value of R_s and drain current (I_D) determines the bias operating point of the circuit. Note the polarity difference in the P-channel circuit.

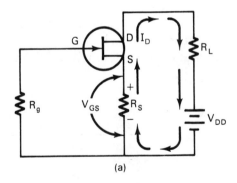

(a)

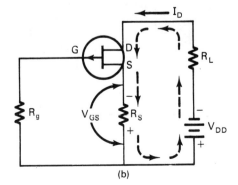

(b)

FIGURE 1-25 Self-biasing of JFETs: (a) N-channel; (b) P-channel.

FET Circuit Configurations

FET amplifiers, like their bipolar counterparts, can be connected in three different circuit configurations. These are described as *common-source, common-gate,* and *common-drain* circuits. One lead is connected to the input with a second lead connected to the output. The third lead is commonly connected to both input and output. This lead is used as a circuit reference point. It is often connected to the circuit ground. This has brought about the terms *grounded source, grounded gate,* and *grounded drain.* The terms common and ground refer to the same type of connection.

Common-source Amplifiers.   The *common-source amplifier* is the most widely used FET circuit configuration. This circuit is similar in many respects to the common-emitter bipolar amplifier. The input signal is applied to the gate-source and the output signal is taken from the drain-source. The source lead is common to both input and output.

A practical common-source amplifier is shown in Figure 1-26. This circuit is essentially the same as the one used in our basic JFET amplifier. It is shown here so that a comparison can be made with the other circuit configurations. The device can be a JFET, a D-MOSFET, or an E-MOSFET. Circuit characteristics are primarily the same for all three devices.

The signal being processed by the common-source amplifier is applied to the gate-source. Self-biasing of the circuit is achieved by the source resistor R_2. This voltage establishes the static operating point. The incoming signal voltage is superimposed on the gate voltage. This causes the gate voltage to vary at an ac rate. This action causes a corresponding change in drain current. The output voltage developed across the source and drain is inverted 180°. Voltage gain A_V is V_{DS}/V_{GS}. Typical A_V values are 5 to 10. The input impedance is extremely high for nearly any signal

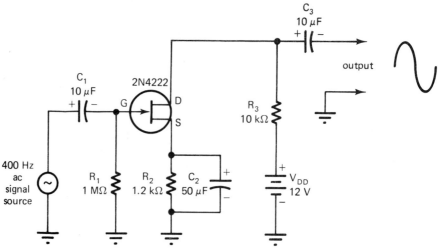

FIGURE 1-26   Common-source JFET amplifier.

source. One to several megohms is common. The output impedance (Z_{out}) is moderately high. Typical values are in the range of
2 to 10 kΩ. Z_{out} is dependent primarily on the value of R_L.

A common-source amplifier has a very good ratio between input and output impedance. Circuits of this type are extremely
valuable as impedance-matching devices. Primarily, common-
source amplifiers are used exclusively as voltage amplifiers. They
respond well in radio-frequency signal applications.

Common-gate Amplifiers.   The *common-gate FET amplifier*
is similar in many respects to the common-base bipolar transistor
circuit. The input signal is applied to the source gate and the output appears across the drain gate. The common-gate amplifier is
capable of voltage gain but has a current gain capability that is
less than 1. JFETs, E-MOSFETs, and D-MOSFETs may all be
used as common-gate amplifiers.

A practical common-gate amplifier is shown in Figure 1-27.
This circuit employs an N-channel JFET. The operating voltages
are the same as those of the common-source circuit. Self-biasing
is achieved by the source resistor R_1. This voltage is used to establish the static operating point. An input signal is applied to R_1
through capacitor C_1. A variation in signal voltage causes a corresponding change in source voltage. Making the source more positive has the same effect on I_D as making the gate more negative.
The positive alternation of the input signal makes the source more
positive. This, in turn, reduces drain current. With less I_D, there
is a smaller voltage drop across the load resistor R_2. The drain or
output voltage therefore swings positive. The negative alternation
of the input reduces the source voltage by an equal amount. This
is the same as making the gate less negative. As a result, I_D is
increased. The voltage drop across R_2 increases similarly. This, in
turn, causes the drain or output to be less positive or negative
going. The input signal is therefore *in phase* with the output signal.

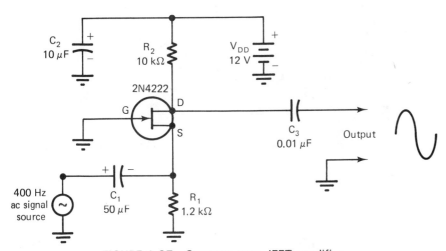

FIGURE 1-27   Common-gate JFET amplifier.

The common-gate amplifier has a number of rather unusual characteristics. Its voltage gain is somewhat less than that of the common-source amplifier. Representative values are 2 to 5. The common-gate circuit has very low input impedance (Z_{in}). The *output impedance* (Z_{out}) is rather moderate. Typical Z_{in} values are 200 to 1500 Ω with Z_{out} being 5 to 15 kΩ. This type of circuit configuration is often used to amplify radio-frequency signals. Amplification levels are very stable for *RF*, without feedback between the input and output.

Common-drain Amplifiers.   A common-drain amplifier has the input signal applied to the gate and the output signal removed from the source. The drain is commonly connected to one side of the input and output. Common-drain amplifiers are also called *source followers*. This circuit has similar characteristics to those of the common-collector bipolar amplifier.

Figure 1-28 shows a practical common-drain amplifier using an N-channel JFET. The input of this amplifier is primarily the same as that of the common source amplifier. The input impedance is therefore very high. Z_{in} is determined largely by the value of R_1. The operating point of our amplifier is determined by the value of R_2. Essentially, this circuit has the same operating point as our other circuit configurations. Resistor R_3 has been switched from the drain to the source in this circuit. Resistors R_2 and R_3 are combined to form the load resistance. The output impedance is based on this value.

When an ac signal is applied to the input of the amplifier, it changes the gate voltage. The dc operating point is established by the value of source resistor R_2. The positive alternation of an input signal makes the gate less negative. This causes an N-channel device to be more conductive. With more current through R_3 and R_2, the source swings positive. The negative alternation of the input then makes the gate more negative. This action causes the channel

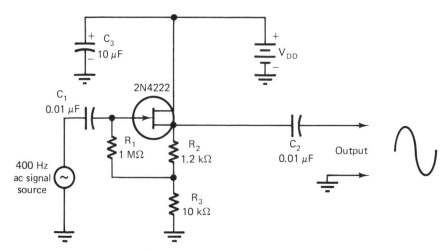

FIGURE 1-28   Common-drain JFET amplifier.

to be less conductive. A smaller current therefore causes the source to swing negative. In effect, this means that the input and output voltage values of the amplifier are in step with each other. These signals are *in phase* in a common-drain amplifier.

Common-drain amplifers are primarily used today as impedance-matching devices. They are used to match a high-impedance device to a low-impedance load. This type of circuit is capable of handling a high input signal level without causing distortion. The input impedance places a minimum load on the signal source. Common-drain amplifiers are frequently used today to match high-impedance devices such as microphones and phonograph pickups to the input of an audio amplifier.

SELF-EXAMINATION

1. A signal to be processed is applied to the _____ of an amplifier and reproduced at the _____.

2. The ratio of output voltage to input voltage of an amplifier is called _____ _____.

3. The primary function of an amplifier is to provide _____, _____, or _____ gain.

4. A bipolar transistor amplifier has its emitter-base junction _____ biased and its collector-base junction _____ biased.

5. The two types of bipolar transistors are _____ and _____.

6. The formula: $V_{CC} \div R_B$ is used to calculate _____.

7. If the output voltage of an amplifier is 5 V p-p and the input voltage is 0.025 V p-p, what is the voltage gain? $A_V = $ _____.

8. If the output current of an amplifier is 8 mA when the input current is 100 μA, what is the current gain? $A_I = $ _____.

9. How much base current is produced by a transistor amplifier when a 120-kΩ resistor is connected to 9 V? $I_B = $ _____.

10. What is the beta of a transistor amplifier having an I_C of 6 mA, an I_B of 120 μA, and I_E of 6.12 mA? $B = $ _____.

11. The two extreme conditions of operation of a bipolar transistor used to develop a load line are _____ and _____.

12. With a sine wave applied to the input of a bipolar amplifier, where must the operating point be adjusted to achieve class A, B, and C amplification?

Class A = _____
Class B = _____
Class C = _____.

13. Draw symbols for the following: (a) JFET, (b) D-MOS-FET, and (c) E-MOSFET.

14. Two types of transistor amplifier biasing methods are _____ and _____.

15. A method of analyzing transistor amplifier circuits is called _____.

16. With a sine wave input and a sine wave output, an amplifier is said to be _____ operation.

17. When a large input signal causes amplifier distortion, the condition is called _____.

18. Three classes of amplifiers are: _____, _____, and _____.

19. Three transistor circuit configurations are _____, _____, and _____.

20. In a CB amplifier, the current gain is called _____.

21. Two types of JFET's are _____ and _____.

22. The three elements of a JFET are: _____, _____, and _____.

23. Two types of MOSFETs are _____ and _____.

24. Biasing methods for FETs include _____, _____, and _____.

25. Three FET amplifier configurations are: _____, _____, and _____.

26. Load-line analysis of an amplifier:

   a. Using the collector family of characteristic curves in Fig. 1-29(a), develop a dc load line for the circuit in Fig. 1-29(b).
   b. The two locating points are _____ and _____.
   c. With a straightedge, connect the locating points with a pencil.

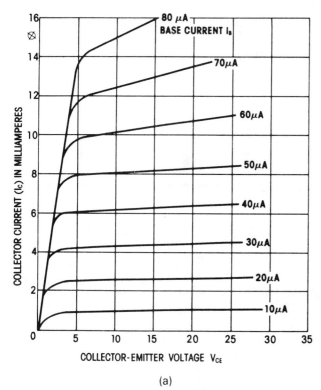

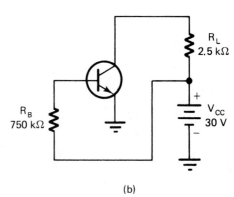

(a)   (b)

FIGURE 1-29   Transistor load-line analysis exercise: (a) I_C-V_{CE} curves; (b) transistor circuit.

d. Determine the Q point for the circuit. Locate this point on the load line.

e. Determine the static operation points for V_{CE} and I_C.

f. If a 0.5-V p-p signal is applied, it causes 20 μA of I_B.

g. Locate these points on the load line. By the projection of lines, determine the ΔI_C, ΔV_{CE}, and ΔI_B:

$\Delta I_C =$ _____ ; $\Delta V_{\text{CE}} =$ _____

_____ ; $\Delta I_B =$ _____ .

h. Calculate the dc beta, ac beta, and voltage gain:

$\beta_{DC} =$ _____ ; $\beta_{AC} =$ _____

$A_V =$ _____ .

ANSWERS

1. Input; output	2. Voltage amplification
3. Voltage; current; power	4. Forward; reverse
5. NPN; PNP	6. Base current (I_B)
7. 200	8. 80
9. 75μA	10. 50
11. Saturation, cutoff	12. A—midway between saturation and cutoff; B—near cutoff; C—below cutoff
13. See Figure 1-18	14. Beta-dependent; beta-independent

15.	Load-line analysis	16.	Linear, or Class A
17.	Overdriving	18.	A; B; C
19.	Common base, common emitter, common collector	20.	Alpha
21.	N-channel; P-channel	22.	Source; drain; gate
23.	D-type; E-type	24.	Fixed; voltage divider; self
25.	Common-source; common-drain; common-gate		

EXPERIMENTAL ACTIVITIES FOR CIRCUITS AND OP-AMPS

The experimental activities that follow stress the practical applications of electronics. They parallel the content of each of the units in this book. The expense of the equipment is kept at a minimum. A few of the activities require no lab equipment.

Each experimental activity is organized in the following way:

1-1

(First activity of Unit 1)

TITLE

(Topic of the activity)

Introduction: One or two paragraphs, the first containing an overview of the activity and practical applications; the other includes the purpose of the activity and suggested observations that should be made.

Objective: Expected learning to take place when the experiment is completed.

Equipment: Necessary equipment and materials to perform the experiment.

Procedure: Logical step-by-step sequence for completing the learning activity; maximum use is made of charts and tables that will aid in the recording of data.

Analysis: Specific related questions and problems which supplement the experimental activity.

The experimental material is presented by using a single-concept approach. Activities organized in this way require only a short time to assemble and make the necessary measurements to facilitate learning.

In this book there are several experimental activities that are used to reinforce the text material. These activities provide a different direction for the learning process. As a rule, the activity is experimentally based. This involves some manipulative activity or hands-on operations. They deal with such activities as constructing circuits, testing operations, calculations, using instruments, and identifying and using components. Through this approach you will become more familiar with electronic components and their uses in a specific circuit application.

Tools and Equipment

A variety of tools and components are needed to perform the experimental activities of this course. These may be obtained from electronic supply houses, mail-order supply houses, and educational vendors. A listing of these sources appears in Appendix C. A specific parts package for this course is available through Heath/ Zenith Company of Benton Harbor, Michigan 49022. The component hardware package number is EB 6104-31.

For the more venturesome learner, these components may be obtained through a variety of electronic supply houses. As a rule, there is a standard part number that must be used in obtaining these components. In many cases, there may be an equivalent component made by another manufacturer. Refer to Appendix C.

IMPORTANT INFORMATION

At this time you may want to turn to the back of the book and review the following information.

Appendix A Electronic Symbols
Appendix D Soldering Techniques
Appendix E Electronic Tools

The information in these sections will help you to perform the experimental activities in this book.

LAB ACTIVITY TROUBLESHOOTING AND TESTING

The lab activities included in this book provide an opportunity to practice troubleshooting and testing for electronic circuits, devices, and systems. This section of the book is used as a comprehensive list of troubleshooting and testing procedures that may be accomplished while performing each lab activity. Emphasis should be placed on *understanding circuit operation* and *understanding proper use of test equipment*. If the technician understands how the circuit, device, or system functions and knows how to use test equipment, troubleshooting and testing are relatively easy to accomplish. This is true even for the simplest type of electronic circuit or for more complex circuits.

Competencies for Troubleshooting and Testing

The specific competencies that should be developed during completion of this book are listed in this section. Remember that this book is one of a series and many other competencies for troubleshooting and testing are included in the other books.

Objectives

Upon completing the activities presented in this book, you should be able to:

1. Outline basic troubleshooting procedures for locating specific trouble with devices and equipment.
2. Find the parts or circuits that are defective by using a "common sense" approach.
3. Test devices and circuits of electronic equipment using correct procedures.

UNIT 1: Amplifier Circuits

Experiment 1-1: Transistor Biasing

1. Construct a common-base transistor amplifier circuit.
2. Measure I_E, I_B, and I_C of a transistor amplifier circuit with a multimeter.
3. Calculate I_B, I_C, and I_E values for a transistor amplifier using values obtained by voltage measurements with a multimeter.
4. Determine proper biasing of a transistor amplifier.

Experiment 1-2: Transistor Gain

1. Construct common-base, common-emitter, and common-collector transistor circuits.
2. Measure voltage, current, and resistance values of common-base, common-emitter, and common-collector transistor circuits with a multimeter.
3. Calculate alpha current gain and voltage gain of transistor circuits by using multimeter measurements.
4. Make circuit adjustments that affect voltage and current gain of a transistor amplifier.
5. Compare current gain of common-base, common-emitter, and common-collector amplifiers.

Experiment 1-3: Amplifier Phase Relationships

1. Construct common-base, common-emitter, and common-collector transistor amplifier circuits.
2. Observe the phase and amplitude of input and output signals of C-B, C-E, and C-C amplifier circuits with an oscilloscope.
3. Determine voltage of an amplifier circuit by making measurements.
4. Connect an oscilloscope for external synchronization.
5. Predict the input and output phase relationships of common-base, common-emitter, and common-collector amplifiers.

Experiment 1-4: Transistor Biasing Methods

1. Construct transistor amplifier circuits with fixed, voltage divider, and self-biasing methods used.
2. Measure and observe the voltage and polarity of the three methods of biasing transistor amplifier circuits using an oscilloscope.
3. Make voltage measurements at test points of a transistor amplifier circuit.

Experiment 1-5: Amplifier Classes

1. Construct Class A, Class B, and Class C transistor amplifier circuits.
2. Adjust the bias operating point of a transistor circuit by changing the relationship of base voltage and emitter voltage.
3. Observe the input and output waveforms of Class A, Class B, and Class C amplifier circuits.

UNIT 2: Amplifying Systems

Experiment 2-1: DC Transistor Amplifiers

1. Construct a common-emitter DC transistor amplifier circuit.
2. Measure input and output voltages with a multimeter and/or oscilloscope.
3. Calculate voltage and current gain using measured values.

Experiment 2-2: AC Transistor Amplifiers

1. Construct an AC transistor amplifier circuit.
2. Measure input and output signals of an AC transistor amplifier that are 180 degrees out-of-phase.

3. Measure voltage and current values of an AC transistor amplifier with a multimeter.
4. Calculate voltage and current gain of an AC transistor amplifier.

Experiment 2-3: JFET Amplifiers

1. Construct a JFET amplifier circuit.
2. Measure voltage, current, and resistance values of a JFET circuit using a multimeter or oscilloscope.
3. Calculate the gain of a JFET amplifier circuit by making measurements.
4. Recognize biasing methods used with JFET amplifier circuits.

Experiment 2-4: *RC* Coupling

1. Construct a two-stage *RC* coupled transistor amplifier circuit.
2. Trace the signal path through an amplifier circuit using an oscilloscope (signal tracing).
3. Test waveforms and phase relationships of an *RC* coupled amplifier.
4. Calculate amplifier voltage gain by using measured values.

Experiment 2-5: Transistor Coupling

1. Construct a two-stage, transformer-coupled transistor amplifier circuit.
2. Perform signal tracing through an amplifier circuit with an oscilloscope.
3. Test waveforms and phase relationships with an oscilloscope.
4. Calculate amplifier voltage gain by using measured values.

Experiment 2-6: Single-ended Power Amplifier

1. Construct a single-ended power amplifier circuit.
2. Measure voltage and current values of a single-ended power amplifier using an oscilloscope and multimeter.
3. Calculate voltage gain, operating current, power, and efficiency values using measured values.

Experiment 2-7: Push-pull Amplifiers

1. Construct a push-pull amplifier circuit.
2. Perform signal tracing through push-pull amplifier circuit with an oscilloscope.

3. Measure signal levels and observe signal phase relationships with an oscilloscope.
4. Calculate voltage gain and power values using measured values.

UNIT 3: Operational Amplifiers

Experiment 3-1: Op-amps

1. Construct an op-amp test circuit.
2. Use an oscilloscope to test the condition of an op-amp.
3. Use an audio test to check the condition of an op-amp.
4. Measure current of an op-amp circuit with a multimeter.

Experiment 3-2: Inverting Op-amps

1. Construct an inverting op-amp circuit.
2. Calculate voltage gain, current gain, and resistance values of an op-amp circuit by making measurements.
3. Design a simple op-amp circuit and determine its gain.
4. Measure voltage, current, and resistance values of an inverting op-amp circuit with a multimeter or oscilloscope.
5. Determine the effect of source resistance and feedback resistance on op-amp voltage gain.

Experiment 3-3: Inverting Op-amps

1. Construct a noninverting op-amp circuit.
2. Calculate amplifier voltage gain by using the resistance method and voltage method.
3. Design a noninverting op-amp circuit.
4. Adjust the gain of a noninverting op-amp circuit by changing resistance values.
5. Measure voltage, current, and resistance values of a noninverting op-amp circuit with a multimeter or oscilloscope.

UNIT 4: Power Supply Circuits

Experiment 4-1: Half-Wave Rectifier

1. Construct a single-phase, half-wave rectifier circuit.
2. Measure ac and dc voltage values for a half-wave rectifier circuit using a multimeter.
3. Make waveform observations of a half-wave rectifier circuit using an oscilloscope.
4. Compare measured and calculated dc output voltage values for a half-wave rectifier circuit.

Experiment 4-2: Full-wave Rectifier

1. Construct a single-phase, full-wave rectifier circuit.
2. Measure ac and dc voltage values for a full-wave rectifier circuit using a multimeter.
3. Make waveform observations of a half-wave rectifier circuit using an oscilloscope.
4. Compare measured and calculated *LC* output voltage values for a full-wave rectifier circuit and compare these values with a half-wave rectifier circuit.

Experiment 4-3: Bridge Rectifier

1. Construct a single-phase bridge rectifier circuit.
2. Measure ac and dc voltage values for a bridge rectifier circuit using a multimeter.
3. Make waveform observations of a bridge rectifier circuit using an oscilloscope.
4. Compare measured and calculated dc output voltage values for a bridge rectifier circuit.

Experiment 4-4: Capacitor Filters

1. Observe the action of a filter capacitor on a bridge rectifier circuit.
2. Calculate dc voltage output, ripple, ripple factor and voltage regulation, and ideal filter capacitor value for a filtered bridge rectifier circuit by making measurements.
3. Measure voltage and current values of a filtered bridge rectifier circuit using a multimeter and oscilloscope.

Experiment 4-5: *RC* and Pi-type Filters

1. Construct *RC* and pi-filter circuits.
2. Observe the action of *RC* and pi-filters on a bridge rectifier circuit.
3. Calculate DC output voltage, ripple, ripple factor and voltage regulation for *RC* and pi-filtered power supply circuits.
4. Measure voltage and current values of a power supply circuit with *RC* and pi-filtering by using an oscilloscope and multimeter.

Experiment 4-6: Zener Diode Voltage Regulator

1. Construct a regulated power supply circuit.
2. Observe the action of a zener diode as a voltage regulator for a power supply circuit.
3. Use a multimeter to measure voltage and current values for a regulated power supply circuit.

4. Calculate and compare values of a regulated power supply as load resistance is varied.

Experiment 4-7: Half-wave Voltage Doubler

1. Construct a half-wave voltage doubler circuit.
2. Use an oscilloscope and multimeter to measure voltages at test points of a half-wave voltage doubler circuit.

Experiment 4-8: Full-wave Voltage Doubler

1. Construct a full-wave voltage doubler circuit.
2. Use an oscilloscope and multimeter to measure voltages at test points of a full-wave voltage doubler circuit.
3. Compare full-wave and half-wave voltage doubler circuits.

UNIT 5: Oscillator Circuits

Experiment 5-1: Hartley Oscillator

1. Construct a Hartley oscillator circuit.
2. Use an oscilloscope to observe.
3. Recognize the construction features and operational characteristics of a Hartley oscillator circuit.
4. Observe the effect of capacitance value on the frequency of a Hartley oscillator circuit.
5. Test the feedback achieved by a Hartley oscillator circuit.

Experiment 5-2: Colpitts Oscillator

1. Construct a Colpitts oscillator circuit.
2. Use an oscilloscope to trace the feedback path of a Colpitts oscillator circuit.
3. Recognize the construction features and operational characteristics of a Colpitts oscillator circuit.

Experiment 5-3: Crystal Oscillator

1. Construct a crystal oscillator circuit.
2. Observe waveforms at key test points of a crystal oscillator circuit with an oscilloscope.
3. Use an AM radio receiver to test the crystal oscillator circuit for oscillation.

Experiment 5-4: Astable Multivibrator

1. Construct an astable multivibrator circuit.
2. Use an oscilloscope to observe waveforms at key test points of an astable multivibrator circuit.

3. Recognize the construction features and operational characteristics of an astable multivibrator circuit.
4. Adjust the switching speed of a astable multivibrator circuit by changing component values.

Experiment 5-5: Triggered Multivibrators

1. Construct bistable and monostable multivibrator circuits.
2. Test stable state and conduction time of a transistor multivibrator.
3. Trigger a multivibrator to cause a state change by applying the proper voltage polarity.
4. Compare monostable and bistable multivibrator circuits.

Experiment 5-6: Function Generators

1. Construct a function generator using a 555 integrated circuit timer.
2. Measure frequency by using a calibrated time-base oscilloscope.
3. Observe square wave and triangular wave outputs with an oscilloscope.

UNIT 6: Communications Circuits

Experiment 6-1: CW Transmitters

1. Construct a continuous wave (CW) transmitter with an RF oscillator circuit.
2. Key information into a transmitter.
3. Observe the output of a transmitter with an oscilloscope.

Experiment 6-2: AM Principles

1. Construct diode modulator and AM transmitter circuits.
2. Observe AF and RF signals and resulting modulated waveforms with an oscilloscope.
3. Alter the level of modulation of an AM signal.

Experiment 6-3: AM Receivers

1. Construct an AM receiver circuit.
2. Observe the principles of signal pickup, tuning, demodulation, and reproduction.
3. Observe the primary functions of an AM receiver with an oscilloscope.
4. Tune in an AM signal and test levels of modulation.
5. Use a block diagram of a superheterodyne receiver to make tests.

Experiment 6-4: FM Principles

1. Construct an FM test circuit.
2. Use an oscilloscope to observe the signal at key test points of an FM circuit.
3. Compare AM and FM transmitters.

Experiment 6-5: FM Receivers

1. Use a block diagram of an FM receiver to make tests.
2. Observe the primary functions of an FM receiver.
3. Calculate specific frequencies that appear at selected points in an FM receiver.
4. Compare AM and FM receivers.

Experiment 6-6: Sound Transducers

1. Test headphones to see how they operate.
2. Test a speaker to see how it operates.
3. Use a signal generator to listen to a signal with headphones and a speaker.

Experiment 6-7: Monochrome TV Receivers

1. Become familiar with the basic controls on a monochrome TV receiver and understand their functions.
2. Describe how a picture is produced on a monochrome TV receiver.

Experiment 6-8: Monochrome TV Controls

1. Adjust basic monochrome TV controls and observe their functions.
2. Use a schematic diagram of a monochrome TV set to observe the functional parts of the TV set.

EXPERIMENT 1-1
TRANSISTOR BIASING

In a transistor circuit, the emitter-base junction is forward biased and the base-collector junction reverse biased. Forward biasing is achieved when the emitter is connected to the positive side of the source and the base is connected to the negative side. Forward biasing normally produces low resistance and high current flow. Normally, the collector of a transistor is reverse biased. A reverse-biased junction normally produces high resistance and a very small amount of current flow.

OBJECTIVES

1. To construct a common-base (C-B) transistor amplifier circuit.
2. To measure I_E, I_B and I_C of the circuit.

EQUIPMENT

Multifunction meter (VOM)
NPN transistor
1.5-V dc batteries (2)
Resistors: 220 Ω (2)
SPST switches (2)

PROCEDURE

1. Construct the transistor circuit of Fig. 1-1A.
2. Prepare the meter to measure a dc current flow of approximately 5 mA. Connect it to measure I_B.

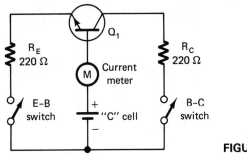

FIGURE 1-1A

53

3. Turn on the emitter-base switch. Emitter-base current flow with the base-collector open is _____ mA.

4. Turn off the emitter-base (E-B) switch and turn on the base-collector (B-C) switch. The base-collector current flow with the emitter-base open is _____ mA. What do steps 3 and 4 demonstrate about transistor operation?

   _____

   _____

5. Turn off both switches and reverse the polarity of the battery and meter.

6. Turn on the emitter-base (E-B) switch. The emitter-base current flow with base-collector (B-C) open is ____ mA.

7. Turn off the emitter-base switch and turn on the base-collector switch. The base-collector current with the emitter base open is _____ mA. What do steps 6 and 7 demonstrate about transistor operation?_____

   _____

8. Turn off both switches and construct the circuit of Fig. 1-1B.

9. Before testing the circuit, how does B_1 bias the emitter-base of transistor?_____

   _____

10. How does B_2 bias the base-collector of transistor?

    _____

    _____

11. Turn on the emitter-base and base-collector switches. Measure and record the voltage across R_E and R_C. V_{R_E} = _____ V; V_{R_C} = _____ V.

12. Calculate emitter current using the value of R_E and measured V_{R_E}: I_E = _____ mA.

13. Using the same procedure, calculate collector current using the value of R_C and measured V_{R_C} voltage: I_C = _____ mA.

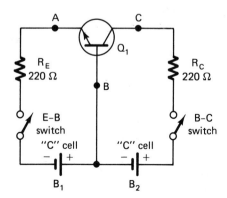

FIGURE 1-1B

ANALYSIS

1. How do you account for the collector current I_C through the reverse-biased base-collector? _____
   _____
   _____

2. Why is I_C less than I_E in steps 12 and 13? _____
   _____
   _____

3. How must the emitter-base and base-collector junctions of a transistor normally be biased? _____
   _____

EXPERIMENT 1-2
TRANSISTOR GAIN

One of the more important characteristics of a transistor is its ability to achieve current gain. The value of this characteristic can be determined by dividing the output current flow by the input current. Since the input and output relationships differ greatly among three different circuit configurations, current gain is based on the specific circuit used. A common-base circuit, for example, has a current gain of less than 1. The current gain of the common emitter and common collector circuits may reach values of 100.

OBJECTIVES

1. To construct common-base (C-B), common-emitter (C-E), and common-collector (C-C) transistor circuits.
2. To measure I_E, I_B and I_C in three different transistor circuit configurations.
3. To calculate the current gains of the three transistor circuit configurations.

EQUIPMENT

Multifunction meter (VOM)
Transistor
1.5-V dc batteries (2)
Potentiometer: 100 KΩ
Resistors: 220 Ω(2)
SPST Switches (2)

PROCEDURE

Part A: Common-Base Circuit

1. Connect the transistor circuit of Fig. 1-2A. Note that test points A, B, and C of this circuit indicate current meter–connection points.
2. Insert a current meter at test point A in Fig. 1-2A.
3. Adjust R_1 to its middle position; then turn on SW_1 (SW_2 remains off). Adjust R_1 to produce a 1.0-mA reading at points A and B.
4. If the circuit is working properly, I_E should equal I_B.

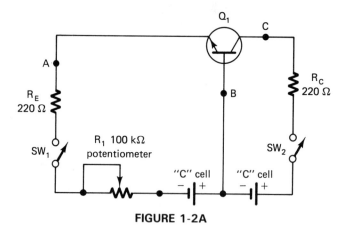

FIGURE 1-2A

Why does I_E equal I_B in this circuit? _____
_____

5. With SW_1 remaining on, turn on SW_2. Record indicated values of I_E and I_B: I_E = _____ mA; I_B = _____ mA. The value of I_B usually is quite small and must be read on a range less than 1 mA.

6. Turn off both SW_1 and SW_2. Insert a current meter at point C.

7. Turn on SW_1 and SW_2. Collector current (I_C) at point C = _____ mA.

8. A check of meter readings should indicate that $I_C + I_B$ = I_E: I_E = _____.

9. Calculate current gain or alpha of this circuit: Alpha = I_C/I_E = _____.

10. Turn off SW_1 and SW_2. Insert the meter at test point A.

11. Turn on SW_1 and SW_2. Both I_E and I_C are to be observed in this circuit. Adjust value of R_2 while observing change in I_E and then I_C. Describe your observations.
_____
_____

12. Assume now that the resistance of the reverse-biased base-collector junction of the transistor circuit is a high resistance value of 100 kΩ or more. Resistance of the forward-biased emitter-base junction is quite low. Assume a typical E-B resistance of 1 kilohm. Voltage gain capability of this transistor can be calculated as

$$A_V = \text{alpha} \times \frac{\text{output resistance}}{\text{input resistance}}$$

A_V for this circuit = _____.

ANALYSIS

1. Why is the alpha of this circuit less than 1?_____

   _____

2. If the value of R_1 of Fig. 1-2A was doubled, what effect would this have on I_C? _____

3. Why does the value of I_B in step 5 decrease when SW_1 and SW_2 are both ON, compared with step 3 where $I_B = I_E$? _____

   _____

Part B: Common-Emitter Circuit

1. Connect transistor circuit of Fig. 1-2B. Test points *A*, *B*, and *C* of circuit are used to indicate milliampere meter connection points. To insert a meter at any of these points, break the circuit and connect the meter in *series* with the two connections. Be sure to observe polarity. Also, after removing the meter from a test point, the circuit path must be reconnected.

2. Connect the meter at test point *B* and then at test point *A*. Turn on SW_1 (SW_2 remains off).

3. Adjust potentiometer R_1 through its range while observing the meters at test point *A* and then at point *B*. This part of the circuit is connected in series.

4. Record the maximum base current (I_B) and emitter current (I_E) readings on the meters: $I_B = $ _____ mA; $I_E = $ _____ mA.

5. Turn on switch SW_2. How does this condition alter values of I_B and I_E? _____

   _____

6. Adjust R_1 to produce an I_B of 0.005 mA or 5μA. Record the value of I_B and I_E: $I_B = $ _____ mA; $I_E = $ _____ mA.

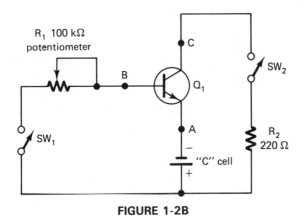

FIGURE 1-2B

7. Turn off SW_1 and SW_2. Remove the meter from test point A and complete the circuit at this point. Insert the meter at test point C.

8. Turn on SW_1 and SW_2. Record the value of I_C: $I_C =$ _____ mA.

9. Current gain or beta of this circuit is determined by dividing output (I_C) by input I_B: beta = _____.

10. Adjust base current to produce a value change of from 5 μA to 4 μA (0.005 mA to 0.004 mA). This change in I_B causes a change in I_C from _____ mA to _____ mA.

11. A changing value of I_B causes a corresponding value change in I_C.

ANALYSIS

1. Why is the beta value higher than the alpha value of the previous circuit (Part A)? _____

_____

2. What are the input and output of this common emitter type of circuit? _____

_____

3. What is meant by the term common emitter amplifier?

_____

_____

Part C: Common-Collector Circuit

1. Connect the transistor ciruit of Fig. 1-2C. Test points A, B, and C of this circuit indicate current meter test points. To insert a meter at these test points, open the circuit path and connect the meter in series.

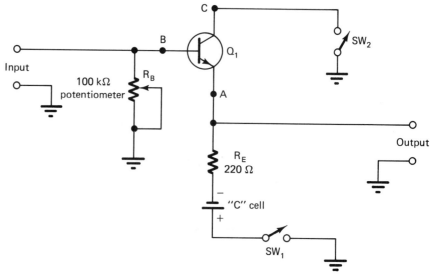

FIGURE 1-2C

2. Prepare the meter to measure 1 mA of current and insert at test point B. This meter will indicate base current (I_B). This meter at test point A will indicate emitter current (I_E).

3. Adjust R_B to its approximate middle position. Turn on SW$_1$ (SW$_2$ remains off). Adjust I_E and I_B to produce a 1-mA reading. Why is the value of I_E and I_B the same?

_____

_____

4. Turn on SW$_2$ (SW$_1$ remains on). If necessary, adjust R_B to produce an I_E of 1.0 mA. It may be necessary to change the meter to a lower current range to measure I_B: I_B = _____ mA.

5. Current gain of this circuit is output current (I_E) divided by input current (I_B): current gain = _____.

6. Carefully decrease the value of I_B by altering the resistance of R_B. Observe the change in I_E that occurs. Describe your findings. _____

_____

7. Return I_B and I_E to their original values. Turn off SW$_1$, and remove the meter from test point B. Insert the meter at test point C.

8. Turn on SW$_1$, and measure collector current (I_C): I_C = _____ mA.

9. Turn off SW$_1$, and remove the meter from test point C.

10. Prepare the meter to measure dc voltage and turn on SW$_1$. Measure the circuit input voltage between base and ground: base input voltage = _____ V.

11. Measure output voltage from emitter to ground with VOM: emitter output voltage = _____ V.

12. Calculate voltage gain of circuit by dividing output voltage by input voltage: voltage gain = _____.

ANALYSIS

1. Using measured values of I_E and I_B from step 4, what is the I_C of this circuit?_____

_____

2. An increase in input (I_B) causes a corresponding _____ in I_E.

3. How does the current gain of the common collector circuit compare with the current gain of common-emitter and common-base circuits? _____

_____

EXPERIMENT 1-3
AMPLIFIER PHASE RELATIONSHIPS

An important characteristic of a transistor amplifier is the phase relationship of the input and output signals. This relationship is based primarily upon the circuit configuration of the amplifier. When you connect a transistor in either a common-base or common-collector configuration, the input and output signals remain in phase. In a common-emitter circuit; however, the input signal is 180° out of phase with the output.

OBJECTIVES

1. To construct common-base (C-B), common-emitter (C-E), and common-collector (C-C) transistor amplifier circuits.
2. To observe the phase and amplitude of the input and output signals of CB, CE, and CC amplifiers and determine voltage gain.
3. To connect the oscilloscope for external synchronization.

EQUIPMENT

Multimeter (VOM)
Oscilloscope
Signal generator
NPN Transistor
CT Transformer
C-type dry cells with holders (2)
Potentiometer: 100 kΩ
Resistors: 220Ω, 1 kΩ
Capacitor: 0.68 μF
SPST switch

PROCEDURE

Part A: Common-Base Phase Relationships

1. Construct the common-base circuit of Figure 1-3A. Ground symbols of diagram must all be connected together in this circuit.
2. Turn on the signal generator and adjust it to 1 kHz. Reduce signal amplitude control to minimum.

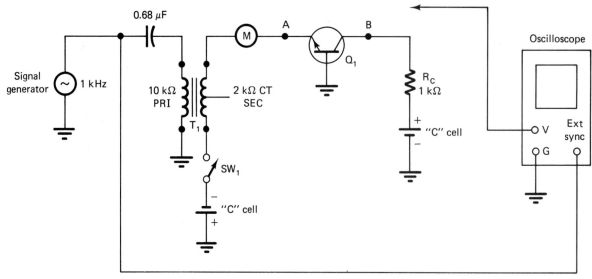

FIGURE 1-3A

3. Turn on the oscilloscope and prepare it for operation. Set the trigger source to the external sync position. Connect the external sync input of the oscilloscope to the signal generator as indicated in Figure 1-3A.

4. Connect the vertical probe of oscilloscope to the output of the transistor at test point *B*. Adjust the signal generator to produce a sine wave of maximum amplitude with a minimum of distortion. Adjust the oscilloscope to display three or four complete sine waves.

5. Adjust the horizontal position control of oscilloscope to display the beginning edge or start of the wave. Make a sketch of the observed waveform in the part of Figure 1-3B marked "output." The sketch should show the phase and peak-to-peak amplitude of the signal.

6. Move the vertical probe of the oscilloscope to the transistor input at test point *A*.

7. Make a sketch of the observed wave in the part of Figure 1-3B marked "input." Indicate peak-to-peak amplitude and phase of the wave.

Test point B _____ (output)

Test point A _____ (input)

B _____ $V_{p\text{-}p}$; _____ Phase

A _____ $V_{p\text{-}p}$; _____ Phase

FIGURE 1-3B

8. Using the peak-to-peak input and output values of Figure 1-3B, calculate voltage gain: voltage gain of the common base circuit = _____.

ANALYSIS

1. Describe the phase relationship of the input and output signals. _____

2. Explain what occurs in the common-base amplifier when the positive alternation is applied to the input and how it produces a resulting output. _____

3. Explain what occurs when the negative alternation of a sine wave is applied to the input of a common-base amplifier, and how it produces a resulting output.

Part B: Common-Emitter Phase Relationships

1. Construct the common-emitter amplifier circuit of Figure 1-3C.
2. Adjust potentiometer R_1 to the center of its range. Adjust R_1 to produce 0.02 mA (20 μA) of base current.
3. Turn on and adjust the signal generator to produce 1 kHz. Adjust the signal amplitude control to minimum.
4. Prepare the oscilloscope for operation. Set the trigger source to the external sync position. Connect the external sync of the oscilloscope to the signal generator as indicated in Figure 1-3C.
5. Connect the vertical probe of the oscilloscope to the out-

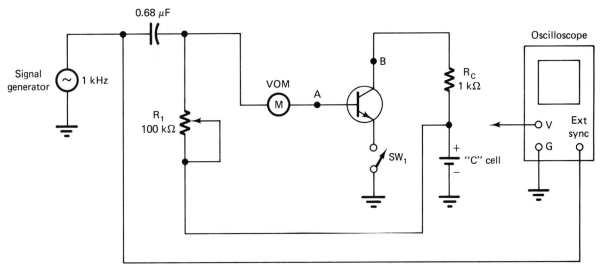

FIGURE 1-3C

put of the transistor at test point *B*. Adjust the signal generator to produce a sine wave of maximum amplitude with minimum distortion. Adjust the oscilloscope to display three or four complete sine waves.

6. Adjust the horizontal position control of the oscilloscope to display the beginning edge or start of wave. Make a sketch of the observed wave in the part of Figure 1-3D marked "output." The sketch should show the phase and peak-to-peak amplitude of the signal.

7. Move the vertical probe of the oscilloscope to the transistor input at test point *A*.

8. Make a sketch of the observed wave in the part of Figure 1-3D marked "input." Indicate the peak-to-peak amplitude and phase of the waveform.

9. Using peak-to-peak input and output values of Figure 1-3D, calculate voltage gain: voltage gain of common-emitter circuit = _____.

ANALYSIS

1. Describe the input and output phase relationship of the common-emitter circuit. _____

2. Explain what occurs in the common-emitter amplifier when the positive alternation is applied to the input and how it produces a resulting output. _____

3. Explain what occurs when the negative alternation of a sine wave is applied to the input of a common-emitter amplifier, and how it produces a resulting output.

Part C: Common Collector Phase Relationships

1. Construct the common collector circuit of Figure 1-3E.

2. Adjust potentiometer R_1 to its center position. Turn on

Test point B (output) _____   Test point A (input) _____

B _____ $V_{p\text{-}p}$; _____ Phase   A _____ $V_{p\text{-}p}$; _____ Phase

FIGURE 1-3D

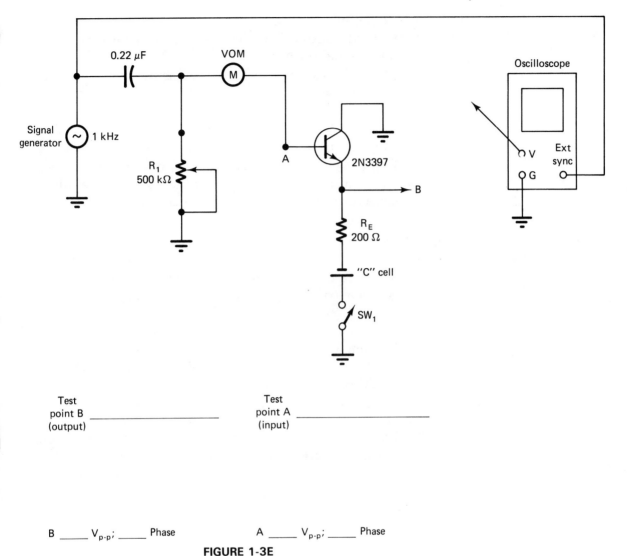

Test
point B _____
(output)

Test
point A _____
(input)

B _____ V_{p-p}; _____ Phase   A _____ V_{p-p}; _____ Phase

FIGURE 1-3E

the circuit switch and adjust R_1 to produce 0.02 mA (20 µA) of base current.

3. Turn on and adjust the signal generator to 1 kHz. Adjust the signal amplitude control to its minimum value.

4. Prepare the oscilloscope for operation. Set the trigger source to the external sync position. Connect the external sync of the oscilloscope to signal generator as indicated in Figure 1-3E.

5. Connect the vertical probe of the oscilloscope to the transistor output at test point *B*. Adjust the signal generator to produce a sine wave of maximum amplitude and minimum distortion. The display should show three or four complete sine waves.

6. Adjust the horizontal position of the oscilloscope to display the beginning edge or start of the wave. Make a

sketch of the observed wave in the part of Figure 1-3E marked "output." Indicate the peak-to-peak signal amplitude and phase.

7. Move the vertical probe of the oscilloscope to the transistor input at point *A*.

8. Make a sketch of the observed wave in the part of Figure 1-3E marked "input." Indicate the peak-to-peak amplitude and phase.

9. Using peak-to-peak input and output values of Figure 1-3E, calculate voltage gain: voltage gain of the common collector circuit = _____.

ANALYSIS

1. Describe the input and output phase relationship of the common-collector circuit. _____

_____

_____

2. Explain what occurs in the common-collector circuit when a positive alternation is applied to the input, and how it produces a resulting output. _____

_____

_____

3. Explain what occurs in common-collector amplifier when a negative alternation of a sine wave is applied to the input and how it develops a resulting output. _____

_____

_____

EXPERIMENT 1-4
TRANSISTOR BIASING METHODS

When a transistor is placed in a circuit, it requires a certain amount of bias voltage to operate. The emitter-base junction must be *forward biased* and the collector *reverse biased*. How this bias voltage is achieved is an important consideration in the operation of an amplifier.

Fixed biasing is accomplished by connecting resistors to the same side of the dc source while the emitter is connected to the opposite side of the source.

The *voltage-divider* method of biasing employs resistors connected in series across the dc source. Appropriate bias voltages are developed according to the ratio of resistance in the divider network.

Emitter biasing develops a constant current through an emitter resistor. The base and collector voltages must be in excess of the emitter voltage in order to be of proper polarity and voltage value.

OBJECTIVES

1. To construct transistor circuits that have fixed, voltage divider, and self-biasing methods used.
2. To measure and observe the voltage and polarity of the three methods of biasing.

EQUIPMENT

Multimeter (VOM)
Variable power supply, 0-15 V
NPN transistor
Resistors: 220 Ω, 1 kΩ, 4.7 kΩ, 10 kΩ
SPST Switch

PROCEDURE

1. Construct the fixed bias circuit of Figure 1-4A.
2. Before turning on the circuit switch, adjust the dc power supply to produce 5 V.
3. Prepare the multimeter to measure dc voltage. With the negative lead connected to ground, measure emitter-base

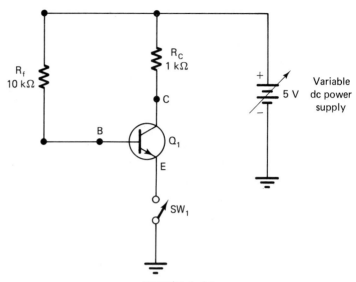

FIGURE 1-4A

(E-B) and emitter (E-C) voltage, noting the value and polarity: E-B = _____ V and
_____ polarity; E-C = _____ V
and _____ polarity.

4. Turn off the circuit switch. Alter the circuit to conform
with the voltage divider biasing method of Figure 1-4B.

5. Turn on the circuit switch and measure E-B and E-C
voltages, noting value and polarity: E-B = _____ V
and _____ polarity; E-C = _____
V and _____ polarity.

6. Turn off the circuit switch. Alter the circuit to conform
with the self bias circuit of Figure 1-4C.

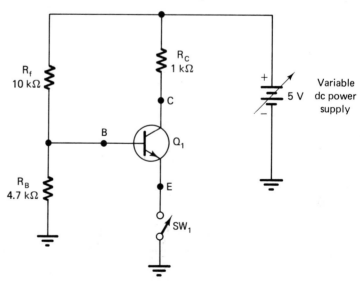

FIGURE 1-4B

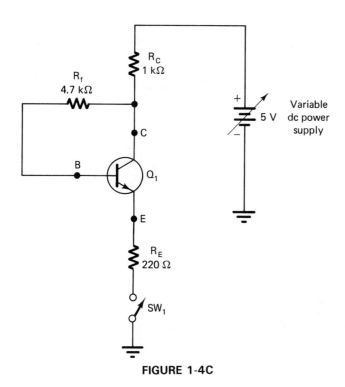

FIGURE 1-4C

7. Turn on the circuit switch and connect the negative meter lead to ground. Measure E-G, B-G, and C-G voltage while noting polarity: E-G = _____ V with a _____ polarity; B-G = _____ V with a _____ polarity; C-G = _____ V with a _____ polarity.

8. E-B voltage of this circuit is determined by subtracting E-G from B-G: E-B voltage = _____ V.

9. E-C voltage of this circuit = C-G minus E-G: E-C voltage = _____ V.

10. To verify these values, connect the negative lead of the meter to the emitter (E) and measure E-B and E-C. Note the polarity of the voltages: B- _____ with respect to *E*; C- _____ with respect to *E*.

ANALYSIS

1. When biasing an NPN transistor, the base and collector must be _____ with respect to the emitter.

2. In the self-bias circuit of Figure 1-4B, how is the **proper**

bias voltage polarity achieved when the E-B-C are all positive? _____

_____

3. How would the three circuits of this experiment change if a PNP transistor were used? _____

_____

EXPERIMENT 1-5
AMPLIFIER CLASSES

Transistor amplifiers are frequently classified according to the relationships of the input and output signals. In this regard, there is class A, B, or C amplification. Amplifier classes are primarily determined by the bias operating point of the circuit.

When an amplifier is operated *class A*, it produces a full sine-wave output when input receives a sine wave: Bias operating point usually is well above the cutoff point. A typical class A operating point is midway between cutoff and saturation points.

Class B amplifiers have a sine-wave input and produce an amplified half-wave output. The bias operating point of this amplifier usually is adjusted to the cutoff point. With no signal applied, there is no current consumed by the transistor.

A *Class C* amplifier has its bias operating point adjusted to a value well below transistor cutoff. In operation only the positive-most part of the input produces an output. A class C amplifier has sine wave input and less than half sine-wave output.

OBJECTIVES

1. To construct class A, B, and C amplifiers circuits.
2. To adjust the bias operating point of a transistor circuit by changing the relationship of the V_B and V_E voltages.
3. To observe input and output waveforms of the three different amplifier classes.

EQUIPMENT

Multimeter (VOM)
Variable power supply, 0–15 V
Signal generator
Oscilloscope
NPN transistor
Capacitor: 100 μF
Resistors: 1 kΩ, 4.7 kΩ, 10 kΩ
Potentiometer: 5 kΩ
C cell with holder
SPST switch

PROCEDURE

1. Connect the common-emitter circuit of Figure 1-5A.
2. Turn on the signal generator. Adjust it to produce a 1-kHz signal with a low amplitude output.
3. Turn on the dc power supply, and adjust it to produce 5 V. Then turn on the circuit switch.
4. Prepare the oscilloscope for operation. Connect the vertical probe to the base (test point B) and the common. Set the horizontal sweep time to 0.1 ms/cm or to a horizontal sweep frequency range of 1 kHz. Set the trigger or sync selector switch to external sync. Connect the sync input to the collector (C). Adjust the display to produce two or three sine waves.
5. Move the vertical probe of the oscilloscope to the collector (test point C).
6. Prepare the meter to measure dc emitter voltage across R_3. Adjust R_3 to produce 0.3 V of emitter voltage with respect to ground.
7. Measure the base voltage across R_2. This value should be approximately 0.9 V dc.
8. V_{BE} voltage is determined by subtracting V_E from V_B:
   V_{BE} = _____ V.
9. Adjust the amplitude control of the signal generator to produce the maximum output signal with least distortion.
10. Make a sketch of the observed output for a class A amplifier in the space of Figure 1-5B.
11. Move the vertical probe of the oscilloscope to the input

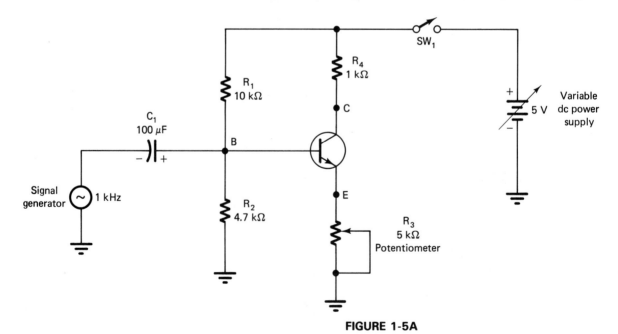

FIGURE 1-5A

Class A
output _____

Signal
input ____ _____
point B

Class C
output _____

Class B
output _____

FIGURE 1-5B

(test point *B*). Make a sketch of the observed input in the space of Fig. 1-5B.

12. Connect the meter across the 5-kΩ potentiometer and adjust the emitter voltage to 0 V. Measure and record base voltage across R_2: V_B = _____ V.

13. Calculate V_{BE} voltage: V_{BE} = _____ V.

14. Connect the vertical probe of the oscilloscope to the collector again and observe the output of a class B amplifier. Make a sketch of the wave form in the space of Fig. 1-5B.

15. Turn off the circuit switch and disconnect the 5-kΩ potentiometer. In its place, connect a C cell with positive attached to emitter and negative to ground.

16. Turn on the circuit switch. Connect the vertical probe to the collector, and adjust the signal-generator amplitude control to produce part of a sine wave. Make a sketch of the observed wave form in the space provided for class C output of Figure 1-5B.

17. With a meter, measure and record V_E and V_B voltage:
    V_E = _____ V; V_B = _____ V;
    V_{BE} = _____V.

ANALYSIS

1. With a sine wave applied, what is the output of a class A amplifier? _____ class B? _____ class C? _____

2. Where must the operating point be adjusted to achieve class A, B, and C amplification?
   Class A- _____
   Class B- _____
   Class C- _____
   Class AB- _____

3. How does adjustment of the emitter resistance change the bias operating point? _____
   _____

UNIT 1 EXAMINATION
AMPLIFIER CIRCUITS

Instructions: For each of the following, circle the answer that most correctly completes the statement.

1. The ability of a transistor amplifier to compensate for temperature changes is known as:
   a. Positive feedback
   b. Regenerative feedback
   c. Thermal instability
   d. Thermal stability

2. When observing an amplifier output on an oscilloscope, you notice that the signal is an exact reproduction of the input. The class of amplifier you are observing is:
   a. A b. AB
   c. B d. C

3. A typical value of input impedance for a common-emitter amplifier:
   a. 1 kΩ
   b. 100 kΩ
   c. 1 MΩ
   d. 100 MΩ

4. The class of amplifier that produces a minimum amount of distortion is:
   a. A b. AB
   c. B d. C

5. The class of amplifier that is biased below cutoff with no input signal applied is:
   a. A b. AB
   c. B d. C

6. The point at which the input signal to a transistor causes collector current to stop flowing is called:
   a. Critical point
   b. Saturation point
   c. Cutoff point
   d. Complementary point

7. If an increase in base current fails to cause an increase in collector current, a transistor amplifier is:
   a. Critical
   b. Saturated

c. Cutoff

d. Complemented

8. The class of amplifier which is cutoff 50% of the time that the input signal applied is:

   a. *A*   b. *AB*
   c. *B*   d. *C*

9. In a transistor amplifier, as temperature rises:

   a. R_L increases
   b. R_L decreases
   c. I_C increases
   d. I_C decreases

10. The relationship that best describes the current flowing through R_1 in Figure E-1-10 is:

    a. $I_1 = I_2 + I_E$
    b. $I_1 = I_2 - I_E$
    c. $I_1 = I_2 + I_B$
    d. $I_1 = I_2 - I_B$

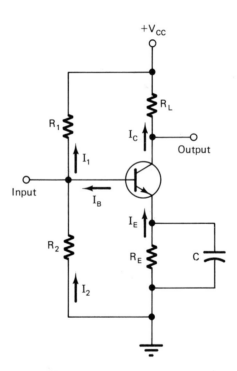

FIGURE E-1-10

11. The emitter-base voltage drop of a forward biased silicon transistor is:

    a. 0.2–0.3 V
    b. 0.4–0.5 V
    c. 0.6–0.7 V
    d. Over 1 V

12. The circuit shown in Fig. E-1-12 is a:

    a. Common-emitter amplifier with voltage-divider bias

b. Common-base amplifier with voltage-divider bias
c. Common-collector amplifier with voltage-divider bias
d. Common-collector amplifier with collector feedback.

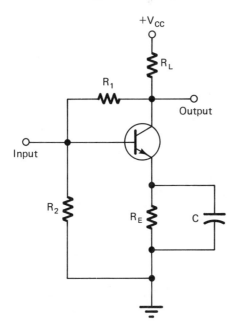

FIGURE E-1-12

13. The circuit shown in Fig. E-1-13 is a:
   a. Common-emitter amplifier with voltage-divider bias
   b. Common-base amplifier with voltage-divider bias
   c. Common-collector amplifier with voltage-divider bias
   d. Common-collector amplifier with collector feedback.

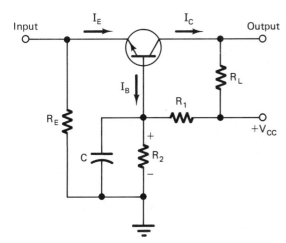

FIGURE E-1-13

14. The circuit shown in Fig. E-1-14 is a:
   a. Common-emitter amplifier with voltage-divider bias
   b. Common-base amplifier with voltage-divider bias
   c. Common-collector amplifier with voltage-divider bias
   d. Common-collector amplifier with collector feedback.

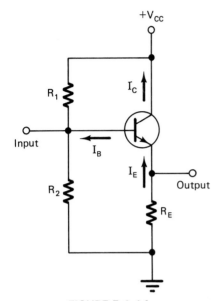

FIGURE E-1-14

FIGURE E-1-15

15. In the circuit of Fig. E-1-15, the formula that would be used to calculate the base voltage is:
    a. $I_{R1} \times R_1$
    b. $I_{R2} \times R_2$
    c. $I_{R3} \times R_3$
    d. $I_{R4} \times R_4$

16. In the circuit of Figure E-1-16, the resistors which determine the transistor's bias are:
    a. R_1 and R_2
    b. R_2 and R_3
    c. R_3 and R_4
    d. R_4 and R_1

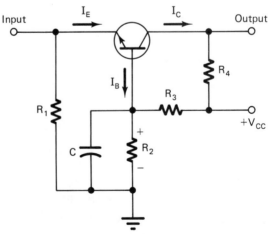

FIGURE E-1-16

17. Normally, a transistor amplifier has its:
    a. Collector and emitter junctions forward biased
    b. Collector and emitter junctions reversed biased
    c. Collector junction forward biased and emitter junction reverse biased
    d. Collector junction reverse biased and emitter junction forward biased

18. A common-emitter amplifier has a:
    a. Low-input impedance and low-output impedance
    b. Low-input impedance and high-output impedance
    c. High-input impedance and high-output impedance
    d. High-input impedance and low-output impedance

19. A common-collector circuit produces a large current gain because it provides:
    a. A high-input and low-output impedance
    b. A low-input and high-output resistance
    c. A low-input and low-output resistance
    d. A high-input and high-output resistance

20. A typical input impedance of a common-collector amplifier is:
    a. 1 Ω
    b. 100 Ω
    c. 1 kΩ
    d. 100 kΩ

21. The type of amplifier configuration that has its input applied to the emitter and base and its output taken across the collector and base is the:
    a. Common-base
    b. Common-emitter
    c. Common-collector
    d. Emitter follower

22. A typical input impedance of a common-base amplifier is:
    a. 5 MΩ
    b. 500 kΩ
    c. 50 kΩ
    d. 50 Ω

23. The type of amplifier configuration that has its input applied to the base and emitter and its output taken across the collector and emitter is the:
    a. Common-base
    b. Common-emitter
    c. Common-collector
    d. Emitter follower

24. The current gain of a common-emitter amplifier is calculated as:
    a. Gain $= \Delta I_B / \Delta I_E$
    b. Gain $= \Delta I_B / \Delta I_C$
    c. Gain $= \Delta I_E / \Delta I_C$
    d. Gain $= \Delta I_C / \Delta I_B$

25. A common-base circuit has a:
    a. Low-input impedance and low-output impedance
    b. Low-input impedance and high-output impedance
    c. High-input impedance and high-output impedance
    d. High-input impedance and low-output impedance

UNIT TWO

Amplifying Systems

UNIT INTRODUCTION

Amplifying systems are widely used in the electronics field today. This particular type of system may be only one part or section of a rather large and complex system. A television receiver is a good example of this application. Several individual amplifying systems are included in a TV receiver. A phonograph, by comparison, has only one amplifying system. Its primary function is to process a sound signal and build it to a level that will drive a speaker. The function of an *amplifier* is basically the same regardless of its application.

In this unit we investigate how amplifying devices are used in an operating system. Transistors and integrated circuits can be used for this type of operation. A number of specific circuit functions are also discussed. These include gain, coupling, load devices, output circuits, transducers, and signal processing. The amplifying devices and their circuit components play an important role in the operation of a system.

UNIT OBJECTIVES

Upon completion of this unit you will be able to:

1. Identify basic coupling techniques and explain advantages and disadvantages of each type.
2. Identify different types of power amplifiers.
3. Explain how volume and tone controls function in audio amplifiers.
4. Construct a dc amplifier.
5. Explain the function and operation of dc, audio, video, power, RF, and IF amplifiers.
6. Explain the operation of a Darlington arrangement.
7. Explain how beta can be increased with the Darlington circuit.

IMPORTANT TERMS

Amplitude modulation (AM). A method of transmitter modulation achieved by varying the strength of the radio-frequency (RF) carrier at an audio signal rate.

Audio frequency (AF). A range of frequency from 15 Hz to 15 kHz, the human range of hearing.

Bel (B). A measurement unit of gain that is equivalent to a 10:1 ratio of power levels.

Cascade. A method of amplifier connection in which the output of one stage is connected to the input of the next amplifier.

Characteristic. The magnitude range of a number of logarithms.

Complementary transistors. PNP and NPN transistors with identical characteristics.

Coupling. A process of connecting active components.

Crossover distortion. The distortion of an output signal at the point where one transistor stops conducting and another conducts, in class B amplifiers.

Crystal microphone. Input transducer that changes sound into electrical energy by the piezoelectric effect.

Darlington amplifier. Two transistor amplifiers cascaded together.

Decibel (dB). One-tenth of a bel; used to express gain or loss.

Dynamic microphone. An input transducer that changes sound into electrical energy by moving a coil through a magnetic field.

Impedance. An opposition to ac measured in ohms.

Impedance ratio. The ac resistance ratio between the input and output of a transformer.

Logarithm. A mathematical expression dealing with exponents showing the power to which a value called the base must be raised in order to produce a given number.

Mantissa. Decimal part of a logarithm.

Midrange. A speaker designed to respond to audio frequency in the middle part of the AF range.

Peak to peak (p-p). An ac value measured from the peak positive to the peak negative alternations.

Piezoelectric effect. The property of a crystal material that produces voltage by changes in shape or pressure.

Push-pull. An amplifier circuit configuration using two active devices with the input signal being of equal amplitude and 180° out of phase.

Read head. A signal pickup transducer that generates a voltage in a tape recorder.

Root mean square (RMS). Effective value of ac, or 0.707 X peak value.

Speaker. A transducer that converts electric current and/or voltage into sound waves. Also called a loudspeaker.

Stage of amplification. A transistor or IC amplifier and all the components needed to achieve amplification.

Stereophonic (stereo). Signals developed from two sources that give the listener the impression of sound coming into both ears from different points.

Stylus. A needlelike point that rides in a phonograph record groove.

Transducer. A device that changes energy from one form to another. A transducer can be on the input or output.

Tweeter. A speaker designed to reproduce only high frequencies.

Voice coil. The moving coil of a speaker.

Voltage divider. A combination of resistors that produce different or multiple voltage values.

Woofer. A speaker designed to reproduce low audio frequencies with high quality.

AMPLIFYING SYSTEM FUNCTIONS

Regardless of its application, an amplifying system has a number of primary functions that must be performed. An understanding of these functions is an extremely important part of operational theory. A block diagram of an amplifying system is shown in Fig. 2-1. The triangular-shaped items of the diagram show where amplification is performed. A *stage* of amplification is represented by each triangle. Three amplifiers are included in this particular system. A stage of amplification consists of an active device and all its associated components. *Small-signal amplifiers* are used in the first three stages of this system. The amplifier on the right side of the diagram is an output stage. A rather large signal is needed to control the output amplifier. An output stage is generally called a *large-signal amplifier,* or a *power amplifier.* In effect, this amplifier is used to control a rather large amount of current and voltage. Remember that power is the product of current and voltage.

In an amplifying system, a signal must be developed and applied to the input. The source of this signal varies a great deal with different systems. In a *stereophonic (stereo)* phonograph, the signal is generated by a phonograph cartridge. Variations in the groove of a record are changed into electrical energy. This signal is then applied to the input of the system. The amplitude level of the signal is increased to a suitable value by the amplifying devices. A variety of different input signal sources may be applied to the input of an amplifying system. In other systems the signal may also be developed by the input. Transducers are responsible for this function. An *input transducer* changes energy of one form into energy of a different type. Microphones, phonograph pickup cartridges, and tape-recorder heads are typical input transducers. Input signals may also be received through the air. Antenna coils and networks serve as the input transducer for this type of sys-

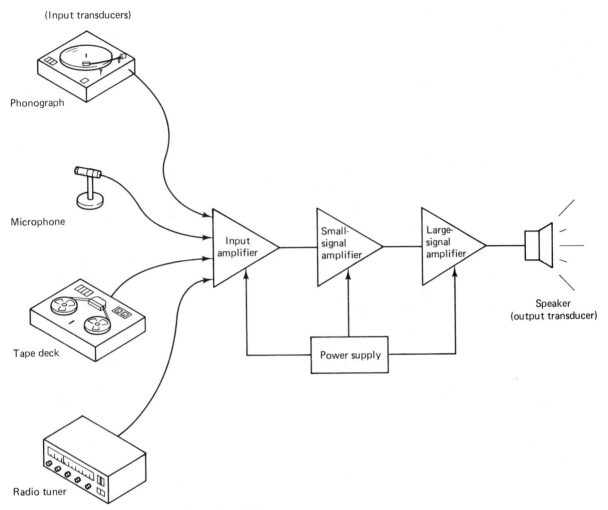

(Input transducers)

Phonograph

Microphone

Tape deck

Radio tuner

Input amplifier

Small-signal amplifier

Large-signal amplifier

Power supply

Speaker (output transducer)

FIGURE 2-1 Amplifying system.

tem. An antenna changes electromagnetic waves into radio-frequency (RF) voltage signals. The signal is then processed through the remainder of the system.

Signals processed by an amplifying system are ultimately applied to an *output transducer.* This type of transducer changes electrical energy into another form of energy. In a sound system, the speaker is an output transducer. It changes electrical energy into sound energy. Work is performed by the speaker when it achieves this function. Lamps, motors, relays, transformers, and inductors are frequently considered to be transducers. An output transducer is also considered to be the load of a system.

For an amplifying system to be operational, it must be supplied with electrical energy. A dc power supply performs this function. A relatively pure form of dc must be supplied to each amplifying device. In most amplifying systems, ac is the primary energy source. Ac is changed into dc, filtered, and—in some systems—regulated before being applied to the amplifiers. The reproduction quality of the amplifier is very dependent on the dc supply voltage. Some portable stereo amplifiers may be energized by batteries.

AMPLIFIER GAIN

The *gain* of an amplifier system can be expressed in a variety of ways. Voltage, current, power, and—in some systems—decibels are expressed as gain. Nearly all input amplifier stages are voltage amplifiers. These amplifiers are designed to increase the voltage level of the signal. Several voltage amplifiers may be used in the front end of an amplifier system. The voltage value of the input signal usually determines the level of amplification being achieved.

A three-stage voltage amplifier is shown in Fig. 2-2. These three amplifiers are connected in *cascade*, which refers to a series of amplifiers where the output of one stage is connected to the input of the next amplifier. The voltage gain of each stage can be observed with an oscilloscope. The waveform shows representative signal levels. Note the voltage-level change in the signal and the amplification factor of each stage.

The first stage has a voltage gain of 5 V. With 0.25 V p-p input, the output is 1.25 V p-p. The second stage also has a voltage gain of 5. With 1.25 V p-p input, the output is 6.25 V p-p. The

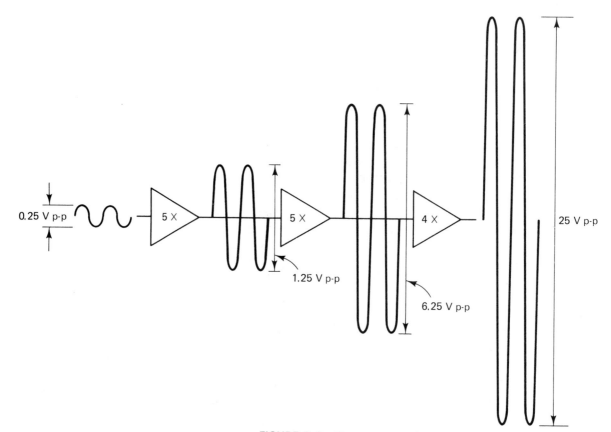

FIGURE 2-2   Three-stage voltage amplifier.

output stage has a gain factor of 4. With 6.25 V p-p input, the output is 25 V p-p.

The total gain of the amplifier is 5 × 5 × 4, or 100. With 0.25 V p-p input, the output is 25 V p-p. Note that output is the product of the individual amplifier gains. It is not just the addition of 5 + 5 + 4. *Voltage gain* (A_V) is an expression of output voltage (V_{out}) divided by input voltage (V_{in}). For the amplifier system, A_V is expressed by the formula

$$A_V = \frac{V_{out}}{V_{in}} = \frac{25 \text{ V p-p}}{0.25 \text{ Vp-p}} = 100$$

Note that the units of voltage cancel each other in the problem. Voltage gain is therefore expressed as a pure number. It is not correct to say that the voltage gain of an amplifier is 100 V p-p. The voltage gain is described by a unitless value such as 100.

Power gain is generally used to describe the operation of the last stage of amplification. Power amplification (A_p) is equal to power output (P_{out}) divided by power input (P_{in}). If the last stage of amplification in Fig. 2-2 were a power amplifier, its gain would be expressed in watts rather than volts. In this case, the gain would be

$$A_P = \frac{P_{out}}{P_{in}} = \frac{25 \text{ W}}{6.25 \text{ W}} = 4$$

Note that the power units of this problem also cancel each other. Power amplification is expressed as a pure number such as 4.

DECIBELS

The human ear does not respond to sound levels in the same manner as an amplifying system. An amplifier, for example, has a linear rise in signal level. An input signal level of 1 V can produce an output of 10 V. The voltage amplification then is 10:1, or 10. The human ear, however, does not respond in a linear manner. It is, essentially, a nonlinear device. As a result of this, sound-amplifying systems are usually evaluated on a *logarithmic* scale. This is an indication of how our ears actually respond to specific signal levels. Gain expressed in logarithms is much more meaningful than linear gain relationships.

The logarithm of a given number is the power to which another number, called the *base,* must be raised to equal the given number. Basically, a logarithm has the same meaning as an *exponent.* A *common logarithm* is expressed in powers of 10. This is illustrated by the following:

$$10^3 = 1000$$
$$10^2 = 100$$
$$10^1 = 10$$
$$10^0 = 1$$

This means that the logarithm of any number between 1000 and 9999 has a *characteristic* value of 3. The characteristic is an expression of the magnitude range of the number. Numbers between 100 and 999 have a characteristic of 2. Numbers between 10 and 99 have a characteristic of 1. Between 1.0 and 9.0 the characteristic is 0. Numbers less than 1.0 have a negative characteristic. It is generally not customary in electronics to use negative characteristic values.

When a number is not an even multiple of 10, its log is a decimal. The decimal part of the logarithm is called the *mantissa*. The log of a number such as 4000 is expressed as 3.6021. The characteristic is 3 because 4000 is between 1000 and 9999. The mantissa of 4000 is 0.6021. The mantissa is found from a table of common logarithms (see Appendix F). To find this value, simply search down the vertical number column (N) to find the number 40. Moving horizontally to the "0" column, the mantissa for 400 is 6021. The log of 4000 is therefore 3.6021. The characteristic is 3 and the mantissa is 0.6021.

The mantissa is always the same for numbers that are identical except for location of the decimal point. For example, the mantissa is the same for 1630, 163.0, 16.3, 1.63. The only difference in these values is the characteristic. The mantissa for 1.63 is 0.2122. The logs of the five values are 3.2122, 2.2122, 1.2122, and 0.2122. Practice changing several number values into logarithms using the common logarithm table.

If an electronic calculator is available with logarithms, the conversion process is very easy. Simply place the number into the calculator. Then press the log button. For example, the log of 1590 is 3.2013971. Try this operation on a calculator. Use a variety of numbers, such as 2503., 603.8, 14.32, 8.134, and 0.2461.

Consider now the gain of a sound system with several stages of amplification. *Gain* is best expressed as a ratio of two signal levels. Specifically, gain is expressed as the output level divided by the input level. This is determined by the expression

$$\text{Power amplification} = \log \frac{P_{out}}{P_{in}} = \text{bels}$$

where the *bel* (B) is the fundamental unit of sound-level gain. For an amplifier with 0.1 W of input and 100 W of output,

$$A_p = \log \frac{100 \text{ W}}{0.1 \text{ W}} = \log 1000 = 3 \text{ bels}$$

As you can see, the bel represents a rather large ratio in sound level. A *decibel* (dB) is a more practical measure of sound level. A decibel is one-tenth of a bel. The bel is named for Alexander Graham Bell, the inventor of the telephone.

The gain of a single stage of amplification within a system can be determined in decibels. A single amplifier stage might have an input of 10 mW and an output of 150 mW. The *power gain* is determined by the formula

$$\text{Power amplification (dB)} = 10 \log \frac{P_{out}}{P_{in}}$$

$$= 10 \log \frac{150 \text{ mW}}{10 \text{ mW}}$$

$$= 10 \log \ 15$$

$$= 10 \times 1.1761$$

$$= 11.761 \text{ dB}$$

The *voltage gain* of an amplifier can also be expressed in decibels. In order to do this the power-level expression must be adapted to accommodate voltage values. The voltage gain formula is

$$\text{voltage amplification (dB)} = 20 \times \log \frac{\text{voltage output}}{\text{voltage input}}$$

$$= 20 \log \ \frac{V_{out}}{V_{in}}$$

Notice that the logarithm of V_{out}/V_{in} is multiplied by 20 in this equation. Power is expressed as V^2/R. Power gain using voltage and resistance values is therefore expressed as

$$\text{dB} = \frac{V_{out}^2 \times R_{out}}{V_{in}^2 \times R_{in}}$$

If the value of R_{in} and R_{out} are equal, the equation is simplified to be

$$\text{dB} = \frac{V_{out}{}^2}{V_{in}{}^2}$$

The squared voltage values can be expressed as two times the log of the voltage value. Decibel *voltage gain* therefore becomes

$$\text{dB} = 2 \times 10 \log \frac{V_{out}}{V_{in}} \quad \text{or} \quad 20 \log \frac{V_{out}}{V_{in}}$$

It is important to remember that decibel voltage gain assumes the values of R_{in} and R_{out} to be equal. Decibel gain is primarily an expression of power levels. Voltage gain is therefore only an adaptation of the power-level expression.

To demonstrate the use of the dB voltage gain equation, let us apply it to the first amplifier of Fig. 2-2. Note that the input voltage is 0.25 V p-p and the output voltage is 1.25 V p-p. The voltage gain in dB is

$$\text{dB} = 20 \log \frac{V_{out}}{V_{in}}$$

$$= 20 \log \frac{1.25 \text{ V p-p}}{0.25 \text{ V p-p}}$$

$$= 20 \log 5$$

$$= 20 \times 0.69897$$

$$= 13.9794, \text{ or } 14$$

Using the procedure just described, determine the gain of the second amplifier stage. Determine the gain of the output amplifier stage. The total voltage gain of the three amplifiers is the sum of the individual decibel voltage values. To check the accuracy of your calculations, determine the total voltage gain of amplifier using 0.25 as V_{in} and 25 as V_{out}. This value should equal that found by adding individual-stage decibel values.

AMPLIFIER COUPLING

In many systems one stage of amplification will not provide the desired level of signal output. Two or more stages of amplification are therefore connected together to increase the overall gain. These stages must be properly coupled for the system to be operational. *Coupling* refers to a method of connecting amplifiers together. The signal being processed is transferred between stages by the coupling component.

Several methods of interstage amplifier coupling are in use today. Three very common methods include *capacitive coupling*, *direct coupling*, and *transformer coupling*. Each type of coupling has specific applications. In most amplifying devices, the output voltage is greater than the input voltage. If the output of one stage is connected directly to the input of the next stage, this voltage difference can cause a problem. Signal distortion and component damage might take place. Proper coupling procedures reduce this type of problem.

Capacitive Coupling

Capacitive coupling is particularly useful when amplifier systems are designed to pass ac signals. A capacitor will pass an ac signal and block dc voltages. The capacitor selected must have low capacitive reactance (X_C) at its lowest operating frequency. This is done to ensure amplification over a wide range of frequency. In general, the capacitive reactance of a good capacitor is extremely high at low frequency. At zero frequency or dc, X_C is nearly infinite. Capacitive coupling has some difficulty in passing low-frequency ac signals. Large capacitance values are selected when good low-frequency response is desired.

A two-stage amplifier employing capacitor coupling is shown in Fig. 2-3. Transistor Q_1 is the input amplifier. Its collector voltage is 8.5 V. At the base of the second amplifier stage (Q_2), the voltage is approximately 1.7 V. The voltage difference across C_1 is 8.5 V − 1.7 V, or 6.8 V. The capacitor isolates these two operating voltages. It must have a dc working voltage value that will withstand this difference in potential.

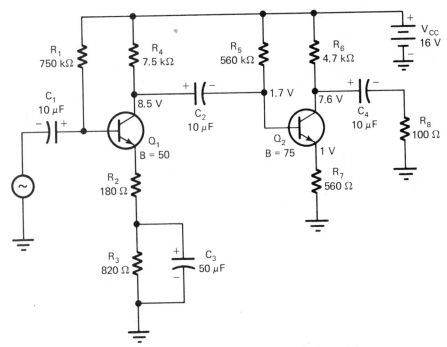

FIGURE 2-3   Two-stage capacitor-coupled amplifier.

The value of a coupling capacitor is a very important circuit consideration. In low-frequency amplifying systems, large values of electrolytic capacitors are normally used. These capacitors respond well to low-frequency ac. In high-frequency amplifier applications, small capacitor values are very common. Selection of a specific coupling capacitor for an amplifier is dependent on the frequency being processed.

Direct Coupling

In *direct coupling* the output of one amplifier is connected directly to the input of the next stage. In a circuit of this type a connecting wire or conductor couples the two stages together. Circuit design must take into account device voltage values. This type of coupling is an outgrowth of the divider method of biasing. A transistor acts as one resistor in a divider network. The output voltage of one amplifier is the same as the input voltage of the second amplifier. When the circuit is designed to take this into account, the two can be connected together without isolation.

Figure 2-4 shows a two-stage direct-coupled transistor amplifier. Notice that the signal passes directly from the collector into the base. The base current (I_B) of transistor Q_2 is developed without a base resistor. Any I_B needed for Q_2 also passes through the load resistor R_3. The collector voltage V_c of Q_1 remains fairly constant when the two transistors are connected. Note also that the emitter bias voltage of Q_2 is quite large (7.2 V). The base-emitter voltage (V_{BE}) of Q_2 is the voltage difference between V_B and V_E. This is 7.8 V_B − 7.2 V_E, or 0.6 V_{BE}. This value of V_{BE}

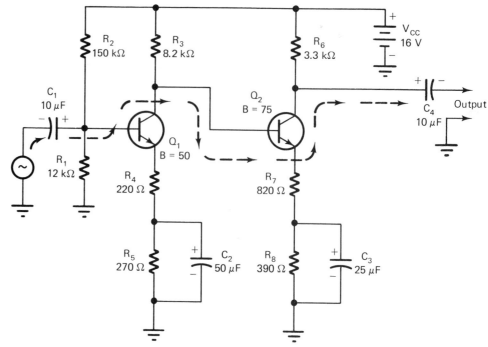

FIGURE 2-4 Two-stage direct-coupled amplifier.

forward-biases the base-emitter junction of Q_2. In direct-coupled amplifiers each stage has a different operating point based on a common voltage source. Several direct-coupled stages supplied by the same source are rather difficult to achieve. As a general rule, only two stages are coupled together by this method in an amplifying system.

Direct-coupled amplifiers are very sensitive to changes in temperature. The beta of a transistor, for example, changes rather significantly with temperature. An increase in temperature causes an increase in beta and leakage current. This tends to shift the operating point of a transistor. All stages that follow amplify according to the operating point shift. Changes in the operating point can cause nonlinear distortion or a lack of stability.

When two transistors are directly coupled, it is often called a *Darlington amplifier.* The transistors are usually called a *Darlington pair.* Two transistors connected in this manner are also manufactured in a single case. This type of unit has three leads. It is generally called a *Darlington transistor.* The gain produced by this device is the product of the two transistor beta values. If each transistor has a beta of 100, the total current gain is 100 × 100, or 10,000. Figure 2-5 shows a Darlington transistor amplifier circuit.

Darlington amplifiers are used in a system where high current gain and high input impedance are needed. Only a small input signal is needed to control the gain of a Darlington amplifier. In effect, this means that the amplifier does not load down the input signal source. The output impedance of this amplifier is quite low.

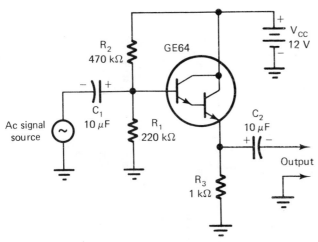

FIGURE 2-5 Darlington transistor amplifier.

It is developed across the emitter resistor R_3. A Darlington amplifier has an emitter-follower output. Several different transistor combinations are used in Darlington amplifiers today.

Transformer Coupling

In transformer coupling, the output of one amplifier is connected to the input of the next amplifier by mutual inductance. Depending on the frequency being amplified, a *coupling transformer* may use a metal core or an air core. The output of one stage is connected to the primary winding, and the input of the next stage is connected to the secondary winding. The number of primary and secondary turns determines the impedance ratio of the respective windings. The input and output impedance of an amplifier stage can be easily matched with a transformer. Ac signals easily pass through the transformer windings. Dc voltages are isolated by the two windings. Figure 2-6 shows some representative coupling transformers.

Figure 2-7 shows a two-stage *transformer-coupled* radio-frequency (RF) amplifier. This particular circuit is used in an *amplitude-modulated (AM)* radio receiver. Circuits of this type operate very well when they are built on a printed circuit board. The operational frequency of this amplifier is 455 kHz.

The operation of a transformer-coupled circuit is very similar to that of a capacitive-coupled amplifier. Biasing for each transistor element is achieved by resistance and transformer impedance. The primary impedance of T_2 serves as the load resistor of Q_1. The secondary winding of the transformer and R_4 serves as the input impedance for Q_2. The low output impedance of Q_1 is matched to the high input impedance of Q_2 by the transformer. The primary tap connection is used to assure proper load impedance of Q_1. The impedance of each winding is dependent on the primary and secondary *turns ratio*. This particular transformer-coupled amplifier is tuned to pass a specific frequency. Selection of this frequency

FIGURE 2-6 Coupling transformers. (Courtesy of TRW/UTC Transformers.)

is achieved by coil inductance (L) and capacitance (C). Tuned transformer-coupled amplifiers are widely used in radio receivers and TV circuits. The dashed line surrounding each transformer indicates that it is housed in a metal can. This is done purposely to isolate the transformers from one another. Several transformers of this type are shown in Fig. 2-8. The smaller units are commonly found in transistor radios.

Transformers are also used to couple an amplifier to a load device. Figure 2-9 shows a transistor circuit with a coupling transformer. In this application the coupling device is called an *output transformer*. The primary transformer winding serves as a collector load for the transistor. The secondary winding couples transistor output to the load device. The transformer serves as an impedance-matching device. Its primary impedance matches the collector load. In a common-emitter amplifier, typical load resistance values are in the range of 1000 Ω. The output device in this application is a speaker. The speaker impedance is 10 Ω. To get maximum transfer of power from the transistor to the speaker, the impedances must match. The primary winding matches the transistor output impedance. The secondary winding matches the speaker impedance. Maximum power is transferred from the transistor to the speaker through the transformer impedance.

Closer examination of the transformer tells a great deal about its impedance. The number of turns of wire on a particular coil determines its impedance. Assume that the output transformer in Fig. 2-9 has 1000 turns on the primary and 100 turns on the secondary. Its turns ratio is

$$\text{Turns ratio} = \frac{\text{primary turns } (N_P)}{\text{secondary turns } (N_S)} = \frac{1000}{100} = 10$$

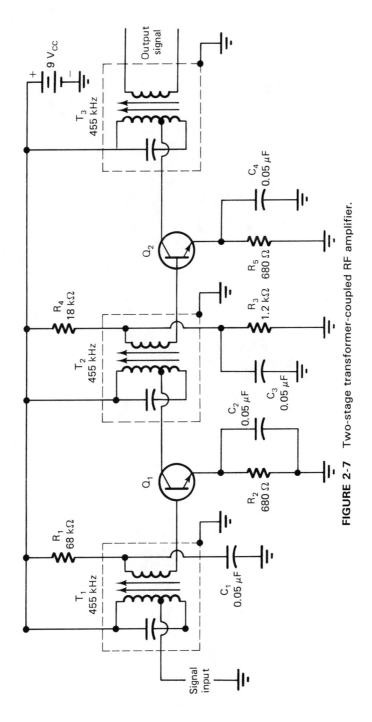

FIGURE 2-7 Two-stage transformer-coupled RF amplifier.

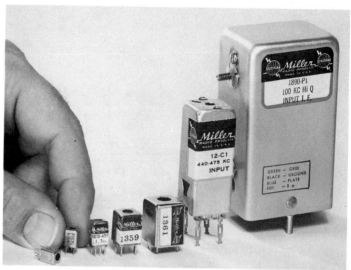

FIGURE 2-8 IF transformers. (Courtesy of J.W. Miller Div./Bell Industries.)

The *impedance ratio* of a transformer is the square of its turns ratio. For the output transformer the impedance ratio is 10^2 Ω, or 100 Ω. This means the impedance of the collector is 100 times more than the impedance of the load device. If the load (speaker) of our circuit has an impedance of 10 Ω, the input impedance is 100 × 10 Ω, or 1000 Ω.

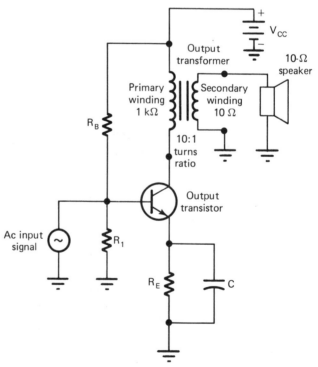

FIGURE 2-9 Transformer-coupled transistor circuit.

Transformer coupling has a number of problems. In sound system applications, the impedance increases with frequency. As a result of this, the high-frequency response of a sound signal is rather poor. Better-quality sound systems do not use transformer coupling. The physical size of a transformer must also be quite large when high-power signals are being amplified. Because of this, low- and medium-power amplifying circuits only use transformer coupling. In addition to this, good-quality transformers are rather expensive. These disadvantages tend to limit the number of applications of transformer coupling today.

POWER AMPLIFIERS

The last stage of an amplifying system is nearly always a *power amplifier*. This particular type of amplifier is generally a low-impedance device. It controls a great deal of current and voltage. Power is the product of current and voltage. Normally, a power amplifier dissipates a great deal more heat than does a comparable signal amplifier. In transistor systems, when 1 W or more of power is handled by a device, it is considered to be a power amplifier. Power amplifiers are generally designed to operate over the entire active region of the device. With careful selection of the device and circuit components, it is possible to achieve power gain with a minimum of distortion.

Single-ended Power Amplifiers

An amplifier system that has only one active device that develops output power is called a *single-ended amplifier*. This type of amplifier is very similar in many respects to a small-signal amplifier. It is normally operated as a class A amplifier. An ac signal applied to the input is fully reproduced in the output. The operating point of the amplifier is near the center of its active region. As a general rule, this type of amplifier operates with a minimum of distortion. The operational efficiency of this amplifier is, however, quite low. Efficiency levels are in the range of 30%. *Efficiency* refers to a ratio of developed output power to dc supply power. The equation for efficiency is

$$\text{Efficiency (\%)} = \frac{\text{ac power output}}{\text{applied dc operating power}}$$

$$= \frac{P_{out}}{P_{in}} \times 100$$

For a class A amplifier it takes 9 W of dc power to develop 3 W of output signal when an amplifier is 33% efficient. Because of their low efficiency rating, signal-ended amplifiers are generally used only to develop low-power output signals.

Figure 2-10 shows a *single-ended audio output amplifier*. The

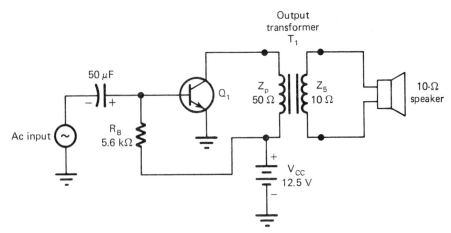

FIGURE 2-10 Single-ended audio amplifier.

term *audio* refers to the range of human hearing. *Audio frequency (AF)* refers to frequencies that are from 15 Hz to 15 kHz. This particular amplifier will respond to signals somewhere within this range.

A family of characteristic curves for the transistor used in the *single-ended power amplifier* is shown in Fig. 2-11. Note that the power dissipation curve of this amplifier is 5 W. The developed load lines for the amplifier are well below the indicated P_D curve. This means that the transistor can operate effectively without becoming overheated.

Investigation of the family of collector curves shows that two load lines are plotted. The static or dc load line is based on the resistance of the transformer primary. For a primary impedance of 50 Ω, the resistance is approximately 10 Ω. The dc load line is therefore based on 12.5 V_{CC} and 1.25 A of I_C. This means that the dc load line will extend almost vertically from the 12.5-V_{CC} point on the curve. The ac load line is determined with the primary winding impedance (50 Ω) and twice the value of V_{CC} (25 V). The ac output signal is, therefore, twice the value of V_{CC}. The operational points are 25V_{CC} and 500 mA of I_C. Note the location of these operating points on the ac load line.

When an ac signal is applied to the input it can cause a swing 12.5 V above and below V_{CC}. The transformer, in this case, accounts for the voltage difference. A transformer is an inductor. When the electromagnetic field of an inductor collapses, a voltage is generated. This voltage is added to the source voltage. In an ac load line, V_{CE} therefore swings to twice the value of the source voltage. This usually occurs only in transformer-coupled amplifier circuits. Note that the operating point (Q point) is near the center of the ac load line.

With the load line of the amplifier established, the base resistance value must be determined. For our circuit the emitter voltage is zero. The emitter-base junction of the transistor, being silicon, has 0.7 V across it when it is in conduction. The base re-

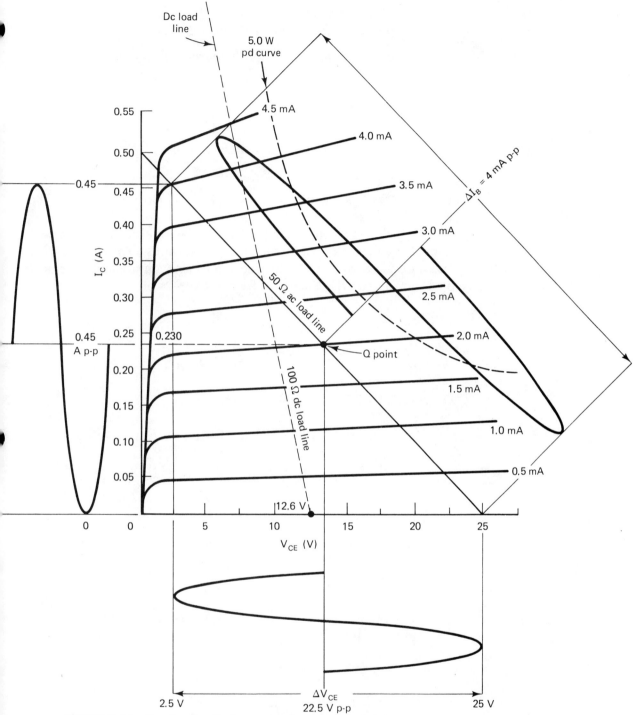

FIGURE 2-11 Family of collector curves for power transistor.

sistor therefore has 12.5 V − 0.7 V, or 11.8 V, across it at the Q point. By Ohm's law, the value of R_B is 11.8 V/2 mA, or 5900 Ω. A standard resistor value of 5.6 kΩ would be selected for R_B. Notice the location of this resistor in the circuit.

With the available data that we have, it is possible to look at the operational efficiency of the single-ended amplifier. The *dc power* supplied to the circuit for operation is

Applied dc operating power = dc source voltage
$\times$ collector current

or

$$P_{in} = V_{CC} \times I_C$$
$$= 12.6 \text{ V} \times 0.230 \text{ A}$$
$$= 2.898 \text{ W}$$

The developed ac output power (P_{out}) is a peak-to-peak value. The V_{CE} is 2.5 to 25 V p-p. The I_C is 0 to 0.45 A, or 0.45 A p-p.

In order to make a comparison in power efficiency, the peak-to-peak values must be changed to rms values. An rms ac value and a dc value do the same effective work. Conversion of peak-to-peak voltage to RMS is done by the expression

$$V_{CE} \text{ (RMS)} = \frac{V_{p\text{-}p}}{2} \times 0.707$$
$$= \frac{22.5 \text{ V p-p}}{2} \times 0.707 = 7.953 \text{ V}$$

For the RMS I_C value, the conversion is

$$I_C \text{ (RMS)} = \frac{I_{p\text{-}p}}{2} \times 0.707$$
$$= \frac{0.45 \text{ A}}{2} \times 0.707 = 0.1591 \text{ A}$$

The developed *ac output power* (P_{out}) is determined by the formula

Ac power output (P_{out}) $= V_{CE} \text{ (RMS)} \times I_C \text{ (RMS)}$
$$= 7.953 \text{ V} \times 0.1591 \text{ A}$$
$$= 1.2652 \text{ W}$$

The operational efficiency of the single-ended amplifier of Fig. 2-10 is determined to be

$$\% \text{ efficiency} = \frac{P_{out}}{P_{in}} = \frac{1.2652 \text{ W}}{2.898 \text{ W}} = 0.44 \text{ or } 44\%$$

This shows that an amplifier of this type is only 44% efficient. It takes 2.898 W of dc supply power to develop 1.2652 W of ac output power. This low operational efficiency does not make the single-ended amplifier very suitable for high-power applications. As a general rule, amplifiers of this type are used only where the output is 10 W or less. The power wasted by a single-ended amplifier appears in the form of heat. Obviously, it would be very desirable to get more sound output and less heat during operation.

Push-pull Amplifiers

Power amplifiers can also be connected in a *push-pull* circuit configuration. Two amplifying devices are needed to achieve this type of output. The operational efficiency of this circuit is much higher than that of a single-ended amplifier. Each transistor handles only one alternation of the sine-wave input signal. Class B amplifier operation is used to accomplish this. A class *B* amplifier has sine-wave input and half-sine-wave output. The bias operating point is at cutoff. This means that no power is consumed unless a signal is applied. The operational efficiency of a push-pull amplifier is approximately 80%.

A simplified *transformer-coupled push-pull transistor amplifier* is shown in Fig. 2-12. The circuit is somewhat like two single-ended amplifiers placed back to back. The input and output transformers both have a common connection. The V_{CC} supply voltage is connected to this common point. For operation, the base voltage must rise slightly above zero to produce conduction. The turn-on voltage is usually about 0.6 V. Base voltage is developed by the incoming signal between the center tap and outside ends of the transformer. The transistor does not conduct until the base voltage rises above 0.6 V. The secondary winding resistance of the input transformer (T_1) is several thousand ohms. The divided transformer winding means that each transistor is alternately fed an input signal. The signals are the same value but 180° out of phase.

A class B push-pull amplifier produces an output signal only when an input signal is applied. For the first alternation of the input, Q_1 will swing positive and Q_2 negative. Only Q_1 will go into conduction with this input. Note the resulting I_C waveform. At the same time, Q_2 is at cutoff. The negative alternation drives it further into cutoff. The resulting I_C from Q_1 flows into the top part

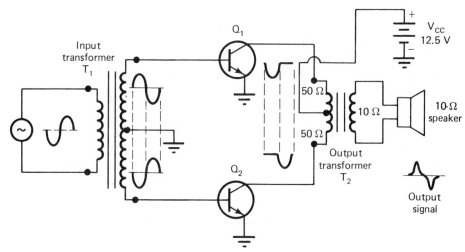

FIGURE 2-12 Transformer-coupled class B push-pull amplifier.

of primary winding of T_2. This causes a corresponding change in the electromagnetic field. Voltage is induced in the secondary winding through this action. Note the polarity of the secondary voltage for this alternation.

For the next alternation of the input signal, the process is reversed. The base of Q_1 swings negative. It becomes nonconductive. Q_2 swings positive at the same point in time. Q_2 goes into conduction. Note the indicated I_C of Q_2 for this alternation. The I_C of Q_2 flows into the lower side of T_2. It, in turn, causes the electromagnetic field to change. Voltage is again induced into the secondary winding. Note the polarity of the secondary voltage for this alternation. It is reversed because the primary current is in an opposite direction to that of the first alternation.

Figure 2-13 shows a combined collector family of characteristic curves for Q_1 and Q_2. These transistors are operated as class B amplifiers. With each transistor operating at cutoff (point Q), only one alternation of output will occur. The entire active region of each transistor can be used in this case to amplify the applied alternation. The combined alternations make a noticable increase in output power.

Using the data of Fig. 2-13, we will analyze the power output and operational efficiency of the push-pull amplifier. The combined change in I_C for both Q_1 and Q_2 is 0.25 A + 0.25 A, or 0.5 A p-p. The combined change in collector voltage is 12 V + 12 V or 24 V p-p. The effective ac current and voltage values are

$$V_{CC} \text{ (RMS)} = \frac{V_{\text{p-p}}}{2} \times 0.707$$

$$= \frac{24 \ V_{\text{p-p}}}{2} \times 0.707 = 8.484 \text{ V}$$

$$I_C \text{ (RMS)} = \frac{I_{\text{p-p}}}{2} \times 0.707$$

$$= \frac{0.5 \text{ A}}{\text{A}} \times 0.707 = 0.17675 \text{ A}$$

The effective *power output* is

$$\text{AC power output } (P_{\text{out}}) = V_{CC} \text{ (RMS)} \times I_C \text{ (RMS)}$$

$$= 8.484 \times 0.17675 \text{ A}$$

$$= 1.4995, \text{ or } 1.5 \text{ W}$$

The operating dc power of our push-pull amplifier must be determined in order to evaluate its operating efficiency. The dc supply voltage V_{CC} is 12.5 V. The I_C changes from 0 to 0.25 A in each transistor. The average value of I_C is

$$\text{Applied } I_C \text{ (average)} = \Delta I_C \times 0.637$$

$$= 0.25 \text{ A} \times 0.637$$

$$= 0.15925 \text{ A}$$

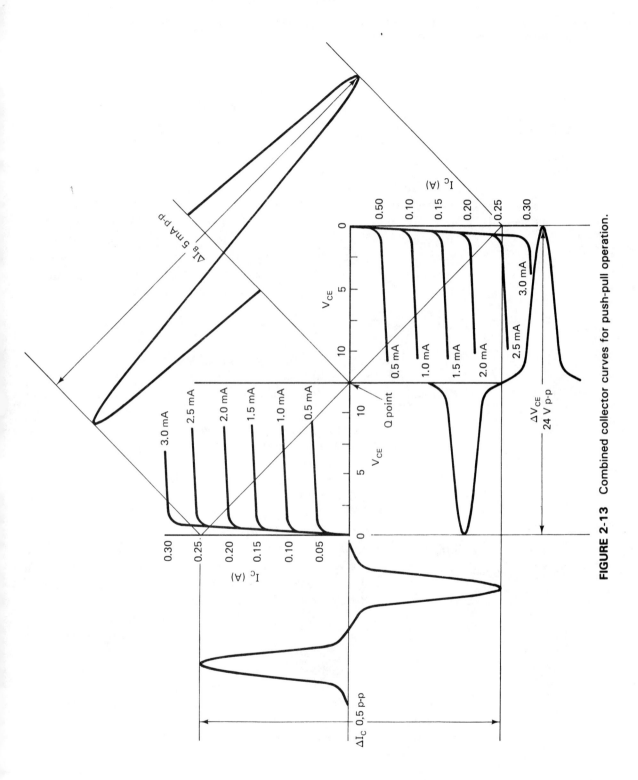

FIGURE 2-13 Combined collector curves for push-pull operation.

The average dc operating power (P_{in}) is equal to the dc voltage times the current flow. This value is

$$P_{in} = \text{dc input V} \times \text{dc input current}$$

$$= 12.5 \text{ V} \times 0.15925 \text{ A}$$

$$= 1.9906 \text{ W}$$

The operational efficiency of our push-pull amplifier is therefore

$$\% \text{ efficiency} = \frac{P_{out}}{P_{in}} = \frac{1.5 \text{ W}}{1.9906 \text{ W}} = 0.753 = 75.3\%$$

This means that it takes nearly 2 W of dc input power to develop 1.5 W of ac output power. This type of operation is very good. The operational efficiency of a push-pull amplifier is much better than that of a comparable class A power amplifier. Its efficiency does not effectively change with the value of the signal-level input. The *power output* of a push-pull amplifier can be nearly four times the output of a single-ended class A power amplifier. Push-pull output circuitry is used primarily to develop high-power signal levels.

Crossover Distortion

Class B push-pull amplification has a very good operating efficiency and does a good job in reproducing high-power signals. It does, however, have a distortion problem when conduction changes back and forth between each transistor. This *distortion* is at the midpoint of the signal. The term *crossover distortion* is used to describe this condition. Figure 2-13 shows crossover distortion near the center of the I_C and V_{CE} waveforms. In sound-amplifying systems the human ear is able to detect crossover distortion.

Crossover distortion is due to the nonlinear characteristic of a transistor when it first goes into conduction. The emitter-base junction of a transistor responds as a diode. A silicon diode does not go into conduction immediately when it is forward biased. It takes approximately 0.6 V to produce conduction. This means that no current will flow between 0 and 0.6 V of base-emitter voltage (V_{BE}). Figure 2-14 shows the input characteristic of a silicon transistor. Notice the nonlinear area between 0 and 0.6 V.

When a transistor is cut off there is no conduction of base current or collector current. When it is forward biased, conduction does not start immediately. There is a slight delay until $0.6 V_{BE}$ is reached. In a push-pull amplifier circuit, this initial delay causes distortion of the input signal. Distortion occurs when the input signal crosses from the conducting transistor to the off transistor. One transistor is turning off and the other starts to conduct. The trailing edge of the off transistor and the leading part of the on transistor both cause some distortion of the waveform. As a general rule, this distortion is very noticeable in small input signals. It is less noticeable in large signals.

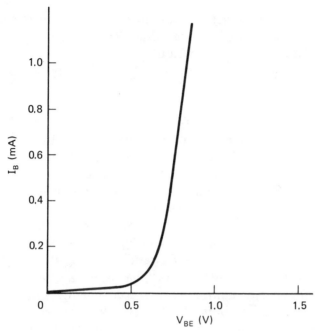

FIGURE 2-14 Input characteristic of a silicon transistor.

Class AB Push-pull Amplifiers

Crossover distortion can be reduced by changing the operating point of a transistor. If each transistor is forward biased by 0.6 V, there will not be any noticeable crossover distortion. A transistor biased slightly above cutoff is considered to be a class AB amplifier. Operation of this type prevents the base-emitter voltage from reaching the nonlinear part of the curve. Figure 2-15 shows the

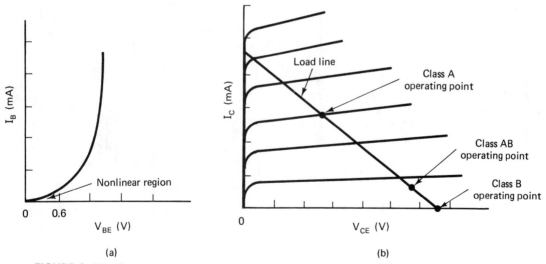

FIGURE 2-15 Transistor characteristics and operating points: (a) input characteristic for a silicon transistor; (b) bias operating points on a load line.

operating point of a class AB amplifier on an input curve and its location on a load line. The load line also shows operating points for class A and class B operation.

The operational efficiency of a class AB amplifier is not quite as good as that of a class B amplifier. There must be some conduction when no signal is applied. The operating point of a class AB amplifier is usually held to an extremely low conduction level. Its operational efficiency is much better than that of a class A amplifier. In a sense, AB operation is a design compromise to minimize distortion at a reasonable efficiency level.

A class AB push-pull amplifier is shown in Fig. 2-16. The circuit is very similar to that of the class B amplifier of Fig. 2-12. A divider network composed of R_{B1} and R_{B2} is used to bias transistors Q_1 and Q_2. The secondary winding resistance of T_1 is rather low compared to the values of R_{B1} and R_{B2}. Resistors R_{E1} and R_{E2} are used to establish emitter bias. The circuit actually has *divider bias* and *emitter bias* combined. Emitter biasing is generally used to provide thermal stability for the two transistors. An amplifier connected in this manner is used for high-power audio systems. Low-power versions of this circuit are widely used in small transistor radios.

Complementary-symmetry Amplifiers

The availability of *complementary transistors* has made the design of transformerless power amplifiers possible. Complementary transistors are PNP and NPN types with similar characteristics.

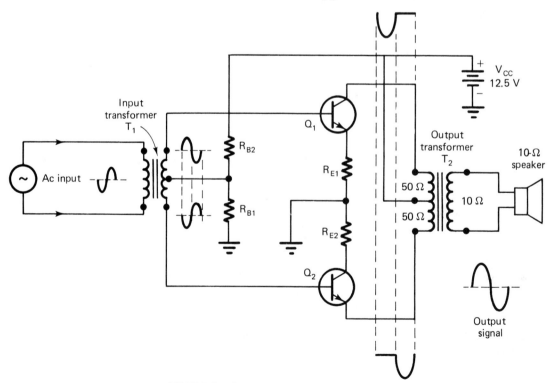

FIGURE 2-16 Transformer-coupled class AB p-p amplifier.

Power amplifiers designed with these transistors have an operating efficiency and linearity that is equal to that of a conventional push-pull circuit. Transformerless amplifiers have the high- and low-frequency response of the output extended. Component cost and weight are also reduced with this type of circuitry.

An elementary *complementary-symmetry amplifier* is shown in Fig. 2-17. This circuit responds as a simple voltage divider. Transistors Q_1 and Q_2 serve as variable resistors with the load resistor (R_L) connected to their common center point. When the positive alternation of the input signal occurs, it is applied to both Q_1 and Q_2. This alternation causes Q_1 to be conductive and Q_2 to be cut off. Conduction of Q_1 causes it to be low resistant. This connects the load resistor to the $+ V_{CC}$ supply. Capacitor C_2 charges in the direction indicated by the solid-line arrows. Note the direction of the resulting current through the load resistor. The polarity of voltage developed across R_L is the same as the input alternation. This means that the input and output signals are in phase. This is typical of a common-collector or emitter-follower amplifier.

When the negative alternation of the input occurs, it goes to both Q_1 and Q_2. The negative alternation reverse-biases Q_1. This causes it to become nonconductive. Q_2, however, becomes conductive. A positive voltage going to the base of an NPN transistor causes it to become low resistant. The positive charge on C_2 from the first alternation forward-biases the emitter of Q_2. With Q_2 low resistant, C_2 discharges through it to ground. The discharge current of C_2 is shown by the dashed arrows. The resulting current flow through R_L causes the negative alternation to occur. The negative alternation of the input therefore causes the same polarity to appear across R_L. Q_2 is also connected as an emitter follower. In effect, conduction of Q_1 causes C_2 to charge through R_L and

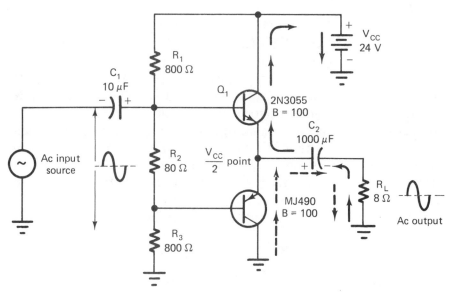

FIGURE 2-17 Complementary-symmetry amplifier.

conduction of Q_2 causes it to discharge. The applied input is reproduced across R_L.

Biasing for Q_1 and Q_2 is achieved by a voltage-divider network consisting of R_1, R_2, and R_3. Each transistor is biased to approximately 0.6 V above cutoff. This biasing must make the emitter junction of Q_1 and Q_2 one-half of the value of V_{CC} when no signal is applied. The voltage drop across R_2 should be 1.2 V if the two transistors are silicon. The voltage drop across R_2 is quite small compared to that across R_1 and R_3. The voltage across R_1 and R_3 must be of an equal value to set the emitter junction at $V_{CC}/2$. To assure proper current output, the value of R_1 and R_3 is based on transistor beta and R_L. This is determined by the formula

$$R_1 \text{ or } R_3 = \text{beta} \times \text{load resistance}$$

or

$$R_1 = R_3 = B \times R_L$$

In Fig. 2-17, R_1 and R_3 are $100 \times 8\ \Omega$, or $800\ \Omega$. The current flow through this network is

$$\text{Network current } (I_n) = \frac{\text{supply voltage } (V_{CC})}{R_1 + R_3}$$

$$I_n = \frac{V_{CC}}{R_1 + R_3} = \frac{24\text{ V}}{800 + 800} = \frac{24\text{ V}}{1600} = 15\text{ mA}$$

The value of R_2 is then determined by the Ohm's law expression:

$$R_2 = \frac{V_2}{I_n} = \frac{1.2\text{ V}}{15\text{ mA}} = 80\ \Omega$$

V_2 is the voltage value used when Q_1 and Q_2 are silicon transistors. The base-emitter voltage V_{BE} is 0.6 V for each transistor. V_2 is therefore 0.6 V_{BE} plus 0.6 V_{BE}, or 1.2 V. The value of V_2 is 0.4 V if germanium transistors were used in the circuit.

IC AMPLIFYING SYSTEMS

A number of manufacturers market amplifying systems in IC packages. A wide range of different ICs are available today. Some of these are designed to perform one specific system function. This includes such things as *preamplifiers, linear signal amplifiers,* and *power amplifiers.* Other ICs may perform as a complete system. Electrical power for operation and only a limited number of external components are needed to complete the system. Low-power audio amplifiers and stereo amplifiers are examples of these devices. Only two of a wide range of IC applications are presented here.

IC *power amplifiers* are frequently used in sound-amplifying systems today. An example of such a circuit is shown in Fig. 2-18. An internal circuit diagram of the IC is shown in part (a). A

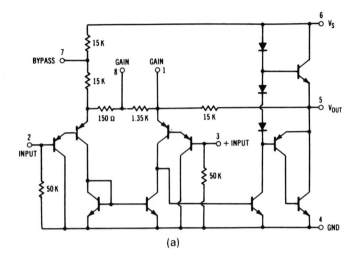

(a)

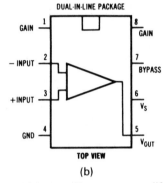

FIGURE 2-18 Low-voltage audio power amplifier IC: (a) internal circuit diagram; (b) pin diagram. (Courtesy of National Semiconductor Corp.)

pin-out diagram of the unit is shown in part (b). This particular chip has a *differential amplifier* input. Darlington pairs are used to increase each input resistance. *Inverting* and *noninverting* inputs are both available. As a general rule, when a signal is applied to one input, the other is grounded. This IC responds only to a difference in input signal levels. It has a a fixed gain of 20 and variable gain capabilities up to 200. Without any external components connected to pins 1 and 8, the device has a *gain* of 20. A 10-mF capacitor connected between these two pins causes a gain of 200. The output of the unit is a Darlington-pair power amplifier. The supply voltage is automatically set to $V_{CC}/2$ at the output transistor common point.

An application of the IC power amplifier is shown in Fig. 2-19. Pins 1 and 8 have a capacitor-resistor combination. This causes the gain to be less than the maximum value of 200. A decoupling capacitor is also connected between pin 7 and ground. R_1 serves as a volume control for the circuit. The amplitude level of the input signal is controlled by this potentiometer. The power

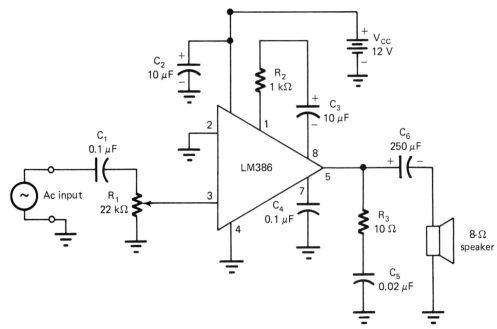

FIGURE 2-19 IC power amplifier. (Courtesy of National Semiconductor Corp.)

output of the IC is 500 mW into a 16-Ω load resistor. Applications include portable AM-FM radio amplifiers, portable tape player amplifiers, and TV sound systems.

A dual 6-W audio amplifier is shown in Fig. 2-20. This particular IC offers high-quality performance for stereo phonographs, tape players, recorders, and AM-FM stereo receivers. The internal circuitry of this IC has 52 transistors, 48 resistors, a zener diode, and two silicon diodes. Two triangle amplifier symbols are used to represent the internal components of this device. This particular amplifier is designed to operate with a minimum of external components. It contains internal bias regulation circuitry for each amplifier section. Overload protection consists of internal current limiting and thermal shutdown circuits.

The *LM 379* has gain capabilities of 90 dB per amplifier. The supply voltage can be from 10 to 35 V. Approximately 0.5 A of current is needed for each amplifier. This particular IC must be mounted to a heat sink. A hole is provided in the housing for a bolted connection to the heat sink. A number of companies manufacture amplifying system ICs today. As a general rule, a person should review the technical data developed by these manufacturers before attempting to use a device.

SPEAKERS

The output of an amplifying system is always applied to a *transducer*. A transducer is a device that changes energy of one type into an entirely different form of energy. Work is performed by a

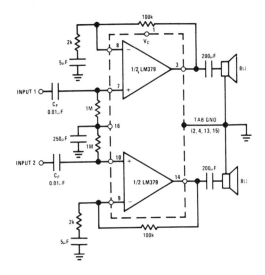

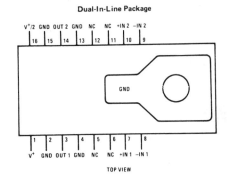

FIGURE 2-20 LM 379 dual 6-W amplifier. (Courtesy of National Semiconductor Corp.)

transducer when energy conversion occurs. The load device of a system is a transducer. Functionally, the load may be resistive, inductive, and in some cases capacitive. Most of our applications have used resistive loads. Resistors generally change electrical energy into heat energy. In an audio amplifying system the load device is a speaker. It changes electrical energy into sound energy. Variations in current cause the mechanical movement of a stiff paper cone. Movement of the cone causes alternate compression and decompression of air molecules. This causes sound waves to be set into motion. The human ear responds to these waves.

The operation of a speaker is based on the interaction of two magnetic fields. One field is usually developed by permanent magnet. The second field is electromagnetic. Permanent-magnet (PM) speakers are of this type. The electromagnetic part of the speaker is generally called a *voice coil*. The voice coil is attached to the cone and suspended around the permanent magnet. See the cross-sectional view of a speaker in Fig. 2-21.

When current flows through the voice coil, it produces an electromagnetic field. The polarity of the field (north or south) depends on the direction of current flow. If ac flows in the voice coil, the field varies in both strength and polarity. The power amplifier of a sound system supplies ac to the voice coil. The changing field

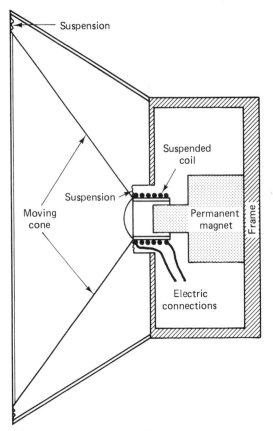

FIGURE 2-21 Cross-sectional view of a speaker.

reacts with the permanent magnetic field. This causes the voice coil to move. With voice coil attached to the cone, the cone also moves. This action causes air molecules to be set into motion. Sound waves are emitted from the speaker cone.

The frequency of the applied ac signal determines how slow or fast the cone of a speaker responds. Operational frequency is based on the rate or speed change of the electromagnetic field. The loudness of the developed sound wave is based on the moving distance of the cone. This depends on the amount of current supplied to the voice coil by the power amplifier. The primary function of an amplifying system is to develop electrical power to drive a speaker.

When selecting a speaker for a specific application, one must take into account a number of considerations. Generally, it takes a large speaker to properly develop low-frequency sounds. Large volumes of air must be set into motion for low-frequency reproduction. Small speakers cannot effectively move enough to produce low tones. Small speakers respond better to high-frequency tones. High-frequency reproduction requires rapid development of air pressure. Small cones can move very rapidly. Large speakers with a big cone and voice coil cannot react quickly enough to produce high-frequency tones. A speaker obviously cannot be large and small at the same time.

In a high-fidelity sound system, at least two speakers are needed to reproduce a typical audio signal. Small speakers are commonly called *tweeters*. The cone of this speaker is generally made of a rather stiff material. Some units employ thin metal cones. Large speakers are called *woofers*. The cone of this speaker is usually quite flexible. Some systems may also employ an *intermediate range,* or *midrange,* speaker. Speakers of this type are designed to respond efficiently to frequencies in the center of the audio range. A great majority of the sound being reproduced falls in the range. These speakers are frequently housed in a wooden enclosure.

INPUT TRANSDUCERS

The input transducer of an amplifying system is primarily responsible for the development of the signal that is being processed by the system. In audio systems, a transducer is used to change sound energy into electrical signal energy. Systems that respond to frequencies above the human range of hearing are called *radiofrequency* (RF) systems. Electromagnetic waves must be changed into electrical signal energy in this type of system. Radio and television receivers respond to RF signals. The type of input transducer being utilized is dependent on the signal being processed. RF systems pick up signals that travel through the air or through an interconnecting cable network. AF systems develop a signal where a system is being operated. Microphones, tape heads, and phonograph pickup cartridges serve as input transducers.

Microphones

The primary function of a *microphone* is to change variations in air pressure into electrical energy. These variations in air pressure are called sound waves. The pressure changes of a sound wave are easily converted into voltage current or resistance. Modern microphones generally change sound energy into a changing voltage signal. *Crystal microphones* and *dynamic microphones* are widely used today.

Crystal microphones generate voltage through a mechanical stress placed on a piece of crystal material. This action is called the *piezoelectric effect.* Rochelle salt crystals produce this effect. Synthetic ceramic materials respond in the same way. Both materials respond to changes in pressure by producing a voltage. Sound waves striking the crystal cause it to bend or squeeze together. Metal plates attached to opposite sides of the crystal develop a potential difference in charge (voltage). The output voltage is then routed away from the crystal by connecting wires. Figure 2-22 shows the construction of a simplified crystal microphone. Part (a) shows a sketch of the internal structure. Part (b) shows an assembled cartridge.

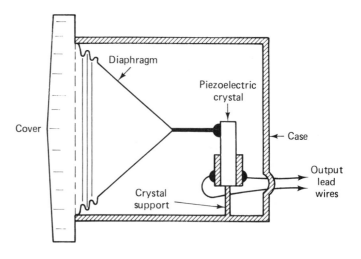

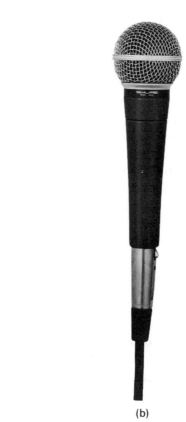

(b)

FIGURE 2-22 Crystal microphone: (a) construction; (b) assembled. (Courtesy of Schure Brothers, Inc.)

Dynamic microphones are considered to be mechanical generators of electrical energy. The mechanical energy of a sound wave causes a small wire coil to move through a magnetic field. This action causes electrons to be set into motion in the coil. A difference in potential charge or voltage is induced in the coil. The resulting output of a dynamic microphone is ac voltage.

A simplified dynamic microphone is shown in Fig. 2-23. Note

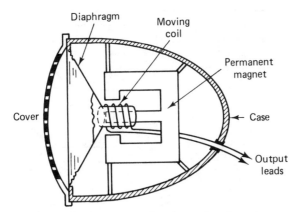

FIGURE 2-23   Structure of a dynamic microphone.

that a lightweight coil of wire is attached to a cone-shaped piece
of material. The cone piece is made of flexible plastic or thin metal.
The cone is generally called a diaphragm. The diaphragm is attached to the outside frame of the microphone. Its center is free
to move in and out when sound waves are applied. Sound waves
striking the diaphragm cause it and the coil to move through a
permanent magnetic field. Through this action sound energy is
changed into voltage. Flexible wires attached to the moving coil
transport signal voltage out of the microphone. This voltage is
applied to the input of a sound system.

Magnetic Tape Input

Magnetic tape is a very common input for many amplifying systems. A long strip of plastic tape is coated with iron oxide. Individual molecules of this material can be easily magnetized. A magnetic tape can therefore store information in the form of minute
charged areas.

Figure 2-24 shows how a sound signal is placed on a magnetic
tape. This is considered to be the recording function. Essentially,
a sound signal is applied to the electromagnetic *head*. Variations
in the applied signal will cause a corresponding change in the electromagnetic field. Tape passing under the air gap of the head will
cause oxide molecules to arrange themselves in a specific order.
In a sense, a changing magnetic field is transferred to the tape.
The recorded signal will remain stored on the tape for a long period
of time. Amplifying systems are used to place information on a
tape. This is achieved by pushing the record button on a tape machine and speaking into a microphone. The signal is then processed by the amplifier. The output of the system supplies signal
current to the electromagnetic recording head. A motor-drive
mechanism is needed to move the tape at a selected recording
speed. The tape head responds as a load device when the system
responds as a recording device.

The recovery of taped information is achieved by a *read head.*

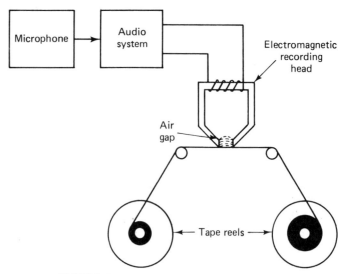

FIGURE 2-24 Tape recording function.

This part of the system is very much like a dynamic microphone. However, the electromagnetic coil is stationary in this type of device. Voltage is induced in the coil by a moving magnetic field. As the magnetic tape moves under the read head, voltage is induced in the coil. Changes in the field cause variations in voltage values. Figure 2-25 shows a simplification of the tape-reading function. The tape head responds as an input transducer when the system responds as a *playback* device.

The head of a tape machine is usually designed to perform both *record* and *playback* functions. A record-playback head is shown in Fig. 2-26. The air gap of a tape head is 6 to 9 mils (0.006 to 0.009 in.) for audio tape recorders. This gap is provided by a

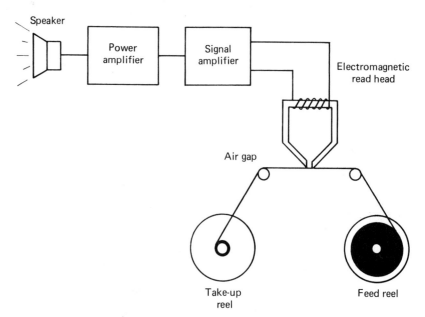

FIGURE 2-25 Tape reading or playback function.

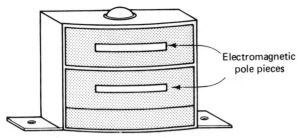

FIGURE 2-26 Record-playback tape head.

thin layer of nonmagnetic material between the poles of an electromagnetic coil. The head must be kept clean and free of foreign material in order to function properly.

Phonograph Pickup Cartridges

The phonograph pickup cartridge is a very important amplifying system input transducer. This device develops an audio signal from the grooves of a record. Phono cartridges are either of the *electromagnetic* or *piezoelectric crystal* type. The output of a phono pickup is much greater than that of a microphone. Typical voltage levels are 5 to 10 mV for magnetic devices and 200 to 1000 mV for crystal types. The resulting frequency output can be 15 to 16,000 Hz.

The construction of an *electromagnetic phono pickup* is shown in Fig. 2-27. This particular cartridge is designed for stereophonic reproduction. The grooves of a record are cut with two audio signals on opposite sides of a 45° angle. As the *stylus* moves the yoke, voltage is induced in each coil. The yoke is a small perma-

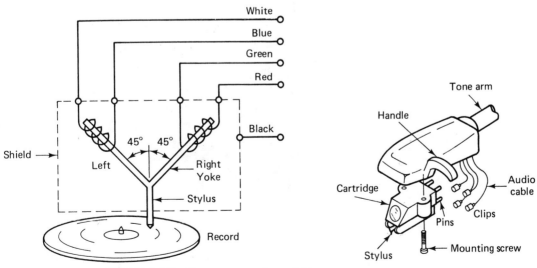

FIGURE 2-27 Electromagnetic cartridges.

nent magnet core. In other cartridges, the magnet may be stationary with the coil being moved by the stylus. In either type of construction, signal voltage is generated by electromagnetic induction.

A *crystal or ceramic stereo cartridge* is shown in Fig. 2-28. This particular design has a Y-shaped yoke structure. A piece of crystal or ceramic material is attached to each side of the yoke. When the stylus moves in the record groove it causes the yoke to twist. This force on the crystal causes a corresponding voltage to be developed. Audio-frequency signal voltage is generated due to the piezoelectric effect. The output from each crystal is developed by wire connections at one end. This particular pickup has a flip-over stylus. With the stylus lever positioned to the right side, it operates in fine-groove records. Positioned to the left it operates in wide-groove records.

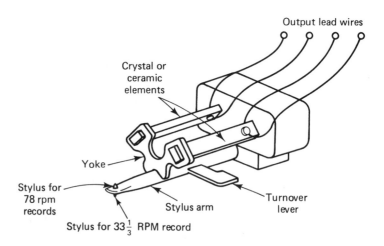

(b)

FIGURE 2-28   Stereo cartridge: (a) crystal or ceramic cartridge structure; (b) assembled phonograph cartridge. [(b) Courtesy of Schure Brothers, Inc.]

SELF-EXAMINATION

1. A _____ changes energy from one form into another.
2. Three examples of input transducers are _____, _____, and _____.
3. The output transducer of a stereo system is a _____.
4. When the output of one amplifier is connected to the input of the next, they are said to be connected in _____.
5. Three common methods of amplifier coupling are _____, _____, and _____.
6. Two directly coupled transistors are referred to as a _____ _____.
7. An output transformer is used for _____ _____.
8. The last stage of an amplifying system is a _____ amplifier.
9. Two types of power amplifiers are _____ and _____.
10. Frequencies from 15 Hz to 15 kHz are called _____ _____.
11. The ratio of an amplifier's power output to its power input is called _____.
12. A transistor biased slightly above cutoff is a class ___ amplifier.
13. An amplifier with a PNP and an NPN transistor coupled together is called a _____ _____.
14. The LM 379 is an _____.
15. The electromagnetic part of a PM speaker is called a _____ _____.
16. Three types of speakers according to frequency range are _____, _____, and _____.
17. Two types of microphones are _____ and _____.
18. Phonograph pickup cartridges are _____ or _____.

ANSWERS

1. Transducer
2. Microphones; phonograph cartridges; tape recorder heads
3. Speaker
4. Cascade

5. Capacitive; direct; transformer
6. Darlington pair
7. Impedance matching
8. Power
9. Single-ended; push-pull
10. Audio
11. Efficiency
12. AB
13. Complementary symmetry amplifier
14. IC power amplifier
15. Voice coil
16. Woofers; tweeters; midrange
17. Crystal; dynamic
18. Electromagnetic; piezoelectric

EXPERIMENT 2-1
DC TRANSISTOR AMPLIFIERS

Transistors are often used to achieve amplification of dc voltages and signals. In this circuit, the signal path between input and output must not be obstructed by capacitors or inductors. Both dc voltage and current can be effectively increased by dc amplifiers.

OBJECTIVES

1. To construct a simple dc amplifier of the common-emitter type.
2. To measure input and output voltages which will be used to calculate the voltage and current gains of the amplifier.

EQUIPMENT

NPN transistor
Resistors: 1 kΩ, 10 kΩ
Potentiometer: 100 kΩ
9-V dc power supply or equivalent battery
Multifunction meter (VOM)
Oscilloscope

PROCEDURE

1. Construct the common-emitter transistor dc amplifier of Figure 2-1A.
2. Before applying V_{CC} power to the circuit, adjust R_1 to midrange and prepare to measure I_B on the 1.0-mA meter range. Turn on the V_{CC} source and adjust I_B to about 50 μA of current.
3. Measure and record the dc emitter-base voltage: V_{BE} = _____ V.
4. Measure and record the dc emitter-collector voltage: V_{CE} = _____ V.
5. With the dc input of an oscilloscope connected across the emitter and collector of the transistor, carefully decrease the I_B current to 20 μA. How does this alter the trace of the oscilloscope? _____
_____
_____

123

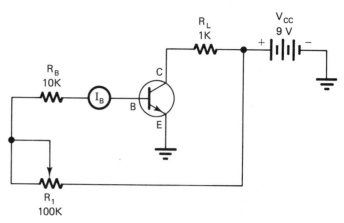

FIGURE 2-1A Dc transistor amplifier.

What does the horizontal trace of the scope actually display about the transistor operation? _____

_____

6. Measure and record the V_{BE} and V_{CE} of the transistor at 20 μA of I_B: V_{BE} = _____ V;
   V_{CE} = _____ V.

7. The total change in V_{BE} between steps 3 and 6 is: V_{BE} (step 3) _____ − V_{BE} (step 6)
   _____ = _____ V.

8. The total change in V_{CE} between steps 4 and 6 is: V_{CE} (step 6) _____ − V_{CE} (step 4)
   _____ = _____ V change.

9. The dc voltage gain of this amplifier is determined by the formula $A_V = \Delta V_{CE}/\Delta V_{BE}$. The Greek letter Δ denotes changing values. Calculate the *dc voltage gain* for the changing voltage values of this amplifier: A_V = _____.

10. Measure and record the voltage that appears across R_L:
    V_{RL} = _____ V.

11. Calculate the current through R_L by using Ohm's law. This value represents the collector current (I_C) of the amplifier: I_C = _____ mA.
    Note: The I_B current setting should remain at 20 μA.

12. Using the formula beta = I_C/I_B, calculate the dc current gain for the circuit: beta = _____.

13. Carefully adjust R_1 to 50 μA, measure V_{RL}, and calculate the circuit I_C: V_{RL} = _____ V; I_C
    = _____ mA.

14. Calculate the dc beta for this operating condition: beta
    = _____.

ANALYSIS

1. What determines the beta of a transistor amplifier?
    _____
    _____

2. If the value of R_L were increased, how would the circuit
    respond? _____
    _____

3. Why does an increase in I_B cause an increase in I_C and a
    reduction in V_{CE}? _____
    _____

4. What is meant by the term *common-emitter* amplifier?
    _____
    _____

EXPERIMENT 2-2
AC TRANSISTOR AMPLIFIERS

Many transistor amplifier circuits are used to increase the strength of alternating current signals. This type of amplifier is somewhat unique when compared with the dc type of transistor amplifier. Essentially dc voltage is used to energize the circuit components of the amplifier. When an ac signal is applied, it causes each dc value to be changed at an ac rate. As a result, this amplifier actually has an ac component along with specific dc component values.

OBJECTIVES

1. To construct a common-emitter amplifier that employs a bias method that is independent of transistor beta.
2. To observe that the input and output signals of a common-emitter amplifier are 180° out of phase.

EQUIPMENT

NPN transistor
Resistors: 100 Ω, 1 kΩ, 2.7 kΩ, 10 kΩ
Capacitors: 10 μF, 33 μF, .01 μF
Dc power supply or 9-V battery
Signal generator
Multifunction meter (VOM)
Oscilloscope
SPST switch

PROCEDURE

1. Construct the ac common-emitter amplifier of Figure 2-2A.
2. Before applying power to the circuit, connect the vertical probe of an oscilloscope to the base of the transistor and the common lead to ground. Turn on the signal generator and adjust it to 400 Hz with approximately 1 V p-p of signal amplitude.
3. Turn on the transistor power supply (V_{CC}) and apply 9 V dc.

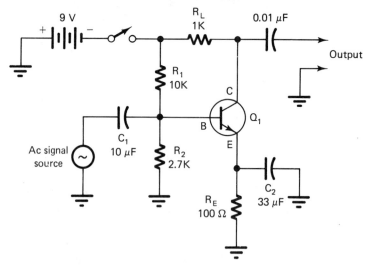

FIGURE 2-2A Common-emitter amplifier circuit.

4. Connect the oscilloscope leads to the output, and adjust the input signal level to produce the *greatest* output signal with a minimum of distortion. Make a sketch of the output wave in Figure 2-2B. Increase the input signal somewhat until clipping occurs. Then reduce the output until a minimum of distortion occurs.
5. Move the vertical probe of the oscilloscope to the base

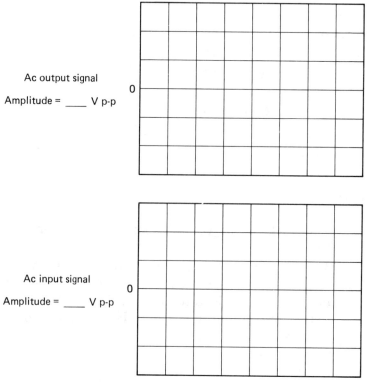

Ac output signal

Amplitude = _____ V p-p

0

Ac input signal

Amplitude = _____ V p-p

0

FIGURE 2-2B Ac input and output signals.

of the transistor amplifier and observe the input waveform. Make a sketch of the waveform in Figure 2-2B.

6. Calculate the ac voltage gain of this amplifier: ac voltage gain = _____.

7. With a multifunction meter, measure and record the following voltages with no signal applied to the amplifier. (*Note:* Do not leave the ground lead of the oscilloscope or signal generator connected to the circuit, or a cross ground will cause the circuit not to operate properly.

V_{R1} = ____ V; V_{R2} = ____ V; V_{RE} = ____ V
V_{RL} = ____ V; V_{BE} = ____ V; V_{CE} = ____ V

8. The emitter-base voltage (V_{BE}) of this circuit is determined by the value of $V_{R2} - V_{RE}$. Calculate this value and compare it with the measured value of V_{BE}: V_{BE} = _____V. _____

9. Since all the transistor current passes through the emitter, calculate the emitter current (I_E) of this circuit using V_{RE} and R_E: I_E = _____ mA.

10. Calculate the collector current (I_C) passing through R_L: I_C = _____ mA.

11. Base current can be determined by the transposing of the $I_E = I_B + I_C$ formula: I_B = _____ mA.

12. Calculate the ac current gain, or *beta*, of the amplifier using the calculated values of I_B and I_C: beta = _____.

13. Measure I_B, I_C, and I_E to compare to your calculations:
I_B = _____ mA; I_C = _____ mA;
I_E = _____ mA.

ANALYSIS

1. Explain how the amplifier of this activity achieves ac amplification with respect to changes in V_{BE}, I_C, I_B, I_E, and V_{CE}. _____

2. How do the measured and calculated values of I_B, I_C, and I_E compare? _____

How do you account for any differences? _____

EXPERIMENT 2-3
JFET AMPLIFIERS

Junction field effect transistors (JFETs) are unique solid-state devices that achieve amplification by controlling the current carriers passing through a semiconductor channel. When a signal is applied to the gate of a JFET, it either adds to or reduces the reverse biasing of the junction. This is the action that controls the flow of current carriers through the conduction channel to the drain (D).

With no gate-source voltage applied, there will be maximum drain current, or I_D. When the gate-source voltage is great enough to stop the flow of I_D, the JFET has reached the "pinch-off voltage." In amplifier operation, the S-G voltage changes somewhere between zero and the pinch-off voltage.

OBJECTIVES

1. To construct a JFET amplifier circuit.
2. To analyze the operation of an JFET amplifier with an oscilloscope and a multimeter.
3. To calculate the gain of a JFET amplifier.

EQUIPMENT

Multimeter (VOM)
Signal generator
Oscilloscope
Variable power supply, 0–15 V
Field effect transistor (General Electric FET-1 or equivalent)
Resistors: 220Ω, 1 kΩ, 47 kΩ, 1 MΩ
Potentiometer: 500 kΩ
Capacitors: 33 μF, 1 μF
SPST switch

PROCEDURE

1. Connect the JFET amplifier circuit of Figure 2-3A.
2. Turn on the variable dc power supply and adjust it to 10 V dc.
3. Turn on the signal generator and adjust it to produce a 1-kHz signal.

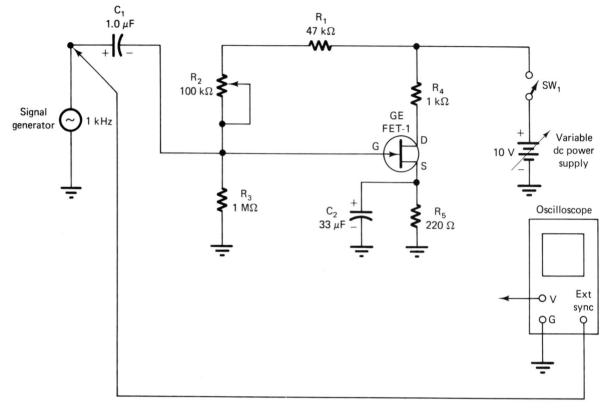

FIGURE 2-3A

4. Turn on the oscilloscope and prepare it for external sync operation. Connect the external sync input to the signal generator as indicated in Figure 2-3A.

5. Connect the vertical probe of the oscilloscope to the signal generator at test point A. Set the sweep frequency to the 1-kHz range or to 0.5-ms/cm sweep rate. A display of two or three complete sine waves should appear on the oscilloscope.

6. Adjust the amplitude control of the signal generator to produce approximately 4 V p-p.

7. Turn on the FET circuit switch and prepare the meter to measure dc voltage at the gate (G). Adjust potentiometer R_2 to produce a gate voltage of approximately 2 V.

8. Move the vertical probe of the oscilloscope to the drain (D) or output of a JFET. Adjust the amplitude control of the signal generator to produce maximum output with a minimum of distortion.

9. Adjust potentiometer R_2 to produce maximum output with least distortion.

10. Make a sketch of the observed output in the provided space of Figure 2-3B. Indicate the peak-to-peak voltage value and phase of the wave form on your sketch.

JFET output _____   Output = _____ $V_{p-p'}$

_____ Phase

JFET input _____   Input _____ $V_{p-p'}$

_____ Phase

FIGURE 2-3B

12. Calculate the voltage gain achieved by the JFET amplifier: voltage gain = _____.

13. With a meter, measure and record the dc source (S), drain (D) and gate (G) voltage of the JFET: V_S = _____ V; V_D = _____ V; V_G = _____ V.

14. Source-gate operating voltage is determined by subtracting V_G from V_S: S-G voltage = _____ V. Note that the value of S-G voltage can be altered by changing R_2.

ANALYSIS

1. What is the bias voltage polarity of the gate with respect to the source voltage? _____
   _____

2. What method of biasing is used by the JFET amplifier of this experiment? _____

3. What type of amplifier configuration is used by this JFET? _____

4. What does transconductance tell about a JFET? _____
   _____
   _____

5. What is dynamic resistance of a JFET? _____
   _____
   _____

6. Why is the JFET considered a voltage operated device?
   _____
   _____

7. Why is the input impedance of the JFET high? _____
   _____
   _____

EXPERIMENT 2-4
RC COUPLING

When two or more transistors are connected together, a great deal more gain can be achieved than by a single amplifier. When resistance-capacitance (RC) coupling is used, the ac part of the signal passes through the capacitor and the dc component will be blocked. Through this method of coupling, the ac part of the signal can be processed by each amplifier without altering the bias voltages.

When an ac signal is applied to the base of a transistor, it causes the collector-emitter voltage to vary above and below its operating point. An increase in V_{CE} causes a coupling capacitor to charge accordingly. A decrease in V_{CE} causes a corresponding discharge of the capacitor voltage. Through this action, changes in V_{CE} voltage can be transferred from the output of one transistor to the base or input of a second transistor. The dc operating voltages of each transistor will remain isolated because they are blocked by the capacitor.

OBJECTIVES

1. To construct a two-stage transistor amplifier using RC coupling.
2. To trace the signal path through the amplifier circuit with an oscilloscope.
3. To observe waveforms and phase relationships and to calculate amplifier voltage gain.

EQUIPMENT

Multimeter (VOM)
Signal generator
Oscilloscope
Variable power supply, 0–15 V
NPN transistors (2)
Resistors: 500Ω, 1 kΩ, 56 kΩ, 100 kΩ
Capacitors: 68 μF, 33 μF, 0.1 μF
SPST switch

PROCEDURE

1. Connect the *RC* coupled amplifier of Fig. 2-4A. To simplify the construction process, parts should be positioned as near as possible to their location in schematic diagram.

2. Turn on the circuit switch, signal generator, power supply and oscilloscope.

3. Adjust the signal generator to produce a 1-kHz signal of approximately .06 *V* p-p. Trace this signal through the amplifier (steps 4 and 5) to see if it is functioning.

4. With the vertical probe of the oscilloscope, start at test point *A* and trace the signal path through the circuit to the output at test point *G*.

5. If the signal passes through the entire circuit, proceed to the next step. If it does not, locate the test point where the signal path stops. Recheck the circuit construction and component values at this point for a construction error.

6. Connect the vertical probe of the oscilloscope to test point *G* and adjust amplitude control of the signal generator to produce maximum output with a minimum of distortion.

7. Peak-to-peak output voltage of the amplifier is _____ V p-p.

8. Move the oscilloscope probe to test point *D* and measure the input signal voltage of Q_2: Q_2 input voltage = _____ V p-p. Using the output volt-

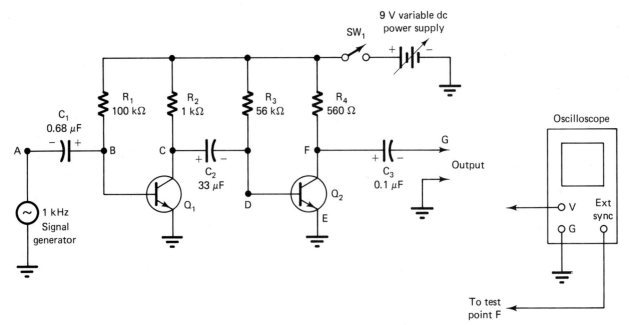

FIGURE 2-4A

age value of step 7, calculate voltage gain of Q_2: Q_2 voltage gain = _____.

9. Using the same procedure, measure the input and output voltage of Q_1 and calculate voltage gain: Q_1 output voltage = _____ V p-p and input voltage = _____ V p-p. Voltage gain of Q_1 = _____.

10. Combined voltage gain of amplifier is the product of Q_1 and Q_2: total amplifier gain = _____.

11. Starting at test point A, trace the signal path to the output while observing phase relationships of the signal. Describe your findings by indicating a plus sign for positive and a minus sign for negative phase observed at each test point.

A _____; B _____;
C _____; D _____;
E _____; F _____;
G _____;

12. Prepare the meter to measure dc voltage at test points B, C, D, E, and F: B = _____ V;
C = _____ V; D = _____ V;
E = _____ V; F = _____ V.

ANALYSIS

1. Why is a capacitor used to couple signals? _____

2. What is the major advantage of RC coupling? _____

3. What limits the frequency passing through an RC-coupled amplifier? _____

EXPERIMENT 2-5
TRANSFORMER COUPLING

When two or more amplifiers are connected together or cascaded, the ac signal must be coupled from one amplifier to the next. The dc voltage values of the first amplifier must also be blocked from passing into the second amplifier. Transformers are commonly used to couple amplifier stages together.

Transformer coupling tends to limit the frequency of signals being transferred through an amplifier. At high frequencies, the inductive reactance of the coils increases and thereby reduces signal gain. Transformer coupling is also somewhat more expensive than other methods of coupling. The advantages of impedance matching must be compared with the disadvantages of cost and poor frequency response when selecting this device as a coupling element.

OBJECTIVES

1. To construct a two-stage, transformer-coupled transistor amplifier.
2. To trace the signal path through the amplifier circuit with an oscilloscope.
3. To measure circuit voltage values and observe wave forms with an oscilloscope.

EQUIPMENT

Multimeter (VOM)
Signal generator
Oscilloscope
Variable power supply, 0–15 V
NPN transistors (2)
Transformer: 10 kΩ to 2 kΩ center-tapped (CT)
Output transformer: 500 Ω CT to 3.2 Ω
Capacitors: 33 μF, 100 μF
Potentiometer: 100 kΩ
Resistors: 220Ω, 4.7 kΩ, 10 kΩ, 100 kΩ
Speaker
Crystal microphone
SPST switch

PROCEDURE

1. Connect the transformer coupled amplifier circuit of Fig. 2-5A.

2. Turn on the variable dc power source. Adjust it to 9 V before turning on the circuit switch.

3. Turn on the signal generator and oscilloscope. Adjust the signal generator to produce a signal of 1 kHz approximately 1 V p-p at test point A. Use the oscilloscope to determine this value. Adjust the oscilloscope to produce a three- or four-cycle display.

4. If the circuit is operating properly, a 1-kHz tone should be heard at speaker. The volume control (potentiometer R_1) may need to be adjusted to produce an output signal. If the amplifier does not work, proceed to step 5. If it does produce a tone, proceed to step 6.

5. Trace the signal path through the amplifier with the oscilloscope. Start at test point G. If a signal does not appear at a specific point, turn off circuit switch and recheck the wiring or component values leading to the test point.

6. Connect the vertical probe of the oscilloscope to test point F. Adjust the amplifier volume control R_1 for a maximum output signal with a minimum of distortion: output signal level = _____ V p-p.

7. Move the vertical probe of the oscilloscope to the amplifier input at test point B: observed input signal = _____ V p-p.

8. Total signal voltage amplification (A_V) between input and output is determined by formula A_V = output divided by input: A_V = _____ .

9. Move the vertical probe of the oscilloscope to test point C and observe the signal amplitude: output voltage of Q_1 = _____ V p-p. Calculate voltage gain A_V of Q_1: A_V of Q_1 = _____ .

10. Using the same procedure, calculate the voltage gain of Q_2 using test points D and F: A_V of Q_2 = _____ .

11. Starting at test point B, observe the phase and amplitude of the signal at test points B, C, D, F, E, and G. Make a sketch of the wave forms in the space of Fig. 2-5B.

12. With a meter, measure dc voltage at B, E, and C of Q_1 and Q_2. Record measured values in the chart of Fig. 2-5C.

13. Disconnect the ac signal source from the input of the amplifier and connect a crystal microphone in its place. Speak into the microphone. Adjust the volume control (R_1) to produce a suitable output signal from the speaker.

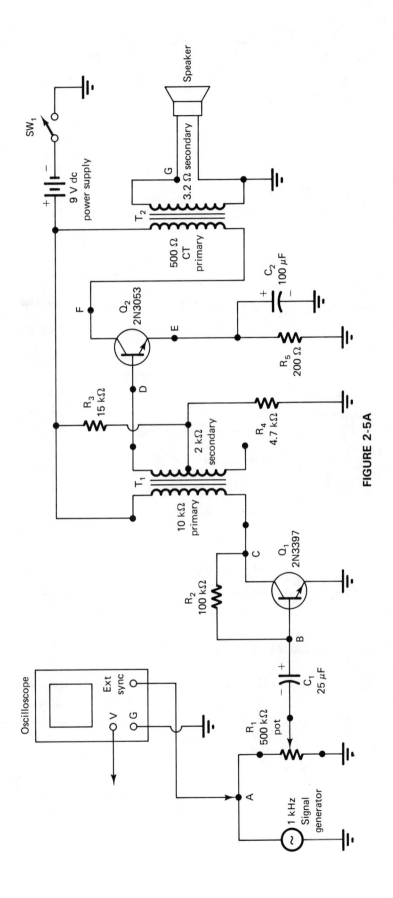

FIGURE 2-5A

141

Q₁ input _____ point B

Q₁ output _____ point C

Q₂ input _____ point D

Q₂ output _____ point F

Q₂ emitter _____ point E

Amplifier output point G _____

B _____ volts p-p, _____ phase

C _____ volts p-p, _____ phase

D _____ volts p-p, _____ phase

F _____ volts p-p, _____ phase

E _____ volts p-p, _____ phase

G _____ volts p-p, _____ phase

FIGURE 2-5B

Element	Q₁	Q₂
Base voltage		
Emitter voltage		
Collector voltage		

FIGURE 2-5C

ANALYSIS

1. How do you account for the decrease in signal amplitude between test points *C* and *D*? _____

2. What are the functions of transformer T_1? _____

3. What are some of the major disadvantages of transformer-coupled amplifiers? _____

EXPERIMENT 2-6
SINGLE-ENDED POWER AMPLIFIER

The output transistor or final stage of amplification is primarily responsible for developing the power needed to drive a speaker or output device. This part of a circuit is generally called a power amplifier or large signal amplifier. Transistors used for this function must be capable of handling large current values and voltages. When one transistor drives the output of an amplifier, it is commonly called a single-ended power amplifier. The output current and voltage applied to this transistor are responsible for developing the output power of the amplifier.

OBJECTIVES

1. To construct a single-ended power amplifier circuit.
2. To measure circuit voltage and current values and to calculate operating power.
3. To use an oscilloscope to observe the operation of a power amplifier circuit.

EQUIPMENT

Multimeter (VOM)
Signal generator
Oscilloscope
Variable power supply, 0–15 V
NPN transistor
Potentiometer: 100 kΩ
Resistors: 220Ω, 4.7 kΩ, 10 kΩ
Capacitors: 33 μF, 100 μF
Speaker
Output transformer: 500 Ω center-tapped (CT) to 3.2 Ω
SPST Switch

PROCEDURE

1. Construct the power amplifier of Fig. 2-6A. Adjust the power supply to produce 9 V dc before turning on the circuit switch.
2. Turn on the signal generator and adjust it to produce a

143

144

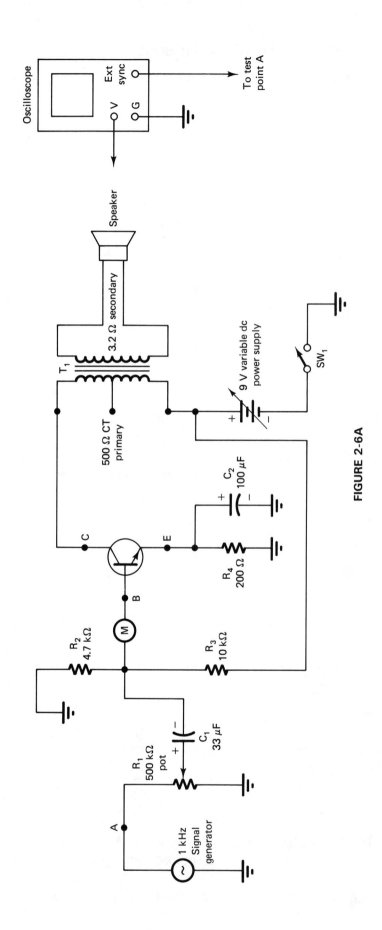

FIGURE 2-6A

1-kHz signal with approximately 5 V p-p output. Use the oscilloscope to measure and observe the signal generator output at test point A.

3. Move the vertical probe of the oscilloscope to test point C. Adjust potentiometer R_1 to produce a maximum output signal with a minimum of distortion: output voltage at test point C = _____ V p-p.

4. Move the oscilloscope probe to the base of transistor at test point B: base signal voltage = _____ V p-p.

5. Calculate the voltage gain of the transistor, using measured input and output voltages: voltage gain = _____.

6. With a meter, measure and record dc emitter, base, and collector voltage: V_E = _____ V; V_B = _____ V; V_C = _____ V.

7. Calculate emitter current flow, using V_C and the value of emitter resistor R_4: I_E = _____ mA.

8. Record the value of base current indicated on the meter: I_B = _____ mA.

9. Using values of I_B and I_E, calculate I_C: I_C = ___ mA.

10. Calculate dc collector power developed by the transistor using I_C and V_C: collector power = _____ mW.

11. Prepare the meter to measure ac (RMS) voltage across the primary winding of the transformer: primary winding voltage = _____ V ac.

12. Using the primary impedance (500 Ω) of the transformer, calculate the primary input power, using the formula $P = V^2/R$: primary power = _____ mW.

13. Prepare a meter to measure total dc current supplied to circuit. A quick way to do this is to place the meter across the circuit switch and turn it *off*: total dc current = _____ mA.

14. Using the dc supply voltage (9 V) and total input current, calculate dc input power to the amplifier: input power = _____ mW.

15. Calculate the efficiency of the amplifier, using ac output power of step 12 and dc input power of step 14. Amplifier efficiency = _____.

ANALYSIS

1. What is the primary function of a single-ended power amplifier? _____

2. What is the function of the output transformer?

3. What does the efficiency of the amplifier indicate?

EXPERIMENT 2-7
PUSH-PULL AMPLIFIERS

A push-pull amplifier is a common type of circuit found in the output of an audio amplifier. Through this type of circuit, it is possible to increase the power output over that developed by a single output transistor.

In order to achieve push-pull amplification, two transistors are connected so that a divided input signal drives each amplifier. The input signals are of equal amplitude but 180° out of phase. The transistors' output signals are combined to produce a relatively pure signal of a higher power level.

OBJECTIVES

1. To construct a transformer-coupled push-pull amplifier.
2. To trace the signal path through the circuit with an oscilloscope.
3. To measure signal levels, calculate transistor voltage gain and observe signal phase relationships.

EQUIPMENT

Multimeter
Oscilloscope
Variable power supply, 0–15 V
Signal generator
Speaker
NPN transistors (3)
Interstage transformer: 10 kΩ to 2 kΩ CT
Output transformer: 500 Ω CT to 3.2 Ω
Potentiometer: 100 kΩ
Resistors: 200Ω, 4.7 kΩ, 10 kΩ, 470 kΩ
Capacitors: 33 μF, 100 μF
Crystal microphone
SPST switch

PROCEDURE

1. Connect the push-pull amplifier of Fig. 2-7A.
2. Turn on the variable dc power supply and adjust it to produce 15 V before turning on circuit switch.

148

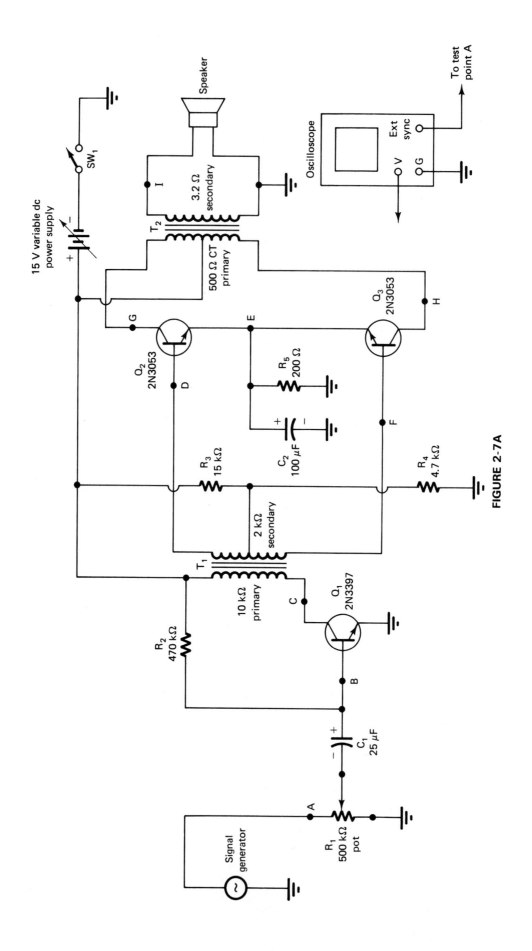

FIGURE 2-7A

3. Turn on the signal generator and oscilloscope. Adjust the signal generator to produce a 1 kHz signal of approximately 2 V p-p at test point *A*. Use the oscilloscope to determine the output. Note that the oscilloscope external sync input is connected to test point *A*.

4. If the amplifier is properly connected, a 1-kHz tone should be heard at the speaker. Volume control R_1 may be used to adjust the signal level of the circuit to produce an output signal. If it is operating properly, proceed to step 6. If an output is not heard at speaker, proceed to step 5.

5. With the oscilloscope, trace the signal path through the circuit from test point *A* through *I*. Points *A* through *C* represent a single path for the signal. Points *D-G* and *F-H* represent two alternate paths. These two signals then combine and appear as one signal at point *I*. A signal not appearing at a test point indicates a problem area. Recheck the circuit wiring or components leading to the inoperative test point.

6. Connect the vertical probe of the oscilloscope to test point *I* and adjust the volume control of the amplifier to produce maximum output with a minimum of distortion. Make a sketch of the observed waveform in the space for *I* in Fig. 2-7B. Indicate the peak-to-peak voltage value and phase of the waveform.

7. Move the vertical probe of the oscilloscope alternately to test points *D, G, B, C, F,* and *H*. Make a sketch of the observed wave forms in the spaces of Figure 2-7B. Indicate the peak-to-peak voltage value and phase of each waveform.

I _____   D _____

G _____ B _____ C _____

F _____   H _____

I ____ volts p-p, ____ phase   D ____ volts p-p, ____ phase

G ____ volts p-p, ____ phase   B ____ volts p-p, ____ phase   C ____ volts p-p, ____ phase

F ____ volts p-p, ____ phase   H ____ volts p-p, ____ phase

FIGURE 2-7B

Element	Q_1	Q_2	Q_3
Base voltage			
Emitter voltage			
Collector voltage			

FIGURE 2-7C

8. Prepare a meter to measure ac signal voltage across the speaker at test point *I:* speaker voltage = _____ V.
9. Calculate the voltage gain of Q_1, Q_2, and Q_3, using measured input and output voltages of Fig. 2-7B: A_V of Q_1 = _____; A_V of Q_2 = _____;
A_V of Q_3 = _____.
10. With a meter, measure dc voltage at *B, E,* and *C* of Q_1, Q_2, and Q_3. Record measured values in the chart of Fig. 2-7C.
11. Disconnect ac signal source from input of amplifier and connect a crystal microphone in its place. Speak into the microphone and test the amplifier output. The volume control may need to be adjusted to produce the best output.

ANALYSIS

1. Using the speaker output voltage of step 8 and the 3.2-Ω speaker impedance, calculate the power output of the amplifier with the $P = V^2/R$ formula: P_{out} = _____.
2. What is the phase relationship of the input signals to Q_2 and Q_3?_____.
3. What class of amplification is achieved by Q_1, Q_2, and Q_3?

UNIT 2 EXAMINATION
AMPLIFYING SYSTEMS

Instructions: For each of the following, circle the answer that most correctly completes the statement.

1. The circuit shown in Fig. E-2-1 is a:
   a. Complementary amplifier
   b. Darlington amplifier
   c. Differential amplifier
   d. Push-pull amplifier

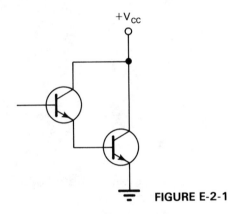

$+V_{CC}$

FIGURE E-2-1

2. The type of arrangement formed when the emitter and collector leads of one transistor are connected to the base and collector leads of a second transistor is called a:
   a. Complementary amplifier
   b. Darlington amplifier
   c. Differential amplifier
   d. Push-pull amplifier
3. The gain of a Darlington amplifier pair is determined by:
   a. Adding the beta values
   b. Subtracting the beta values
   c. Dividing the beta values
   d. Multiplying the beta values
4. The type of coupling used by a Darlington arrangement is:
   a. Direct
   b. Transformer
   c. Impedance
   d. Resistance-capacitance

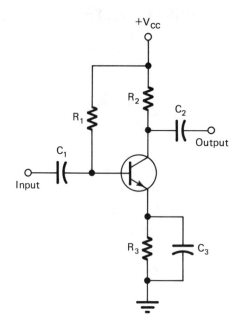

FIGURE E-2-5

5. The type of coupling used in the audio amplifier shown in Fig. E-2-5 is:
   a. *RC*
   b. *LC*
   c. *RL*
   d. Direct

6. The following type of coupling which would be used in the design of a dc amplifier is:
   a. *RC*
   b. *LC*
   c. Direct
   d. Transformer

7. Transformer coupling is frequently used in audio amplifier design instead of *RC* coupling due to:
   a. Increased frequency response
   b. Increased efficiency
   c. Increased loading effect
   d. Increased miller effect

8. The type of feedback used in Fig. E-2-8 is:
   a. Collector feedback
   b. Emitter feedback
   c. Base feedback
   d. Resistor feedback

9. The type of feedback used in Fig. E-2-9 is:
   a. Collector feedback
   b. Emitter feedback

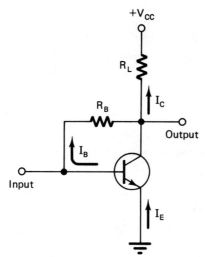

FIGURE E-2-8

c. Base feedback
d. Resistor feedback

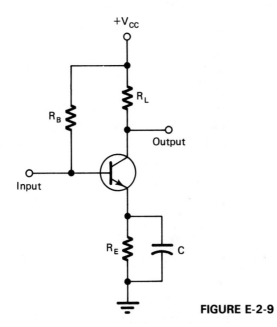

FIGURE E-2-9

10. The type of feedback which tends to stabilize the gain of a transistor amplifier is:
    a. Regenerative feedback
    b. Degenerative feedback
    c. Generative feedback
    d. Positive feedback

11. The method of coupling which would *never* be used in a dc amplifier is:
    a. Diode
    b. Direct

    c. Resistor

    d. Transformer

12. For maximum power transfer with distortion at a minimum, the load impedance should:

    a. Match the collector resistance of the transistor

    b. Be twice the collector resistance of the transistor

    c. Be half the collector resistance of the transistor

    d. Be ten or more times the collector resistance of the transistor

13. Reduced collector current in a power amplifier could be caused by:

    a. A shorted collector decoupling capacitor

    b. An open emitter bypass capacitor

    c. Excessive reverse bias of the collector

    d. Loss of emitter bias voltage

    e. Reduced source voltage

14. In a carbon microphone:

    a. Resistance varies directly with pressure

    b. Resistance varies inversely with pressure

    c. Pressure varies inversely with resistance

    d. Pressure varies directly with resistance

    e. Resistance remains constant

15. The input voltage to an amplifier is 0.5 V and the output voltage is 50 V. The gain of the amplifier is:

    a. 0 to 10 dB

    b. 11 to 20 dB

    c. 21 to 30 dB

    d. 31 to 40 dB

    e. 41 dB or higher

16. An unbypassed emitter resistor in an AF amplifier causes:

    a. Distortion

    b. Reduced gain

    c. Regenerative feedback

    d. No change in frequency response

    e. Increased amplification

17. The main advantage of class B push-pull amplification over class A push-pull amplification is:

    a. Greater power output can be obtained

    b. Harmonic reduction

    c. Output transformer saturation reduction

    d. Reduced distortion of signal

    e. Better frequency response

18. The main disadvantage of class B push-pull amplification over class AB amplification is:
    a. Frequency response reduction
    b. Harmonic content reduction
    c. Crossover distortion
    d. Reduction of transformer saturation distortion
    e. Reduced collector current swing

19. The best frequency response of a transistor is achieved with:
    a. Transformer coupling
    b. Impedance coupling
    c. Direct coupling
    d. RC coupling
    e. LC coupling

20. The term *load* of a transistor refers to the:
    a. Applied collector-emitter voltage
    b. Total power consumed from the source
    c. Resistance or impedance into which the output is fed
    d. Internal resistance between E-C
    e. Heat dissipated

UNIT THREE

Operational Amplifiers

UNIT INTRODUCTION

An *operational amplifier,* or op-amp, is a high-performance, directly coupled amplifier circuit containing several transistor devices. The entire assembly is built on a small silicon substrate and packaged as an integrated circuit. ICs of this type are capable of high-gain signal amplification from dc to several million hertz. An *op-amp* is a modular, multistage amplifying device.

Operational amplifiers have several important properties. They have open-loop gain capabilities in the range of 200,000 with an input impedance of approximately 2 MΩ. The output impedance is rather low, with values in the range of 50 Ω or less. Their bandwidth, or ability to amplify different frequencies, is rather good. The gain does, however, have a tendency to drop, or roll off, as the frequency increases.

UNIT OBJECTIVES

Upon completion of this unit, you will be able to:

1. Define common mode rejection ratio, input resistance, output resistance, offset voltage, offset current, bias current, slew rate, and other op-amp characteristics.

2. Analyze op-amp circuits.
3. Design basic inverting and noninverting op-amp circuits.
4. Explain the purpose and characteristics of a voltage follower.
5. Explain the operation of summing and difference amplifiers.
6. Explain how a differential amplifier operates.
7. Investigate the characteristics of a differential amplifier circuit.
8. Construct an op-amp using a 741 integrated circuit.
9. Calculate the input and output resistances of an operational amplifier circuit.
10. Construct inverting and noninverting operational amplifiers, observe their operation, and compute their gain.

IMPORTANT TERMS

Closed-loop gain. An op-amp that describes the gain or amplification achieved by an amplifier that has negative feedback between the input and output.

Common mode (CM). Signals that are identical in phase and amplitude that appear at the inputs of an operational amplifier simultaneously. The abbreviation CM is used to identify this term.

Common-mode rejection ratio (CMRR). The ability of a differential amplifier of an operational amplifier to cancel a common mode signal.

Comparator. An op-amp function that compares voltage applied to one input to that applied to another input that is of a predetermined value or reference.

Complementary symmetry. Two transistors that have similar characteristics and ratings but are of opposite construction polarity, such as NPN and PNP bipolar transistors. A signal is applied to a common input that sends the correct polarity to the transistor that will cause conduction. Used as the output of some operational amplifiers.

Darlington connection or pair. A circuit configuration in which two similar transistors are connected together so that they act as a single transistor. The overall current gain is the product of the two individual current gains.

Differential amplifier. An amplifier having a high common-mode rejection capability and whose output is proportional to the difference of its two input signals. This type of amplifier is also called a difference amplifier.

Inverting input. One of two inputs of an operational amplifier or voltage comparator. The inverting input changes the signal 180° in the output and identified by a − sign on an amplifier symbol.

Noninverting input. One of two inputs of an operational amplifier or voltage comparator. This input accepts a signal and causes it to appear in the output without a change in phase and is identified by a + sign on an amplifier symbol.

Op-amp. An abbreviation for the term operational amplifier.

Open-loop gain. The voltage gain of an amplifier connected to a load that does not have a feedback path between the output and input.

Operational amplifier. A high-gain DC amplifier with high input impedance and low output impedance. It was originally designed to achieve math functions such as summing, difference, integration, and differentiation.

Slew rate. An op-amp characteristic expressed in volts per microseconds. This is a measure of the amplifiers switching speed. It is defined as the maximum time rate of change between the input and output voltage signals of an amplifier with a gain factor of one.

Summing amplifier. An op-amp function that takes the instantaneous algebraic sum of two or more input signals.

INSIDE THE OP-AMP

The internal construction of an op-amp is quite complex and usually contains a large number of discrete components. A person working with an op-amp does not ordinarily need to be concerned with its internal construction. It is helpful, however, to have some general understanding of what the internal circuitry accomplishes. This permits the user to see how the device performs and indicates some of its limitations as a functioning unit.

The internal circuitry of an op-amp can be divided into three functional units. Figure 3-1 shows a simplified diagram of the internal functions of an op-amp. Notice that each function is enclosed in a triangle. Electronic schematics use a triangle to denote the amplification function. This diagram shows that the op-amp has three basic amplification functions. These functions are generally called *stages* of amplification. A stage of amplification contains one or more active devices and all the associated components needed to achieve amplification.

The first stage, or input, of an op-amp is usually a *differential amplifier.* This amplifier has two inputs, which are labeled V_1 and V_2. It provides high gain of the signal difference supplied to the two inputs and low gain for common signals applied to both inputs simultaneously. The input impedance is high to any applied signal. The output of the amplifier is generally two signals of equal amplitude and 180° out of phase. This could be described as a push-pull input and output.

One or more intermediate stages of amplification follow the differential amplifier. Figure 3-1 shows an op-amp with only one intermediate stage. This amplifier is designed to shift the operating point to a zero level at the output and has high current and voltage gain capabilities. Increased gain is needed to drive the output stage without loading down the input. The intermediate stage generally has two inputs and a single-ended output.

The *output stage* of an op-amp has a rather low output impedance and is responsible for developing the current needed to drive an external load. Its input impedance must be great enough that

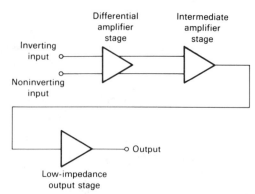

FIGURE 3-1 Op-amp diagram.

it does not load down the output of the intermediate amplifier. The output stage can be an emitter-follower amplifier or two transistors connected in a complementary-symmetry configuration. Voltage gain is rather low in this stage with a sizable amount of current gain.

Differential Amplifier Stage

A *differential amplifier* is the key or operational basis of most op-amps. This amplifier is best described as having two identical or balanced transistors sharing a single emitter resistor. Each transistor has an input and an output. A schematic diagram of a simplified differential amplifier is shown in Fig. 3-2. Notice that the circuit is energized by a dual-polarity, or *split-power, supply*. The source leads are labeled $+V_{CC}$ and $-V_{CC}$ and are measured with respect to a common ground lead.

Operation of a differential amplifier is based on its response to input signals applied to the base. Grounding one base and applying an input signal to the other base produces two output signals. These signals have the same amplitude but are *inverted* 180°. This type of input causes the amplifier to respond in its *differential mode* of operation.

When two signals of equal amplitude and polarity are applied to each base at the same time, the resulting output is zero. This type of input causes a difference or canceling voltage to appear across the commonly connected emitter resistor. In a sense, the differential amplifier responds as a balanced bridge circuit to identical input signals. There is no output when the circuit is balanced and output when it is unbalanced. This is called the *common-mode* (CM) condition of operation. A differential amplifier is designed to reject signals common to both inputs. The term *common-mode rejection ratio* (CMRR) is used to describe this action of the ampli-

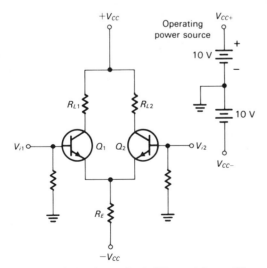

FIGURE 3-2 Simplified differential amplifier.

fier. CMRR is a unique characteristic of the differential amplifier. Undesirable noise, interference, or ac hum can be rejected by this operating condition.

Figure 3-3 shows a simplified schematic of a *differential amplifier* connected for differential mode operation. In this circuit an input signal is applied to the base of Q_1, and the base of Q_2 is left open, or in a *floating* state. This condition causes signals to be developed at both outputs and across the emitter resistor. The emitter signal, as indicated, is in phase with the input. The two output signals are out of phase with each other and have a substantial degree of amplification. Output V_{o1} is out of phase with the input and V_{o2} is in phase.

A differential amplifier will produce two output signals when only one input signal is applied. Coupling of the input signal from Q_1 to Q_2 is accomplished through the emitter resistor. The positive alternation of the input signal, for example, causes increased forward bias of Q_1. This causes an increase in the conduction of Q_1. With more I_E, there is a greater voltage developed across the emitter resistor. This in turn causes both emitters to be less negative. The conduction of Q_1 is not appreciably influenced by this voltage because it has an external signal applied to its input. Q_2 is, however, directly influenced by the reduced negative voltage to its emitter. This causes the conduction of Q_2 to be reduced. Reduction in current through Q_2 causes less voltage drop across R_2 and the collector voltage to swing in the positive direction. In effect, an input signal applied to the base of Q_1 reduces the V_E of Q_2, which in turn increases the value of the output voltage V_2. An input signal is therefore coupled to Q_2 by the commonly connected emitter resistor.

The negative alternation of the input signal causes a reversal of the action just described. Q_1, for example, will be less conductive and Q_2 will have increased conduction. This action causes a reduction of I_C through Q_1 and an increase swing in the value of V_1. *Increased conduction of Q_2 causes a corresponding reduction*

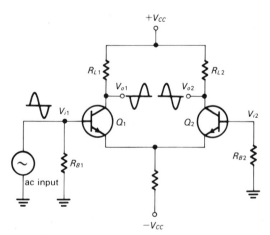

FIGURE 3-3 Ac differential amplifier.

in V_2. The two output signals continue to be 180° out of phase. In effect, both alternations of the input appear in the output. A differential amplifier connected in the *differential mode* will develop two output signals that are reflective of the entire input signal.

Differential amplifiers respond in primarily the same way when the inputs are reversed. In this case an input signal is applied to the base of Q_2 with the base of Q_1 open or floating. V_{o2} is out of phase with the input, and V_{o1} is in phase. The amplitude, or output signal level, of the amplifier is still based on the signal difference between the two inputs. With only one signal applied to the input at a time, the amplifier sees a very large differential input and develops a sizable output voltage.

The differential amplifier of Fig. 3-3 is rarely used in op-amp construction today. Ordinarily, the resistance of R_E needs to be quite large in order to have good coupling and common-mode rejection capabilities. Large resistance values are rather difficult to fabricate in IC construction. R_E can, however, be replaced with a transistor. This transistor and its associated components are called a constant-current source.

Figure 3-4 shows a differential amplifier with a *constant-current source* in the emitter circuit. Transistor Q_3, in this case, has a fixed or constant bias voltage. This voltage maintains the internal resistance of Q_3 at a rather high value. In some op-amps the constant-current source can be altered by an external base bias voltage. This is achieved by having an external lead connection to the base of Q_3. The current source of the differential amplifier can then be altered to some extent. The resistance or impedance of Q_3 can be adjusted to meet the design parameters of a specific circuit application. When the impedance of Q_3 is high, the *common-mode*

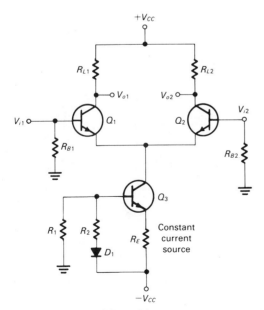

FIGURE 3-4   Differential amplifier with constant-current source in the emitter.

gain of the amplifier is very low and signal coupling is good. The constant-current source of an input differential amplifier is an important part of op-amp construction.

The differential amplifier of an op-amp can be made with other transistor devices and connected in a variety of different configurations. Two rather common configurations are shown in Fig. 3-5. The amplifier of Fig. 3-5 (a) is achieved with bipolar junction transistors connected in a Darlington-pair configuration. The input impedance of this circuit is increased by a factor of 1.5 times the beta. The beta of each transistor branch is also increased by the same value. Darlington-pair devices respond at high speed to a wide range of frequencies.

The differential amplifier of Fig. 3-5 (b) employs JFETs instead of bipolar junction transistors (BJTs) in its input. The *JFET*

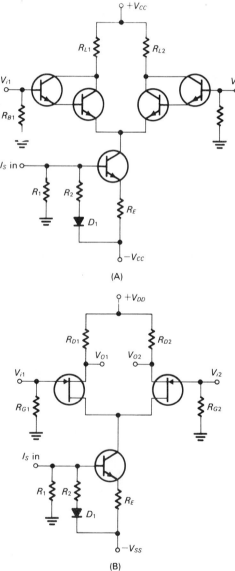

FIGURE 3-5   Differential amplifiers: (a) Darlington transistor; (b) JFETs.

input is extremely high resistant when compared with the BJT. Typical input impedance values are 10 MΩ for this type of differential input. Op-amps employing the JFET input are widely used in process control, medical instrumentation, and other applications requiring very low input current values.

Intermediate Amplifier Stage

The *intermediate amplifier* stage of an op-amp follows the input differential amplifier stage and feeds the output stage. This part of the op-amp is primarily responsible for additional *gain* and dc *voltage stabilization*. Figure 3-6 shows a simplification of the internal circuitry of the op-amp. Note that the intermediate amplifier stage is the center of the circuit. It contains a differential amplifier with a constant-current source transistor. The amplifier has a push-pull differential input and a single-ended output. Transistors Q_3, Q_4, and Q_7 and diode D_2 make up this part of circuit. As a rule, the intermediate differential amplifier has rather high gain capabilities compared with the input differential amplifier.

The intermediate amplifier stage of Fig. 3-6 is also equipped with a *voltage-stabilizing* transistor, Q_5. This transistor evaluates the signal at the emitters of Q_3 and Q_4. Since the intermediate differential amplifier is driven by a push-pull signal, there should be a zero reference level at its input when the first differential amplifier is operating properly. If an operational error occurs, a *correcting voltage* is developed by Q_5 across R_5 of the input amplifier. Q_5 also feeds the *error bias voltage* to the constant-current transistor Q_7. This voltage further reduces the error and improves the common-mode rejection capabilities of the amplifier. The output of the intermediate amplifier stage is zero-voltage-level corrected and appears at the collector of Q_4.

Output Stage

The output stage of an op-amp is designed primarily to develop the power needed to drive an external load device. In accomplishing this function, there must be a maximum output voltage signal developed across a low-impedance load. The output of an op-amp may be single-ended or two transistors connected in a complementary-symmetry power amplifier.

The output stage of Fig. 3-6 is a *single-ended emitter-follower amplifier*. This part of the op-amp is on the right side of the schematic diagram. Transistors Q_8, Q_9, and Q_{10} make up the output stage. Q_8 is an emitter-follower driver transistor. Q_9 serves as a constant-current source for the output transistor Q_{10}. The single-ended output of the intermediate differential amplifier stage drives the base of Q_8. This transistor matches the output impedance of Q_4 to the low output impedance of Q_{10}. *Maximum signal transfer* is accomplished through Q_8.

The signal *gain* of a single-ended output stage must be con-

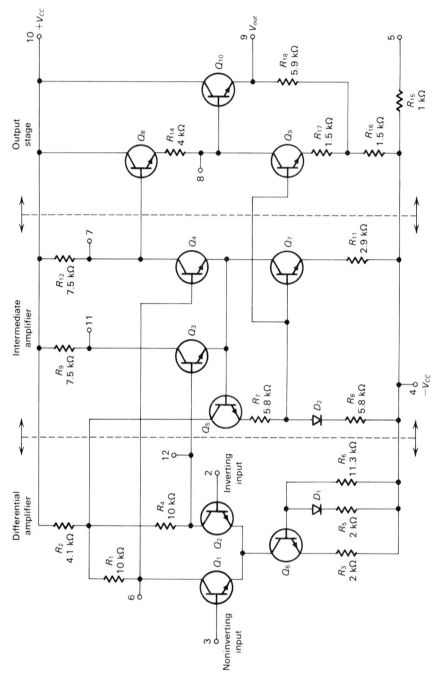

FIGURE 3-6 Internal circuitry of a general-purpose op-amp.

trolled or limited to some extent in order to provide good *stability*. Transistor Q_9 performs this function in the output stage of Figure 3-6. This transistor has a dual role. It responds as a *constant-current source* for the driver transistor Q_8 and as a *feedback regulator* for the output transistor. Feedback is regulated by maintaining the current through R_{16} at a rather constant level by conduction of Q_9. The constant-current function of Q_9 maintains the emitter of Q_8 at a consistent level to reduce level shifting and improve *common-mode rejection*.

Many op-amps employ two transistors in a *complementary-symmetry* output circuit instead of the single-ended emitter-follower circuit. The emitter resistor of a single-ended emitter-follower output generally consumes a great deal of power at high current levels. The complementary-symmetry output circuit overcomes this problem by using two transistors. The transistors are NPN and PNP complements that have the same or symmetrical characteristics. Figure 3-7 shows a simplification of the complementary-symmetry power amplifier used in many op-amps today.

The output of the complementary-symmetry amplifier of Figure 3-7 is developed across an external load connected to pin 8. Note that this terminal is connected directly to the emitters of Q_{10} and Q_{11}. These transistors are connected in a common-collector circuit configuration has a low-impedance output.

A resistive divider network is connected across $+V_{CC}$ and $-V_{CC}$. The base of each transistor is biased at cutoff or slightly above cutoff by this network. A signal applied to the input is fed directly to the base of each transistor. The positive alternation of an ac signal causes the NPN transistor Q_{10} to be conductive. Conduction current flows from ground through R_L, Q_{10}, and to $+V_{CC}$ for this alternation. Q_{11} is driven further into cutoff by this alter-

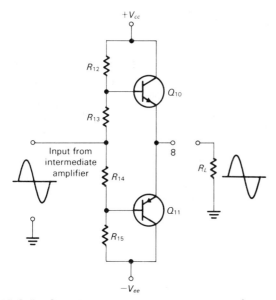

FIGURE 3-7   Complementary-symmetry output of an op-amp.

nation and has no output. For the negative alternation, Q_{11} is forward biased and goes into conduction. Q_{10} is reverse biased by this alternation and goes into a nonconductive state. Conduction of Q_{11} if from $-V_{CC}$ through Q_{11} and R_L to ground. The negative alternation develops output across R_L. This means that each transistor goes into conduction for one alternation. The output current through R_L is a combination of the current flow produced by each transistor. This type of amplifier can develop large amounts of current with good power gain and low output impedance.

Op-Amp Schematic Symbol

An op-amp has at least five terminals or connections in its construction. Two of these are for the power supply voltage, two for differential input, and one for the output. There may be other terminals in the makeup of this device, depending on its internal construction or intended function. Each terminal is generally attached to a schematic symbol at some convenient location. Numbers located near each terminal of the symbol indicate pin designations.

The schematic symbol of an op-amp is generally displayed as a triangular-shaped wedge. The triangle symbol in this case denotes the amplification function. Figure 3-8 shows a typical symbol with its terminals labeled. The point or apex of the triangle identifies the output. The two leads labeled − and + identify the differential input terminals. The − sign indicates *inverting input* and the + sign denotes *noninverting input.* A signal applied to the − input is inverted 180° at the output. Standard op-amp symbols usually have the inverting input located in the upper-left corner. A signal applied to the + input is not inverted and remains in phase with the input. The + input is located in the lower-left corner of the symbol. In all cases, the two inputs are clearly indentified as + and − inside the triangle symbol.

Connections or terminals on the sides of the triangle symbol are used to identify a variety of functions. The most significant of these are the two power supply terminals. Normally, the *positive voltage terminal* (+V) is positioned on the top side and the *negative voltage* (−V) is on the bottom side. In practice, most op-amps are supplied by a *split, or divided, power supply.* This supply has +V, ground, and −V terminals. It is important that the cor-

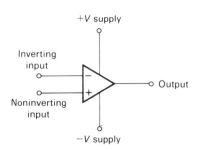

FIGURE 3-8 Op-amp schematic symbol.

rect voltage polarity be supplied to the appropriate terminals or the device may be permanently damaged. A good rule to follow for most op-amps is not to connect the ground lead of the power supply to − V. An exception to this rule is the *current-differencing amplifier*, or *CDA*, op-amp. These op-amps are made to be compatible with digital logic ICs and are supplied by a straight 5-V voltage source.

The actual schematic symbol of an op-amp generally has its terminals numbered and the element names omitted. There is a *pin-out* key to identify the name of each terminal. The manufacturer of the device supplies information sheets that identify the pin-out and operating data. Figure 3-9 shows part of the data sheet for a uA741 *linear op-amp*. Note the package styles, symbol location, and pin-out key for the device. This sheet also shows absolute maximum ratings, electrical characteristics, and typical performance curves.

Figure 3-10 shows the schematic symbol of a uA741 connected as an open-loop amplifier. The power supply voltage terminals are labeled + V, − V, and ground. Typical supply voltages are ±15, ±12, and ±6 V. The absolute maximum voltage that can be applied to this particular op-amp is 36 V between + V and − V, or ±18 V. The output of an op-amp circuit is generally connected through an external load resistor to ground. R_L, in this case, serves as an external load for the uA741. All the output voltage developed by the op-amp appears across R_L. This type of connection is described as a *single-ended output*. The operational load current for this device is a value between 10 and 15 mA. The maximum output, or saturation voltage, that can be developed is approximately 90% of the supply voltage. For a + 15 V supply, the saturation voltage is + 13.5 V. An ac output is + 13.5 and − 13.5, or 27 V p-p. The current and voltage limits of the output restrict the load resistance to a value of approximately 1 kΩ. Most op-amps have a built-in load *current-limiting* feature. The uA741 is limited to a maximum short-circuit current of 25 mA. This feature prevents the device from being destroyed in the event of a direct short in the load.

The inputs of an op-amp are labeled minus (−) for *inverting* function and plus (+) for the *noninverting* function. This is also used to denote that the inputs are differentially related. Essentially, this means that the polarity of the developed output is based on the voltage difference between the two inputs. In Fig. 3-10(a), the output voltage is positive with respect to ground. This is the result of the inverting input being made negative with respect to the noninverting input. Reversing the input voltage, as in Fig. 3-10(b), causes the output to be negative with respect to ground. For this connection, the noninverting input is negative when the inverting input is positive. This condition of operation holds true for both input possibilities. In some applications one input may be grounded. The difference voltage is then made with respect to ground. The inputs of Fig. 3-10 are connected in a *floating* config-

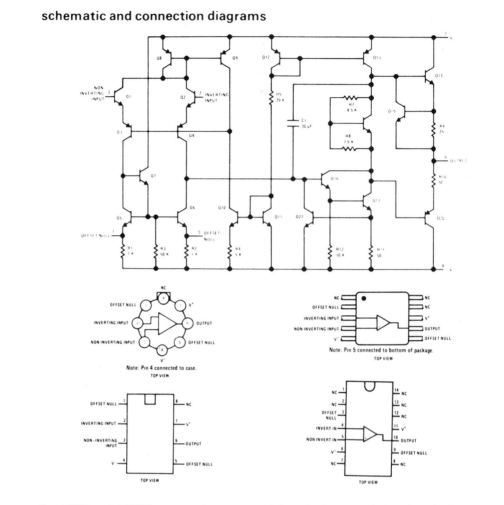

schematic and connection diagrams

The LM741 and LM741C are general purpose operational amplifiers which feature improved performance over industry standards like the LM709. They are direct, plug-in replacements for the 709C, LM201, MC1439 and 748 in most applications.

The offset voltage and offset current are guaranteed over the entire common mode range. The amplifiers also offer many features which make their application nearly foolproof: overload protection on the input and output, no latch-up when the common mode range is exceeded, as well as freedom from oscillations.

The LM741C is identical to the LM741 except that the LM741C has its performance guaranteed over a 0°C to 70°C temperature range, instead of −55°C to 125°C.

FIGURE 3-9 Schematic diagram of an op-amp. (Courtesy of National Semiconductor Corp.)

uration that does not employ a ground. For either type of input, the resulting output polarity is always based on the voltage difference. This characteristic of the input makes the op-amp an extremely versatile amplifying device.

Open-loop Voltage Gain

The *open-loop voltage gain* characteristic of an op-amp refers to an output that is developed when only a difference voltage is applied to the input. The two uA751C op-amps of Fig. 3-10 are con-

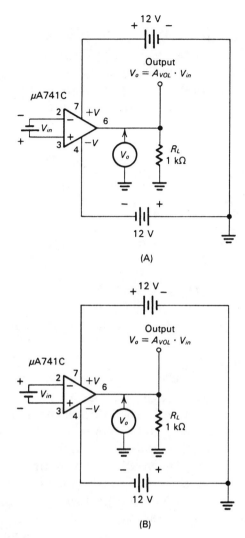

FIGURE 3-10 Inverting dc amplifiers.

nected in an open-loop circuit configuration. The open-loop voltage gain (A_{VOL}) is a ratio of the output voltage (V_{out}) divided by the differential input voltage (V_{in}). This is expressed by the formula

$$A_{\text{VOL}} = \frac{V_{\text{out}}}{V_{\text{in}}}$$

The open-loop voltage gain of an op-amp is usually quite large. Typical values are in the range of 10,000 to 200,000. When the differential input voltage is zero, the output voltage is also zero. If a slight difference in input voltage occurs, the output voltage increases accordingly. The output voltage, however, cannot exceed 90% of the source voltage. When the output voltage reaches this level, the amplifier is said to be *saturated*. Due to the high A_{VOL} of an op-amp, only a few millivolts of V_{in} are needed to cause it to go into saturation. As a rule, op-amps are rarely used in an open-loop circuit configuration.

To see how an op-amp responds in the open-loop mode of operation, refer to the following example.

Assume that the op-amps of Fig. 3-10 have an A_{VOL} of

100,000. The supply voltage is ± 15 V, as indicated. What value of V_{in} is needed to cause this amplifier to reach *saturation*?

For a representative op-amp, saturation occurs when the output voltage (V_{out}) reaches approximately 90% of the source voltage. For this circuit, the saturation voltage is

$$V_{sat} = 0.90 \times + 15 \text{ V} = + 13.5 \text{ V}$$

Transposing the previous A_{VOL} formula for input voltage (V_{in}) causes it to be:

$$V_{in} = \frac{V_{sat}}{A_{VOL}}$$

The value of V_{in} needed to cause saturation of an op-amp is approximately:

$$V_{in} = \frac{+ 13.5 \text{ V}}{100,000}$$

$$= 135 \text{ }\mu\text{V, or } 135 \times 10^{-6} \text{ V}$$

Input Offset Voltage

Ideally, the output voltage (V_{out}) of an op-amp should be 0 V when the input voltage is 0 V. In the internal construction of an op-amp, it is extremely difficult to develop perfectly balanced differential amplifiers that will eliminate this problem. Thus there is actually some output voltage even when the input is zero. Values range from microvolts to millivolts in a typical circuit.

An *input offsetting voltage* is used to overcome the unwanted output voltage of an op-amp. The net effect is 0 V at the output when no voltage is applied to the input. This voltage is called the input offset voltage (V_{io}). Typical V_{io} values are in the range from a few microvolts to several millivolts. The uA741C op-amp of Fig. 3-10 has a V_{io} of 1 mV. As a general rule, the lower the input offset voltage, the better the quality of the op-amp.

Input Bias Current

The internal construction of an op-amp generally causes input current values of the differential amplifier to be unequal to some extent. *Input bias current* (I_b) is the average current flowing into or out of the two inputs. This parameter describes the relationship of the two input current values. In general, smaller values of I_b are used with better-quality op-amps.

The type of transistor device used in the construction of the input differential amplifier has a great deal to do with its input bias current. Op-amps with JFET inputs have lower I_b values than BJT inputs. Typical I_b values for the uA741 are in the range of 80 nA. For a JFET input op-amp such as the LH0042, I_b values drop down to the 5- to 10-pA range. Input bias current values are often used when comparing the quality of different devices.

Input Impedance

The *input impedance* of an op-amp is the equivalent resistance that an input source sees when connected to the differential input terminals. Normally, the input impedance (Z_{in}) of a conventional op-amp is quite high. Typical values are in the range from 10 kΩ to 1 MΩ. If the Z_{in} of a specific device is unknown, an estimate of the value would be 250 kΩ. Impedance in this case refers to the opposition encountered by either an ac or dc signal voltage.

Input Current

The *input current* (I_{in}) of a conventional op-amp is usually quite small. A voltage source connected to the differential input sees an extremely high input impedance. Typical I_{in} values are in the range of a few nanoamperes. A nanoampere is one-billionth, or 1×10^{-9}, ampere. To see how a representative input current of this value occurs, consider the following example.

Using one of the op-amps of Fig. 3-10, determine the *input current* for circuit operation at or near the point of saturation. Assume that the Z_{in} is estimated to be 250 kΩ. The input voltage that produces saturation was previously determined to be 135 μV.

The solution is

$$I_{in} = \frac{V_{in}}{Z_{in}}$$

$$= \frac{135 \ \mu V}{250 \ k\Omega}$$

$$= 0.54 \ nA, \text{ or } 0.00054 \ A$$

With input current values being this small, they are often considered to be negligible in most op-amp circuit calculations.

Slew Rate

The *slew rate* of an op-amp refers to the rate at which its output changes from one voltage to another in a given time. This parameter is extremely important at high frequencies because it indicates how the output responds to a rapidly changing input signal. Slew rate depends on such things as amplifier gain, compensating capacitance, and the polarity of the output voltage. The worse case, or slowest slew rate, occurs when the gain is at 1, or unity. As a rule, slew rate is generally indicated for unity amplification. Mathematically, slew rate is expressed as:

$$\text{Slew rate (SR)} = \frac{\text{maximum change in output voltage } (\Delta V_{out})}{\text{change in time } (\Delta t)}$$

Slew rate is primarily an indication of how an op-amp responds to different frequencies. A slew rate of 0.5 V/μs means that the output has time to rise only 0.5 V in 1 μs. Since frequency is

time-dependent, this indicates where certain input frequencies do
not have enough time to produce a corresponding output signal.
The resulting output is a distorted version of the input. The maximum frequency at which we can obtain an undistorted output is
determined by the expression:

$$\text{Frequency }(f) = \frac{\text{slew rate}}{6.28 \times \text{peak output voltage }(V_{out})}$$

For the uA741 op-amp, the slew rate is 0.5V/μs. This is used to
determine the undistorted output frequency of the op-amp. Consider the following example.

At what maximum frequency can you get an *undistorted* output voltage of 1 V with the uA741 op-amp?

$$f = \frac{\text{SR}}{6.28} \times V_{out}$$

$$= \frac{1}{6.28} \times \frac{0.5\text{ V}}{1\ \mu\text{s}}$$

$$= 0.15924 \times \frac{0.5\text{ V}}{1 \times 10^{-6}}$$

$$= 0.15924 \times 0.5\text{ V} \times 10^6$$

$$= 70.620\text{ Hz}$$

The slew rate of an op-amp is primarily determined by a compensating capacitor that is either internal or externally connected. Essentially, it takes a certain period of time for the capacitor to develop a charge voltage. The value of the capacitor and the charging
current determine the response of the capacitor. Since the op-amp
has internal constant-current sources to limit the current, the capacitor can only charge at a specified rate. The input signal must
occur at a slower rate than the capacitor charge time, or some other
type of distortion will then occur. When the input frequency is
higher than the slew rate limit, the square wave appears as a slope
and the sine wave becomes a triangle.

DIFFERENTIAL VOLTAGE COMPARATORS

Op-amps are rarely used in an *open-loop* circuit configuration. The
A_{VOL} is generally so high that it is rather difficult to prevent the
output (V_{out}) from being driven into saturation. One application of
an op-amp in an open-loop circuit configuration is the *differential
voltage comparator*. Used in this manner, the op-amp simply compares voltage values applied to its two inputs and indicates which
is greater.

In Fig. 3-11(a), the output voltage swings into full saturation
for an extremely small change in input voltage (V_{in}). Note that the
noninverting input is referenced at ground, or 0 V. If V_{in} swings
slightly positive, V_{out} goes into negative saturation ($-V_{sat}$). A

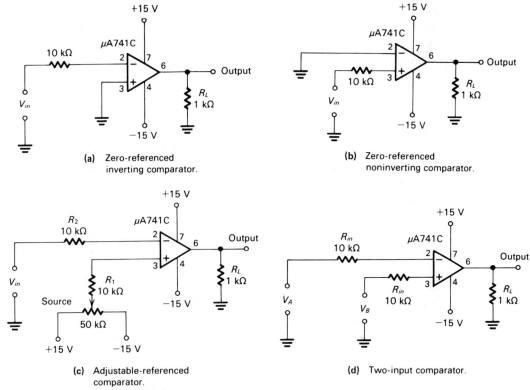

(a) Zero-referenced inverting comparator.

(b) Zero-referenced noninverting comparator.

(c) Adjustable-referenced comparator.

(d) Two-input comparator.

FIGURE 3-11 Op-amp differential comparators.

small change in the value of V_{in} in the negative direction causes the output to go into positive saturation $(+V_{sat})$. Used in this manner, the op-amp compares the voltage difference between its *inverting input* and a zero reference value. An ac voltage applied to the same circuit configuration causes a square wave to appear in the output. The output is inverted, or *180° out of phase* with the ac input. A comparator circuit of this type is often called a *waveshaper.*

Figure 3-11(b) shows a *zero voltage referenced comparator* for noninverting output. In this type of circuit the V_{out} detects, or indicates, the polarity of the input voltage. When V_{in} is positive, the output swings into $-V_{sat}$. A negative value of V_{in} likewise causes the output to reach $-V_{sat}$. An ac input causes an in-phase square wave to appear in the output.

In Fig. 3-11(c), V_{in} is compared with a *variable reference* voltage value. The output voltage goes positive only if V_{in} is made more positive than the voltage level setting of R_3. It also swings negative if V_{in} is more negative than the voltage value adjusted by R_3. A comparator of this type has an adjustable reference level. This circuit is designed to produce an output when the input exceeds its referenced level adjustment value.

The circuit of Fig. 3-11(d) compares the voltage value of signals applied to *both inputs* of an op-amp. If V_A is more positive than V_B at the same time, the output is negative. In the same manner, the output is positive if V_A is more negative than V_B. A similar response occurs if V_B is greater than V_A. In this case, however, the output is of the same polarity as the input to V_B. A comparator of this type is generally used to detect specific voltage values.

An interesting characteristic of the *two-input comparator* of Figure 3-11(d) is its *common-mode rejection ratio* (CMRR). If a signal of the same voltage value and polarity is applied to both inputs at the same time, the output is null, or zero. This type of signal is therefore attenuated and will not appear in the output. The CMRR is used to reject noise or to reduce unwanted ac signal voltage values.

INVERTING CONFIGURATION

One of the most widely used op-amp circuit configurations is the *inverting amplifier.* An amplifier of this type is defined as a circuit that receives a signal voltage at its input and delivers a large undistorted version of the signal at its output. The phase or polarity of the output signal is an inversion of the input. Operation does not ordinarily permit the output to reach saturation. The level of amplification is controlled by a feedback resistor connected between the output and inverting input. This causes the amplifier to have *negative feedback.* The addition of a feedback resistor permits the amplifier to have a controlled level of amplification. Performance is no longer dependent on the open-loop gain (A_{VOL}) of the device. Closed-loop voltage gain (A_{VCL}), or simply A_V, can be controlled by altering the value of feedback resistor (Rf).

A typical *inverting op-amp circuit* is shown in Fig. 3-12. This basic circuit consists of an op-amp and three resistors. The noninverting input is connected to ground. Input to the amplifier is applied to the inverting input through resistor R_{in}. The output signal is developed across the load resistor (R_L) and ground. A portion of the output signal is also returned to the inverting input

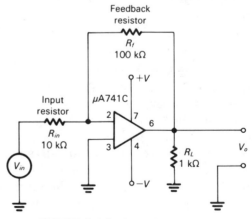

FIGURE 3-12   Inverting op-amp.

through the feedback resistor (R_f). The value of the inverting input signal is therefore determined by a combination of V_{in} and the output signal fed back through R_f.

To assess the operation of an inverting amplifier, we will describe a number of events that occur rather quickly when it is placed into operation. Assume, for example, that the op-amp of Fig. 3-12 is energized by ±15 V. Let us also assume that no signal is initially applied to the input. In this operational state, with no differential input signal applied, the output will be zero, or show no output voltage. This represents the *quiescent* condition, or steady state of operation.

Assume now that an input signal of $+1$ V is applied to the inverting input of our op-amp. With this voltage applied, the inverting input immediately goes positive. This action causes V_{out} to swing immediately in the negative direction. At the same instant, a negative-going voltage is fed back to the inverting input through R_f. This immediately reduces the original $+1$ V applied to the inverting input. The feedback signal does not completely cancel the V_{in} signal. It simply reduces the value. The $+1$ V signal is immediately changed to a value of only a few microvolts. This means that the inverting input is now controlled or limited to a rather low voltage value. As a general rule, the input is considered to be at approximately zero. Through the feedback loop, the inverting input voltage is held to approximately zero regardless of the value of V_{in}.

To see how an inverting op-amp responds to an input signal, we must consider the virtual ground concept. A *virtual ground* is the point of a circuit which is at zero potential (0 V) but is not actually connected to ground. In an inverting op-amp circuit, a virtual ground appears at the inverting input terminal. With the noninverting input grounded, the voltage at the inverting input is never greater than a fraction of a millivolt. V_{in}, V_{out}, R_{in}, and R_f all tend to hold the voltage of the inverting input to practically zero. With this condition existing, the inverting input responds as if it were grounded. It is a common practice to refer to this point of an op-amp as a virtual ground.

Refer now to the *inverting op-amp* and its equivalent circuit in Fig. 3-13. Keep in mind that the voltage at the inverting input terminal is nearly zero and its input impedance is approximately 1 MΩ. Assume now that an input of $+1$ V is applied to V_{in}. This condition causes 1 mA of current to flow through R_{in}. Note the I_{in} calculations near the input of the equivalent circuit. This shows that a 1-V drop will appear across R_{in}. The inverting input continues to remain at zero or virtual ground. The 1 mA of current entering the equivalent circuit at the V_G point must therefore all flow through the feedback resistor (R_f). Very little input control flows into the inverting input. This means that practically all the 1 mA flows through the 10-kΩ feedback resistor. The voltage drop across R_f will be 1 mA times 10 kΩ, or 10 V. Note the calculation near R_f of the equivalent circuit. As indicated, the polarity of the output voltage is negative with respect to the V_G point. This shows the inverting characteristic of the op-amp.

At this point it may be asked, How does an op-amp control a -10-V output signal if the inverting input voltage and current are both zero? It should be pointed out that the inverting input terminal is considered to be at a virtual zero. In a circuit, a signal applied to V_{in} always causes some voltage and current to appear at the inverting input terminal. The actual value of it is so small that it is considered to be zero. In practice, these values are usually not measurable. In effect, this means that the right side of R_{in} is considered to be zero and the left side is V_{in}. Across the feedback resistor, the right side is V_{out} and the left side is zero.

The *closed-loop voltage amplification* of an inverting amplifier is determined primarily by the value of R_{in} and R_f. For the op-amp of Fig. 3-13, this is based on the formula:

$$A_{VOL} = \frac{-R_f}{R_{in}}$$

$$= \frac{-10\ k\Omega}{1\ k\Omega}$$

$$= -10\ V$$

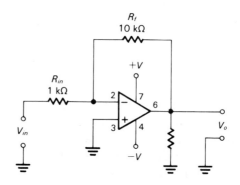

(a) Inverting operational amplifier.

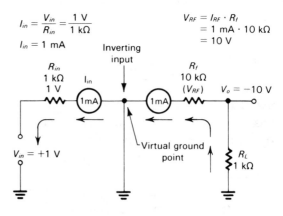

(b) Equivalent circuit.

FIGURE 3-13 Inverting op-amp.

The negative sign of the formula denotes the inversion function of the op-amp. Typical applications of the inverting amplifier have values that range from -1 to $-10,000$.

SUMMING OPERATIONS

A slight modification of the inverting op-amp permits it to achieve the mathematical operation of *addition*. Used in this manner, it can add dc voltages or ac wave forms. Adding or summing operations are very useful in analog computers and in analog-to-digital conversion functions.

    Summing can be achieved when two or more voltage values are applied to the input of an op-amp. Representative schematic diagram of an *adder* is shown in Figure 3-14. Note that the circuit is an inverting op-amp with two inputs. The resistor values are all 1 kΩ. The resulting output voltage that appears across R_1 is the sum of the input voltages applied to V_1 and V_2. For simplicity, the circuit shows the addition of dc voltage values. Keep in mind that it can add ac voltage values equally well.

    The *voltage gain* of our summing op-amp is 1. This is based on the resistance values of R_{in} and R_f. The output, however, is -1 due to the inversion characteristic of the op-amp. If a positive value is desired, the output of the adder can be followed by an op-amp with a gain of 1.

    The voltage applied to each input of a summing amplifier responds as an independent source. Input V_1 does not alter or change the input of V_2. This is primarily due to the virtual ground appearing at the inverting input terminal. In Fig. 3-14, note that input V_1 causes 3 mA of current to flow through input resistor R_1. With only this voltage applied, the feedback resistor R_f would have

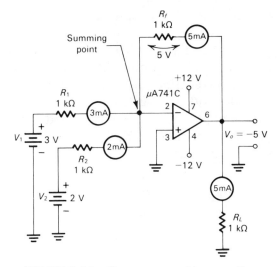

FIGURE 3-14 Op-amp summing operation.

a resulting 3 mA through it. The load resistor R_1 would likewise have 3 mA and develop an output voltage of 3 V. An applied input voltage of + 3 V produces a corresponding − 3 V output. Therefore, 3 V + 0 V equals 3 V.

Applying + 3 V to V_1 and + 2 V to V_2 causes 3 mA and 2 mA to flow through the respective resistors. Current flow at the inverting input will be the sum of these two values or 5 mA. This point of the circuit is usually called the *summing point* instead of the virtual ground. In effect, this common point isolates V_1 from V_2. The combined current from this point must all pass through the feedback resistor R_f. Thus R_f has 5 mA of current. The voltage drop across R_f will be 5 V, with V_{out} being − 5 V due to the inverting characteristic. A similar current value flows through the load resistor R_1. In effect, an input of +3 V and +2 V causes the output to increase to − 5 V. The summing function is expressed by the formula

$$V_{out} = -(V_1 + V_2)$$

This formula is used only when the input resistors are of the same value. Any number of input voltages may be added by this circuit. The input resistors must be the same value for each input voltage.

Figure 3-15 shows three variations of the *summing circuit*. Circuit *A* shows the standard *adding function*. This particular circuit simply adds the input voltages and produces the sum at the output. All resistors must be of an equal value in order for the circuit to respond correctly.

The op-amp of Fig. 3-15(b) achieves both *summing and gain*. The input resistors are of the same value. The feedback resistor is a multiple of the input resistors. The sum of V_1, V_2, and V_3 in this case is multiplied by a factor of 5. Note the formula for this mathematical operation.

Figure 3-15(c) is a *scaling adder*. The input to this circuit sees different weighting factors in the input resistors. An input of +1 V applied to V_1 produces a 1-V output. In the same manner, V_2 is 2 V, V_3 is 4 V, and V_4 is 8 V. This means that the inputs are scaled down by the powers of 2. The resulting output of V_4 carries eight times more weight than V_1. Scaling adders are frequently used in digital-to-analog converter functions. The formula for determining the output of a scaled adder notes the difference in the weighting factor. It should be apparent that voltages fed into the smaller-valued input resistors are more heavily scaled because they produce a larger voltage output.

NONINVERTING OP-AMPS

A *noninverting op-amp* can provide controlled voltage gain with high input impedance and no inversion of the input-output signals. Voltage gain is dependent on the input voltage and feedback resistors. An unusual feature of the noninverting op-amp circuit con-

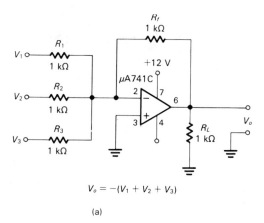

$$V_o = -(V_1 + V_2 + V_3)$$

(a)

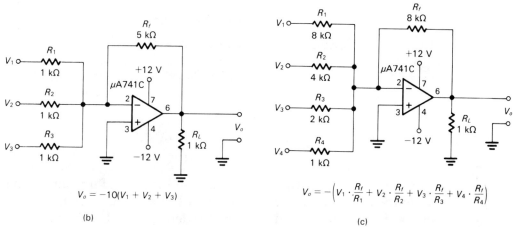

$$V_o = -10(V_1 + V_2 + V_3)$$

(b)

$$V_o = -\left(V_1 \cdot \frac{R_f}{R_1} + V_2 \cdot \frac{R_f}{R_2} + V_3 \cdot \frac{R_f}{R_3} + V_4 \cdot \frac{R_f}{R_4}\right)$$

(c)

FIGURE 3-15 Summing amplifiers: (a) standard adder; (b) adder with gain; (c) scaling adder.

figuration is the placement of the feedback resistor network. It is placed in the *inverting input* with a resistor connected to ground. The input voltage (V_{in}) is applied to the *noninverting input*.

A representative *noninverting amplifier* is shown in Fig. 3-16. In this circuit the signal voltage (V_{in}) to be amplified is applied to the noninverting input (+). A fraction of the output voltage (V_{out}) is returned to the inverting input (−) through a voltage-divider network composed of R_f and R_1. In theory, we again assume that very little current flows into the + and − inputs due to the virtual ground concept. This means that the *differential input voltage* (V_{di}) is essentially zero. The voltage (V_1) developed across resistor R_1 is therefore equal to the input voltage (V_{in}). This means that the current (I_{R1}) passing through R_1 is equal to either

$$I_{R1} = \frac{V_{in}}{R_1} \quad \text{or} \quad I_{R1} = \frac{V_1}{R_1}$$

The feedback current (I_F) is also considered to be the same as I_{R1}. This means that:

$$V_{out} = I_F R_1 + R_f$$

or

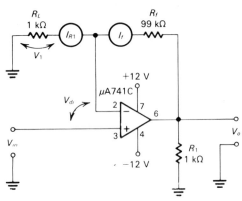

FIGURE 3-16 Noninverting op-amps.

$$V_{\text{out}} = I_{R1}R_1 + R_f$$

In terms of voltage gain, we can then say that:

$$A_V = \frac{V_{\text{out}}}{V_{\text{in}}}$$

or

$$\frac{V_{\text{out}}}{V_{\text{in}}} = \frac{R_1 + R_f}{R_1}$$

In an actual circuit, the closed-loop voltage gain of a noninverting amplifier depends almost entirely on the external circuit components. In this regard, A_V is usually determined by the values of R_1 and R_f. The standard voltage gain formula for a noninverting circuit configuration is expressed as:

$$A_V = \frac{R_1 - R_f}{R_1} \quad \text{or} \quad \frac{R_f}{R_1} + 1$$

For example, to determine the voltage-amplifying capabilities of the noninverting amplifier of Fig. 3-16 by the resistor method:

$$A_V = \frac{R_f}{R_1} + 1$$

$$= \frac{99 \text{ k}\Omega}{1 \text{ k}\Omega} + 1$$

$$= 100$$

SELF-EXAMINATION

1. The input of an op-amp is usually a _____ amplifier.
2. Op-amps are rarely used in _____ configurations.

3. If the feedback resistor for a noninverting circuit is 78 kΩ and the input resistor is 2 kΩ, $A_V = $ _____.

4. An inverting op-amp has a 10-kΩ R_f and a 2-kΩ R_{in}. The potential A_v is $A_V = $ _____.

5. Three stages of an op-amp circuit are _____ _____, _____, and _____.

6. The property of a differential amplifier to reject signals common to both of its inputs is called _____ _____ _____.

7. Dc voltage stabilization of an op-amp is caused by the _____ _____ stage.

8. The output stage of an op-amp may be _____ _____ or _____ _____.

9. Five terminals of an op-amp are _____, _____, _____, _____, and _____.

10. The ratio of output voltage to differential input voltage of an op-amp is called _____ _____.

11. An _____ _____ _____ is used to overcome unwanted output voltage of an op-amp.

12. The rate at which an op-amp's output voltage changes is called _____ _____.

13. One application of an op-amp in an open-loop arrangement is a _____ _____ _____.

14. The type of feedback of an inverting op-amp circuit is _____.

15. An inverting op-amp circuit may be used to accomplish the mathematical function of _____.

16. An op-amp has an open-loop voltage gain of 200,000 with a supply voltage of $\pm$ 12 V. The value of input needed to cause saturation is input = _____ V.

17. The controlled gain of an op-amp using an R_{in} of 4.7 kΩ and an R_f of 100 kΩ is gain = _____.

18. A noninverting op-amp has an R_{in} of 10 kΩ and an R_f of 120 kΩ. With a +0.6-V p-p input voltage, $V_{out} = $ _____.

19. An inverting op-amp has an R_{in} of 22 kΩ and an R_f of 68 kΩ. With a +0.5-V p-p input signal, $V_{out} = $ _____.

20. Refer to the summing amplifier of Fig. 4-15(a), if $V_1 = $ +2 V, $V_2 = $ +3 V, and $V_3 = $ −1 V, $V_{out} = $ _____.

ANSWERS

1. Differential
2. Open-loop
3. $A_V = 40$
4. $A_V = -5$
5. Differential amplifier; intermediate amplifier; low-impedance output stage
6. Common mode rejection ratio
7. Intermediate amplifier
8. Single-ended; complementary symmetry
9. $+V$; $-V$; inverting input; noninverting input; output
10. Open-loop voltage gain (A_{VOL})
11. Input offsetting voltage
12. Slew rate
13. Differential voltage comparator
14. Negative
15. Addition
16. $0.54V = 54mV$
17. -21.277
18. $7.8V$
19. $15.45V$
20. -4

EXPERIMENT 3-1
OP-AMPS

An op-amp is a linear integrated circuit that is capable of high gain and operates over a wide range of voltages. Its internal structure is small and somewhat complex. As a general rule, outside component values attached to the op-amp determine its operating capabilities. Gain ranges from a high of 100,000 for a single op-amp in an open-loop circuit to 1 when it is used as a voltage follower.

OBJECTIVE

To construct a circuit that is commonly used to test the operation of an op-amp. This test circuit is a simple astable multivibrator or free-running oscillator application of the op-amp. The circuit connections are designed primarily for a μA741C op-amp of the 8-pin mini-dual-in-line package type. The same circuit could also be used to test similar op-amps with a slight modification of the pin connections.

EQUIPMENT

Resistors: 2 kΩ, 200 kΩ(2), 560 kΩ, 390 Ω
0.01-μF, 100-V dc capacitor
8-Ω voice-coil speaker
μA741C op-amp
Oscilloscope
Multifunction meter
SPST switch
Power supply: $\pm$9 V dc or $\pm$15 V dc (variable)

PROCEDURE

1. Construct the op-amp test circuit of Fig. 3-1A.
2. The power source for this circuit is developed from a split supply made of two 9-V batteries or two variable dc power supplies connected as indicated in Fig. 3-1B. If variable power supplies are used, the IC should operate at + 15 V and − 15 V for optimum results.
3. Before applying power, prepare a multifunction meter to measure current between pin 7 and the + V_{CC} source.

185

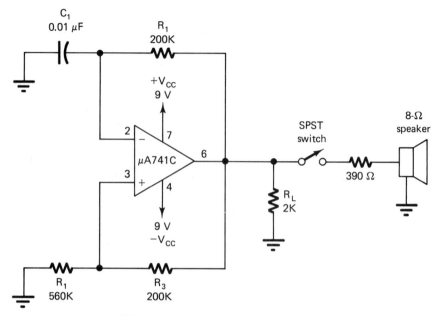

FIGURE 3-1A   Op-amp tester circuit.

Turn on the power supply. Record the current. With a 2-kΩ load, a good IC should show from 1 to 4 mA on the meter. Record your findings.

4. If an oscilloscope is connected across load resistor R_L, a square wave will indicate a good IC. Oscillation in this case is indicative of amplification, which is necessary for a good IC.

5. Make a sketch of the waveform observed across R_L in Fig. 3-1C. Indicate the peak-to-peak voltage change appearing in the output.

6. A quick and easy audio test for the IC is achieved by closing the output switch. A good IC should be capable of producing approximately 25 mA of output current under load. Turn on the speaker for an audio test.

7. The audio test will provide more of a load on the IC than the 2 kΩ load resistor. Measure and record the load cur-

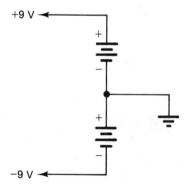

FIGURE 3-1B   Split-supply power source.

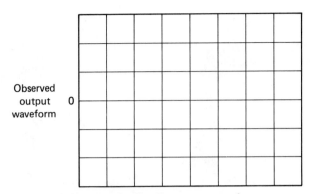

FIGURE 3-1C Oscilloscope waveform.

rent for the load resistor and then do the same for the speaker: R_L _____ mA; speaker _____ mA.

8. When the oscillator is in operation, touch one end of the resistor R_1, R_2, or R_3 with your fingers. How does this alter the circuit? _____ _____

9. Observe the output of the circuit on the oscilloscope while touching resistor R_L. What occurs? _____ _____

10. How does the audio output waveform compare with the 2-kΩ R_L waveform of Fig. 3-1C? _____ _____

ANALYSIS

1. What is an operational amplifier? _____ _____ _____

2. What are three indications of a good op-amp from this test circuit? _____ _____ _____

3. How would a faulty IC respond to the current test? _____ _____

4. What causes the change in step 8? _____ _____ _____

EXPERIMENT 3-2
INVERTING OP-AMPS

Applications of the op-amp are very common in control circuits. Integrated circuits of this type are used primarily as high-gain amplifiers. The versatility of this amplifier makes it very useful in automatic control systems and instrumentation applications. With no external feedback circuit, an op-amp can achieve an open-loop gain that is around 100,000. With a feedback circuit, gains can be achieved from 1 to 100,000 according to the selected resistance values.

OBJECTIVES

1. To construct a simple inverting op-amp, including both ac and dc signals.
2. To employ calculations that show voltage gain, current gain, and resistance value selection.
3. To be able to design a simple operational amplifier that will achieve a specific amount of gain.

EQUIPMENT

μA741C op-amp
Resistors: 10 kΩ, 200 kΩ, 2 kΩ
Split power supply or two 9-V batteries
Variable low-voltage dc source
AC signal generator
4-μF, 25-V dc capacitor
Multifunction meter
Oscilloscope

PROCEDURE

1. Construct the inverting dc op-amp of Fig. 3-2A. The power source for the μA741C is derived from a split power supply of two 9-V batteries (Fig. 3-2B) or variable dc supplies. Figure 3-2C shows the top view of the IC package.
2. Adjust the dc input source to produce 0.1 V dc.
3. Energize the IC by turning on or connecting the split

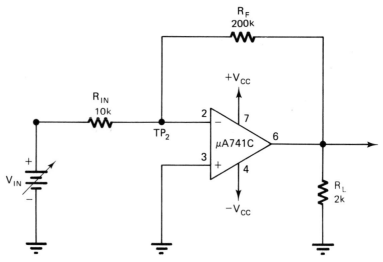

FIGURE 3-2A Dc op-amp.

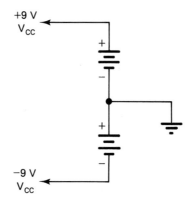

FIGURE 3-2B Split power supply.

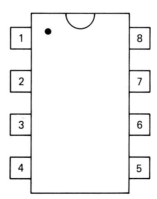

FIGURE 3-2C Top view of *V* package *IC*.

power supply. Measure and record the output voltage across R_L: _____ V output.

4. How does the polarity of the output compare with that of the input? _____

5. Carefully increase the V_{in} voltage to 0.3 V; then measure and record the output: output = _____ V_{dc}.

6. The dc voltage gain of an operational amplifier is determined by the formula $A_V = -R_F/R_{in}$. The negative sign of R_f denotes the inversion function.

7. Calculate the gain factor for 0.3 V input: voltage gain = _____.

8. Test the measured value of steps 3 and 5 with calculated values. How do they compare? How do you account for any differences between calculated and measured values? _____

9. Using the measured voltage across R_L in step 5, calculate the load current passing through R_L: I_L calculated = _____ mA.

10. Measure the load current and record your findings: I_L measured _____ mA.

11. How do the measured and calculated values of I_L compare? _____

12. Measure the input current at test point 1. Then calculate the input current with the formula $I_{in} = V_{in}/(R_S + R_{in})$. We will assume that the resistance of a regulated power supply R_S is quite small and not applicable in this case. How does the measured value compare with the calculated value? _____

13. The current gain (A_i) of an inverting dc amplifier is determined by the formula $A_i = I_{out}/I_{in}$. Calculate this value: A_i = _____.

14. Turn off the power supply and disconnect the dc input source V_{in}. Connect a 4-μF capacitor in series with R_{in} and an ac signal source of 400 Hz. Adjust the output of the signal source to its lowest value. Connect an oscilloscope across output resistor R_L.

15. Turn on the power supply and signal source. Carefully adjust the signal source input until it produces its maximum *undistorted* output. Record the peak-to-peak output voltage: output voltage = _____ V p-p.

16. Measure and record the input voltage at test point 1: input voltage = _____ V p-p.

17. Calculate the ac voltage gain of the amplifier using the formula $A_V = R_F/R_{in}$: voltage gain A_V = _____.

ANALYSIS

1. If the source resistance R_S were 500 Ω, how would it alter I_{in} in Step 10? _____

2. If R_F were increased to 250 kΩ, how would it alter the voltage gain (A_V) of step 7? _____

3. If R_F were changed to 50 kΩ, how would it alter the voltage gain (A_V) Step 7? _____

4. How does the current gain (A_i) compare with voltage gain (A_V) in steps 11 and 7? _____

EXPERIMENT 3-3
NONINVERTING OP-AMPS

Circuits frequently need amplifiers that develop high voltage and current gains without signal inversion. The noninverting op-amp is specifically designed for this type of application. Both ac and dc signals may be amplified by this circuit, with the input and output remaining in phase. The input resistance or impedance of this amplifier is quite high, with typical values exceeding 100 MegΩ. With the input signal applied to the noninverting input terminal, voltage gain is dependent upon the values of R_{in} and R_f. The output signal of this amplifier is developed across the load resistor, R_L. Typical R_L values are 35 to 50Ω.

OBJECTIVES

1. To connect a simple noninverting op-amp and observe its operation.
2. To calculate amplifier gain using both the resistance method and the voltage method.
3. To be able to design a simple noninverting op-amp and alter its gain by changing the resistance values.

EQUIPMENT

μA741C op-amp IC
Resistors: 200 kΩ (2), 10 kΩ, 2 kΩ
Split power supply or two 9-V batteries
0.01-μF, 100-V dc capacitor
Variable dc V_{in} source or a 1.5-V C cell with a potentiometer
Multifunction meter
Oscilloscope
Ac signal generator

PROCEDURE

1. Construct the noninverting op-amp of Fig. 3-3A.
2. Adjust the voltage input source to 0.1 V input and connect it to point A. Turn on the IC power source, and measure and record the output voltage developed across R_L:
   V_{out} = _____V dc.

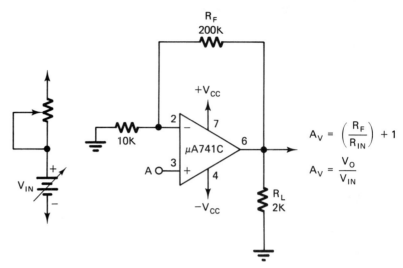

FIGURE 3-3A Noninverting op-amp dc circuit.

3. Using the voltage amplification formula, calculate the A_V for the circuit using measured voltage values: $A_V =$ _____.

4. Using the A_V formula for resistance, calculate the gain using the circuit values for R_{in} and R_F: $A_V =$ _____.

5. If there is a significant difference in values of these two voltage calculations, measure the actual value resistance of R_{in} and R_F. Then calculate A_V using these actual values.

6. Connect two 200-kΩ resistors in series to make a 400-kΩ R_F. Repeat steps 2, 3, and 4 for the new value of R_F. Record the measured value and calculated values: $V_{out} =$ _____ V; A_V by voltage = _____; A_V by resistance = _____.

7. Connect the two 200-kΩ resistors in parallel and repeat step 6 for V_{in} of 0.3 V. Record the measured and calculated values: $V_{out} =$ _____ V; A_V by voltage = _____; A_V by resistance = _____.

8. Alter the dc amplifier of Fig. 3-3A to conform with the ac amplifier of Fig. 3-3B. Resistors R_{in} and R_1 should be of equal value.

9. Turn on the ac signal generator and adjust it to produce a 400-Hz signal of 0.1 V p-p. Turn on the IC power supply.

10. Measure and record the output voltage appearing across R_L: $V_{out} =$ _____ V p-p.

11. Using the voltage formula calculate the voltage gain: A_V = _____.

12. Calculate the voltage gain with the resistor formula: resistor $A_V =$ _____.

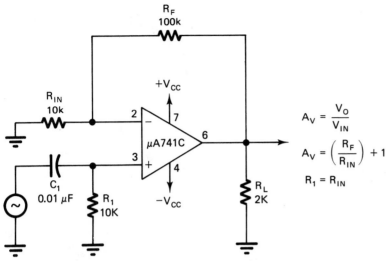

FIGURE 3-3B   Noninverting op-amp ac circuit.

13. Observe the phase relationship of the input signal while observing the output signal across R_L. At what input value does the output become distorted? _____

_____

ANALYSIS

1. What is the primary difference between an ac and a dc noninverting op-amp? _____

_____

2. Why is a capacitor used on the input of the ac amplifier?

_____

3. How would the value of the capacitor limit the amplifying frequency of the amplifier? _____

_____

_____

4. Why does R_F influence voltage gain? _____

_____

_____

UNIT 3 EXAMINATION
OPERATIONAL AMPLIFIER CIRCUITS

Instructions: For each of the following, circle the answer that most correctly completes the statement.

1. The correct method of finding the gain of an operational amplifier circuit is:

    a. $A = \dfrac{R_1}{R_2}$

    b. $A = \dfrac{R_2}{R_1}$

    c. $A = \dfrac{R_1}{R_2} + 1$

    d. $A = \dfrac{R_2}{R_1} + 1$

2. In the circuit shown in Fig. E-3-2, if R_1 and R_2 were both 1000-Ω resistors, the gain of the operational amplifier circuit would be:

    a. 1.0   b. 1.5

    c. 2.0   d. 2.5

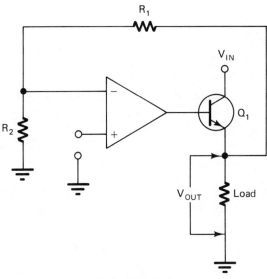

FIGURE E-3-2

3. The gain of the operational amplifier in Fig. E-3-3 is:

    a. $+20$ b. -20

    c. $+21$ d. -21

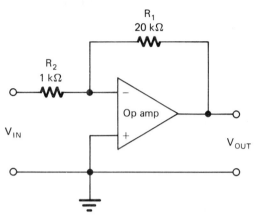

FIGURE E-3-3

4. The voltage gain figure listed below which is typical of a voltage follower is:

   a. 1   **b.** 10

   c. 100   **d.** 1000

5. The characteristic that would describe an operational amplifier used as a voltage follower is:

   a. Inverting, high gain

   b. Inverting, low gain

   c. Noninverting, high gain

   d. Noninverting, low gain

6. Impedance characteristics that describe a voltage follower are:

   a. Low input, low output

   b. Low input, high output

   c. High input, low output

   d. High input, high output

7. An inverting op-amp has a 20-kΩ input resistor and a 50-kΩ feedback resistor. With a +2-V input signal applied, the output voltage would be:

   a. +2 V   **b.** −2 V

   c. +5 V   **d.** −5 V

8. An operational amplifier that has an output of −5 V with an input of +10 V is considered:

   a. Inverting, with a gain of 2

   b. Noninverting, with a gain of 2

   c. Inverting, with a gain of 0.5

   d. Noninverting, with a gain of 0.5

9. The gain of a *noninverting* operational amplifier may be calculated by the formula:

   a. $A = \dfrac{R_f}{R_{in}} + 1$

b. $A = \dfrac{R_f}{R_{in}}$

c. $A = \dfrac{R_{in}}{R_f} + 1$

d. $A = -\dfrac{R_{in}}{R_f}$

10. The gain of an *inverting* op-amp may be calculated by the formula:

 a. $A = \dfrac{R_f}{R_{in}} + 1$

 b. $A = -\dfrac{R_f}{R_{in}}$

 c. $A = \dfrac{R_{in}}{R_f} + 1$

 d. $A = -\dfrac{R_{in}}{R_f}$

11. The gain of an operational amplifier with a bandwidth of 10 kHz and a gain bandwidth product of 100,000 is:

 a. 1   **b.** 10

 c. 100   **d.** 1000

12. A noninverting operational amplifier has a +20 V output when the input is +5 V. If the input resistor is 5 kΩ, the value of the feedback resistor is:

 a. 10 kΩ **b.** 15 kΩ

 c. 20 kΩ **d.** 30 kΩ

13. A noninverting operational amplifier has a 10-kΩ feedback resistor and a 5-kΩ input resistor. The input voltage required to obtain a +10 V output is:

 a. 3.3 V   **b.** 5.0 V

 c. 20 V   **d.** 30 V

14. The voltage gain of an operational amplifier may be calculated by the formula:

 a. $A = V_{in} \times V_{out}$

 b. $A = \dfrac{V_{in}}{V_{out}}$

 c. $A = \dfrac{V_{out}}{V_{in}}$

 d. None of the above

15. The minimum number of transistors needed to make a discrete component differential amplifier is:

 a. One   **b.** Two

 c. Three   **d.** Four

16. The input stage of almost all operational amplifiers is a(n):

 a. NPN stage

b. PNP stage
c. Ground-base stage
d. Differential amplifier

17. Common-mode rejection ratio is described as:
 a. Differential gain × common-mode gain
 b. $\dfrac{\text{Differential gain}}{\text{Common-mode gain}}$
 c. $\dfrac{\text{Common-mode gain}}{\text{Differential gain}}$
 d. $\dfrac{\text{Differential gain}}{\text{Differential gain-common mode gain}}$

18. The circuit shown in Fig. E-3-18 is a:
 a. Ramp generator
 b. Constant current source
 c. Nonsinusoidal oscillator
 d. Basic differential amplifier

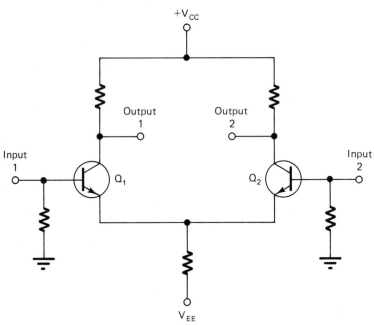

FIGURE E-3-18

19. The output of the circuit of Fig. E-3-19 when the inputs are +2 V, −2 V, and +1 V is:
 a. +1 V   b. −1 V
 c. +10 V   d. −10 V

20. The output of the circuit of Fig. E-3-19 when the inputs are +1 V, +2 V, and +3 V is:
 a. +3 V   b. +6 V
 c. −6 V   d. −60 V

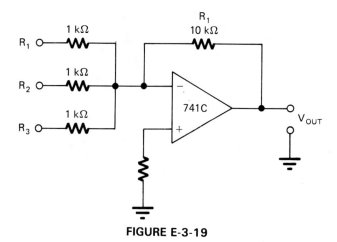

FIGURE E-3-19

UNIT FOUR

Power-Supply Circuits

UNIT INTRODUCTION

All electronic systems require certain voltage and current values for operation. The energy source of the system is primarily responsible for this function. As a general rule, the source is more commonly called a *power supply*. In some systems, the power supply may be a simple battery or a single dry cell. Portable radios, tape recorders, and calculators are examples of systems that employ this type of supply. Television receivers, computer terminals, stereo amplifier systems, and electronic instruments usually derive their energy from an ac power line. This part of the system is called an *electronic power supply*. Electronic devices are used in the power supply to develop the required output energy.

The energy source of an electronic power supply is alternating current. In most systems this is supplied by the ac power line. Most small electronic systems use 120-V, single-phase, 60-Hz ac as their energy source. This energy is readily available in homes, buildings, and industrial facilities.

UNIT OBJECTIVES

Upon completion of this unit, you will be able to:

1. Explain the operation and characteristics of half-wave, full-wave, and bridge-rectifier circuits.
2. Explain the effect of a filter capacitor on the output voltage and ripple voltage of a rectifier circuit.
3. Describe the characteristics of capacitor, resistor-capacitor, and pi filters.
4. Explain the operation of voltage doubler circuits.
5. Explain zener regulator circuits.
6. Understand simple transistor and op-amp regulator circuits.
7. Explain the operation of series and shunt transistor regulators.
8. Calculate percent ripple of a power supply circuit.
9. Calculate percent regulation for a power supply.
10. Construct half-wave, full-wave, and bridge-rectifier circuits.
11. Draw typical output waveforms before and after adding a filter for each type of rectifier circuit.

IMPORTANT TERMS

Alternation. The positive or negative part of an ac sine wave.

Anode. The positive terminal or electrode of an electronic device such as a solid-state diode.

Average value. An ac voltage or current value based on the average of all instantaneous values for one alternation. $V_A = 0.637 \times V_{max}$.

Capacitance input filter. A filter circuit employing a capacitor as the first component of its input.

Cathode. A negative terminal of an electronic device such as a solid-state diode.

Center tap. An electrical connection point at the center of a wire coil or transformer.

Counter EMF. The voltage induced in a conductor through a magnetic field that opposes the original source voltage.

Forward bias. A voltage connection method that produces current conduction in a P-N junction.

Inductance. The property of an electric circuit that opposes a change in current flow.

Peak value. The maximum voltage or current value of an ac sine wave. $V_P = 1.414 \times$ RMS.

Pi filter. A filter with an input capacitor connected to an inductor-capacitor filter, forming the shape of the Greek letter pi.

Pulsating dc. A voltage or current value that rises and falls at a rate or frequency with current flow always in the same direction.

Rectification. The process of changing ac into pulsating dc.

Reverse bias. A voltage connection method producing very little or no current through a P-N junction.

Time constant (RC). The amount of time required for the voltage of a capacitor in an RC circuit to increase 63.2% of the value or decrease to 36.7% of the value.

Turns ratio. The ratio of the number of turns in the primary winding to the number of turns in the secondary winding.

POWER-SUPPLY FUNCTIONS

All electronic power supplies have a number of functions that must be performed in order to operate. Some of these functions are achieved by all power supplies. Others are somewhat optional and are dependent primarily on the parts being supplied. A block diagram of a general-purpose *electronic power supply* is shown in Fig. 4-1. Each block represents a specific power-supply function.

Transformer

Electronic power supplies rarely operate today with ac obtained directly from the power line. A *transformer* is commonly used to step the line voltage up or down to a desired value. A schematic

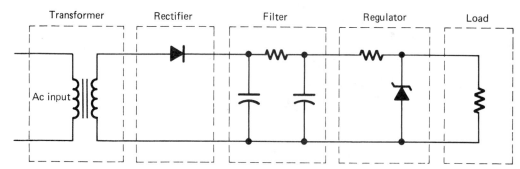

FIGURE 4-1 Functional block diagram of an electronic power supply.

symbol and a simplification of a power supply transformer are shown in Fig. 4-2.

The coil on the left of the symbol is called the *primary wind-ing*. The ac *input* is applied to this winding. The coil on the right side is the *secondary winding*. The *output* of the transformer is developed by the secondary winding. The parallel lines near the center of the symbol are representative of the *core*. In a trans-former of this type, the core is usually made of laminated soft steel. Both windings are placed on the same core. Power-supply trans-formers may have more than one primary and secondary winding on the same core, which permits them to accommodate different line voltages and to develop alternate output values.

The *output voltage* of a power supply transformer may have the same polarity as the input voltage or it may be reversed. This depends on the winding direction of the coils. A dot is often used on the symbol to indicate the *polarity* of a winding. The polarity of the input and output are the same where the dots are indicated.

The output voltage developed by a transformer is primarily dependent on the *turns ratio* of its windings. If the primary wind-ing has 500 turns of wire, and the secondary has 1000 turns, the transformer has a 1:2 turns ratio. This type of transformer will *step up* the input voltage by a factor of 2. With 120 V applied from

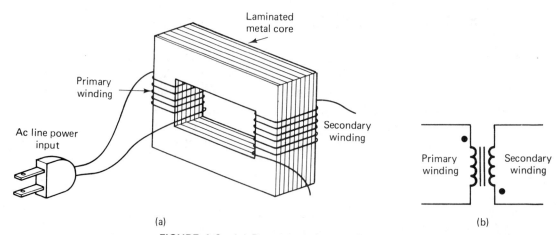

FIGURE 4-2 (a) Power supply transformer; (b) schematic symbol.

the power line, the output voltage will be approximately 240 V. It is interesting to note that the current capability of a transformer is the reverse of its turns ratio. A 1:2 turns ratio steps up the voltage and *steps down* the current capability. One ampere of primary current is capable of producing only 0.5 A of secondary current. The resistance of the wire is largely responsible for this condition. A 1:2 turns ratio has twice as much resistance in the secondary as it has in the primary winding. An increase in resistance therefore causes a decrease current. This is all based on the wire size of the two windings being the same. Figure 4-3 shows some representative transformer types that are widely used today.

Power supplies for solid-state systems generally develop low voltage values. Transformers for this type of supply are designed to *step down* the line voltage. A turns ratio of 10:1 is very common. With 120 V applied to the input, the output will be approximately 12 V. The output current capabilities of this transformer are increased by a ratio of 1:10. This means that a step-down transformer has a low voltage output with a high current capacity. Supplies of this type are well suited for solid-state applications.

Rectification

The primary function of a power supply is to develop dc for the operation of electronic devices. Most power supplies are initially energized by ac. This energy must be changed into dc before it can be used. The process of changing ac into dc is called *rectification*. All electronic power supplies have a rectification function. Figure 4-4 shows a graphic representation of the rectification function.

FIGURE 4-3 Representative transformers. (Courtesy of TRW/UTC Transformers.)

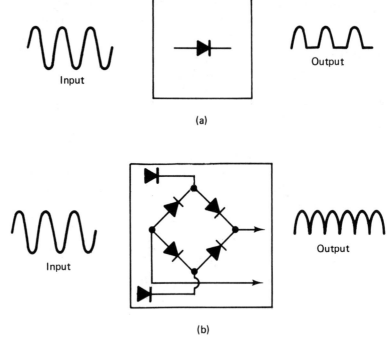

FIGURE 4-4 Rectifier: (a) half wave; (b) full wave.

Notice that rectification can be either half-wave or full-wave. Half-wave rectification uses one alternation of the ac input, whereas full-wave uses both alternations.

Half-wave Rectification. In a half-wave rectification, only one alternation of the ac input appears in the output. The negative alternation of Fig. 4-5(a) has been eliminated by the diode rectifier. The resulting output of this circuit is called *pulsating dc.* A half-wave rectifier can remove either the positive or negative alternation, depending on its polarity in the circuit. When the output has a positive alternation, the resulting dc is positive with respect to the common point or ground. A negative output occurs when the positive alternation is removed. Figure 4-5(b) shows an example of a rectifier with a negative output.

The rectification function of a modern power supply is performed by *solid-state diodes.* This type of device has two electrodes, known as the *anode* and *cathode.* Operation is based on its ability to conduct easily in one direction but not in the other direction. As a result of this, only one alternation of the applied ac appears in the output. This occurs only when the diode is *forward biased.* The anode must be positive and the cathode negative. Reverse *biasing* does not permit conduction through a diode.

A simplification of the rectification process is shown in Fig. 4-6. Figure 4-6(a) shows how the diode will respond during the *positive alternation.* This alternation causes the diode to be forward biased. Note that the anode is positive and the cathode is negative. The resulting current flow is shown by arrows. The output of the

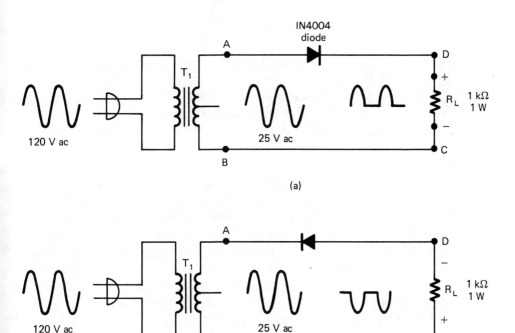

FIGURE 4-5 Half-wave power supplies: (a) positive output; (b) negative output.

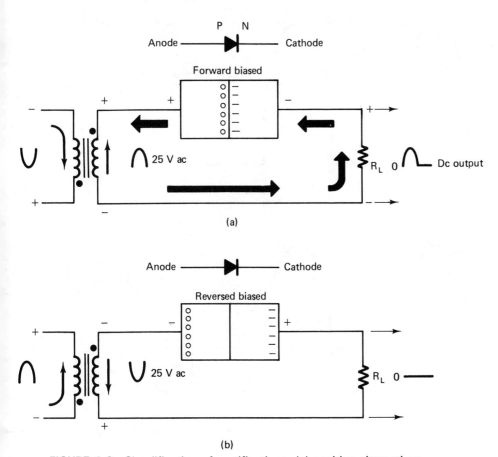

FIGURE 4-6 Simplification of rectifications: (a) positive alternation; (b) negative alternation.

rectifier appears across R_L. It is positive at the top of R_L and negative at the bottom.

Figure 4-6(b) shows how the diode will respond during the *negative alternation*. This alternation causes the diode to be reverse biased. The anode is now negative and the cathode is positive. No current flows during this alternation. The output across R_L is zero for this alternation.

The dc output of a half-wave rectifier appears as a series of pulses across R_L. A dc voltmeter or ammeter would therefore indicate the average value of these pulses over a period of time. Figure 4-7(a) shows an example of the *pulsating dc output*. The average value of one pulse is 0.637 of the peak value. Since the second alternation of the output is zero, the composite average value must take into account the time of both alternations. The composite output is therefore 0.637 ÷ 2, or 0.318 of the peak value. Figure 4-7(b) shows a graphic example of this value.

The equivalent dc output of a half-wave rectifier without filtering can be calculated when the value of the ac input is known. The peak value of one ac alternation is determined by first multiplying the RMS value by 1.414. The composite dc output is then determined by multiplying the peak ac value by 0.318.

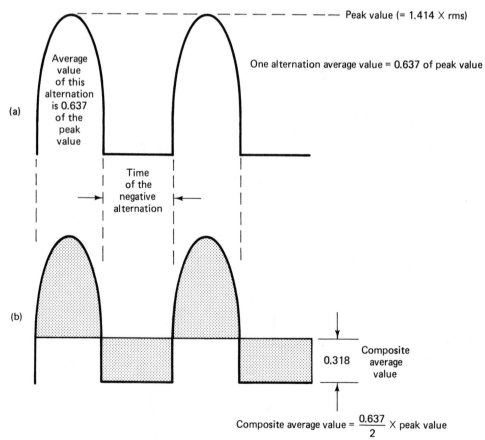

FIGURE 4-7 Half-wave rectifier output: (a) pulsating dc output; (b) composite output.

A simplified method of calculating the dc output of a half-wave rectifier is to combine different values. In this case, $1.414 \times 0.637 \div 2$ equals 0.45. The composite dc output of a half-wave rectifier is therefore 45%, or 0.45, of the RMS input voltage. In an actual circuit we must also take into account the voltage drop across the diode. When a silicon diode is used, this would be 0.6 V. The dc output of an unfiltered half-wave rectifier would therefore be $0.45 \times \text{RMS} - 0.6$ V. Using this procedure you may determine equivalent dc voltage of the half-wave rectifier.

The frequencies of the output pulses of a half-wave rectifier occur at the same rate as the applied ac input. With 60 Hz input, the pulse rate, or *ripple frequency*, of a half-wave rectifier is 60 Hz. Only one pulse appears in the output for each complete sine-wave input. Potentially, this means that only 45% of the ac input is transformed into a usable dc output. Half-wave rectification is therefore only 45% efficient. Due to its low efficiency rating, half-wave rectification is not widely used today.

Full-wave Rectification.   A full-wave rectifier responds as two half-wave rectifiers that conduct on opposite alternations of the input. Both alternations of the input are changed into pulsating dc output. Full-wave rectification can be achieved by two diodes and a *center-tapped transformer* or by four diodes in a *bridge circuit*. Both types of rectifiers are widely used in solid-state power supplied today.

A schematic diagram of a two-diode full-wave rectifier is shown in Fig. 4-8. Note that the anodes of each diode are connected to opposite ends of the transformer secondary winding. The cathode of each diode is then connected to form a common positive output. The *load* of the power supply (R_L) is connected between the common cathode point and the center-tap connection of the transformer. A complete path for current is formed by the transformer, two diodes, and the load resistor.

When ac is applied to the primary winding of the transformer, it steps the voltage down in the secondary winding. The *center tap* serves as the electrical neutral or center of the secondary winding. Half of the secondary voltage will appear between points *CT* and *A* and the other half between *CT* and *B*. These two voltage

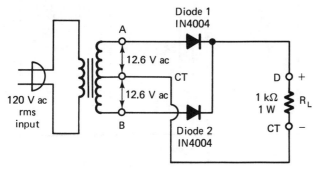

FIGURE 4-8   Two-diode full-wave rectifier.

values are equal and will always be 180° out of phase with respect to point *CT.*

Assume now that the 60-Hz input is slowed down so that we can see how one complete cycle responds. In Figure 4-9 note the polarity of the secondary winding voltage. For the first alternation point *A* is positive and point *B* is negative with respect to the center tap. The second alternation causes point *A* to be negative and point *B* to be positive with respect to *CT.* The center tap will therefore always be negative with respect to the positive end of the winding. This point then serves as the *negative output* of the power supply.

Conduction of a specific diode in a full-wave rectifier is based on the polarity of the applied voltage during each alternation. In Fig. 4-9 conduction is made with respect to point *CT.* For the first alternation, point *A* is positive and *B* is negative. This polarity causes D_1 to be forward biased and D_2 to be reverse biased. Current flow for this alternation is shown by the solid arrows. Starting at *CT*, electrons will flow through a conductor to R_L through D_1 and return to point *A*. This current flow causes a pulse of dc to appear across R_L for the first alternation.

For the second alternation, point *A* become negative and point *B* is positive. This polarity forward-biases D_2 and reverse-biases D_1. Current flow for this alternation is shown by the dashed arrows. Starting at point *CT*, electrons flow through a conductor to R_L, through D_2, and return to point *B*. This causes a pulse of dc to appear across R_L for the negative alternation.

It should be obvious at this point that current flow through R_L is in the same direction for each alternation of the input. This

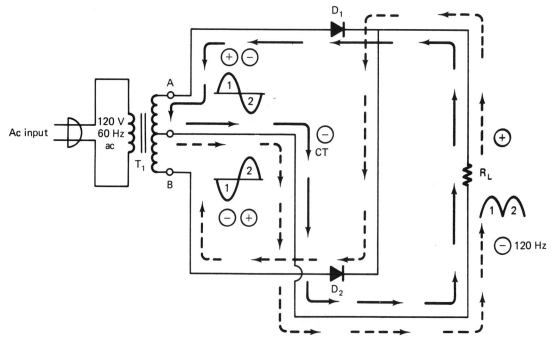

FIGURE 4-9   Conduction of a full-wave rectifier.

means that each alternation is transposed into a pulsating dc output. Full-wave rectification therefore changes the entire ac input into dc output.

The resulting dc output of a full-wave rectifier is 90%, or 0.90, of the ac voltage between the center tap and the outer ends of the transformer. This voltage value is determined by calculating the peak value of the RMS input voltage and then multiplying it by the average value. Since 1.414 × 0.637 equals 0.90, 90% of the RMS value is the equivalent dc output. The output of a full-wave rectifier is 50% more efficient than that of an equivalent halfwave rectifier.

The actual dc output of a full-wave rectifier is slightly less than 0.90 of the RMS input just described. Each diode, for example, reduces the dc voltage by 0.6 V. A more practical calculation is dc output = RMS × 0.90 − 0.6 V. In a low-voltage power supply a reduction of the dc by 0.6 V may be quite significant.

The ripple frequency of a full-wave rectifier is somewhat different from that of a half-wave rectifier. Each alternation of the input, for example, produces a pulse of output current. The ripple frequency is therefore twice the alternation frequency. For a 60-Hz input the ripple frequency of a full-wave rectifier is 120 Hz. As a general rule, a 120-Hz ripple frequency is easier to filter than the 60 Hz of a half-wave rectifier.

Full-wave Bridge Rectifiers.   A bridge rectifier requires four diodes to achieve full-wave rectification. In this configuration, two diodes will conduct during the positive alternation and two will conduct during the negative alternation. A center-tapped transformer is not required to achieve full-wave rectification with a bridge circuit.

The component parts of a full-wave bridge rectifier are shown in Fig. 4-10. In this circuit, one side of the ac input is applied to the junction of diodes D_1 and D_2. The alternate side of the input

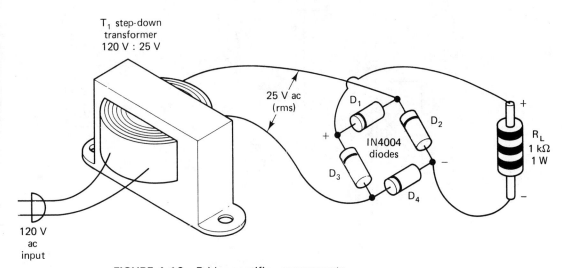

FIGURE 4-10   Bridge rectifier components.

is applied to diodes D_3 and D_2. The alternate side of the input is applied to diodes D_3 and D_4. The diodes at each input are reversed with respect to each other. Output of the bridge occurs at the other two junctions. The cathodes of D_1 and D_3 serve as the positive output. Negative output appears at the anode junction of D_2 and D_4. The *load* of the power supply is connected across the common anode- and common cathode-connection points.

When ac is applied to the primary winding of a power-supply transformer, it can be either stepped up or down, depending on the desired dc output. In Fig. 4-10, the 120-V input is stepped down to 25 V RMS. In normal operation, one alternation will cause the top of the transformer to be positive and the bottom to be negative. The next alternation will cause the bottom to be positive and the top to be negative. Opposite ends of the secondary winding will always be 180° out of phase with each other.

Assume now that the ac input causes point A to be positive and B to be negative for one alternation. The schematic diagram of a bridge in Fig. 4-11 shows this condition of operation. With the indicated polarity, diode D_1 is forward biased and D_2 is reverse biased at the top junction. At the bottom junction D_4 is forward biased and D_3 is reverse biased. When this occurs, electrons will flow from point B through D_4, up through R_L, through D_1, and return to point A. Solid arrows are used to show this current path. The load resistor sees one pulse of dc across it for this alternation.

With the next alternation, point A of the schematic diagram becomes negative and point B becomes positive. When this occurs, the bottom diode junction becomes positive and the top junction goes negative. This condition forward biases diodes D_2 and D_3 while reverse biasing D_1 and D_4. The resulting current flow is indicated by the dashed arrows starting at point A. Electrons flow

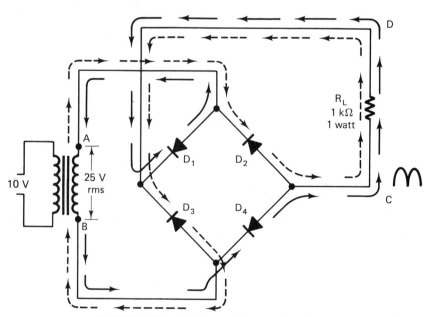

FIGURE 4-11 Bridge rectifier conduction.

from A through D_2, up through R_L, through D_3 and return to the transformer at point B. Dashed arrows are used to show this current path. The load resistor also sees a pulse of dc across it for this alternation.

The *output* of the bridge rectifier has current flow through R_L in the same direction for each alternation of the input. Ac is therefore changed into a pulsating dc output by conduction through a bridge network of diodes. The dc output, in this case, has a *ripple frequency* of 120 Hz. Each alternation produces a resulting output pulse. With 60 Hz applied, the output is 2 × 60 Hz, or 120 Hz.

The dc output voltage of a bridge circuit is slightly less than 90% of the RMS input. Each diode, for example, reduces the output by 0.6 V. With two diodes conducting during each alternation, the dc output voltage is reduced by 0.6 × 2 V, or 1.2 V. In the circuit of Fig. 4-11, the dc output is 25 V × 0.9 − 1.2 V, or 21.3 V. The output of a bridge circuit is slightly less than that of an equivalent two-diode full-wave rectifier. The voltage drop of the second diode accounts for this difference.

The four diodes of a bridge rectifier can be obtained today in a single package. Figure 4-12 shows several different package types. As a general rule, there are two ac input terminals and two dc output terminals. These devices are generally rated according to their current-handling capability and peak reverse voltage rating. Typical current ratings are from 0.5 to 50 A in single-phase ac units. P-R-V ratings are 50, 200, 400, 800 and 1000 V. A bridge package usually takes less space than four single diodes. Nearly all bridge power supplies employ the single package assembly today.

Dual Power Supplies. A rather new variation of the power supply is now being used in some systems. This supply has both negative and positive output with respect to ground. *Dual or split power supplies*, as they are known, have been developed as a voltage source for integrated circuits. The secondary winding of the

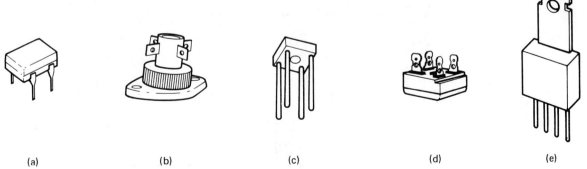

(a)   (b)   (c)   (d)   (e)

FIGURE 4-12 Packaged bridge rectifier assemblies: (a) dual-in-line package; (b) sink-mount package; (c) epoxy package; (d) tab-pack enclosure; (e) therm-tab package.

input transformer is divided into two parts. The center tap or neutral serves as a common ground connection for the two outside windings. Each half of the winding has a complete full-wave rectifier.

A *dual power supply* with two full-wave rectifiers is shown in Fig. 4-13. Notice that the output of this supply is +10.7 and −10.7 V.

Diodes D_1 and D_2 are rectifiers for the positive supply. They are connected to opposite ends of transformer T_1. Diodes D_3 and D_4 are rectifiers for the negative supply. They are connected in a reverse direction to the opposite ends of the transformer. The positive and negative output is with respect to the center tap of the transformer.

For one alternation assume that the top of the transformer is positive and the bottom negative. Current flows out of point B through D_4, R_{L2}, R_{L1}, D_1 and returns to terminal A of the transformer. The top of R_{L1} becomes positive, with the bottom of R_{L2} being negative. Solid arrows show the path of current flow for this alternation.

The next alternation makes the top of the transformer negative and the bottom positive. Current flows out of point A through D_3, R_{L2}, R_{L1}, D_2 and returns to terminal B of the transformer. The top of R_{L1} continues to be positive, whereas the bottom of R_{L2} is negative. The direction of current flow through R_{L1} and R_{L2} is the same for each alternation. The resulting output volt-

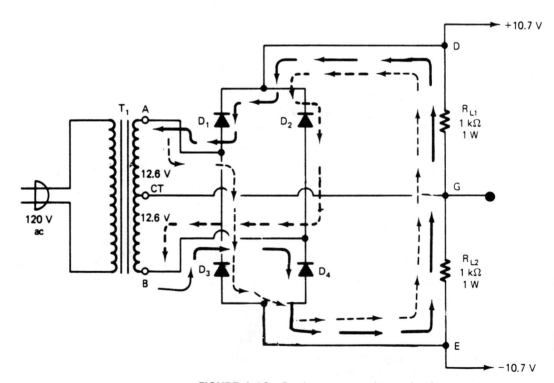

FIGURE 4-13 Dual power supply conduction.

age of the supply appears at the top and the bottom of R_{L1} and R_{L2}. Dashed arrows show the direction of current flow for the second alternation.

Filtering

The output of a half- or full-wave rectifier is *pulsating direct current*. This type of output is generally not usable for most electronic circuits. A rather pure form of dc is usually required. The *filter* section of a power supply is designed to change pulsating dc into a rather pure form of dc. Filtering takes place between the output of the rectifier and the input to the load device. Power supplies discussed up to this point have not employed a filter circuit.

The pulsating dc output of a rectifier contains two components. One of these deals with the dc part of the output. This component is based on the combined average value of each pulse. The second part of the output refers to its ac component. Pulsating dc, for example, occurs at 60 Hz or 120 Hz, depending on the rectifier being employed. This part of the output has a definite ripple frequency. Ripple must be minimized before the output of a power supply can be used by most electronic devices.

Power-supply filters fall into two general classes according to the type of component used in its input. If filtering is first achieved by a capacitor, it is classified as a *capacitor, or C-input, filter*. When a coil of wire or inductor is used as the first component, it is classified as an *inductive, or L-input, filter*. *C*-input filters develop a higher value of dc output voltage than do L filters. The output voltage of a *C* filter usually drops in value when the load increases. *L* filters, by comparison, tend to keep the output voltage at a rather constant value. This is particularly important when large changes in the load occur. However, the output voltage of an *L* filter is somewhat lower than that of a *C* filter. Figure 4-14 shows a graphic comparison of output voltage and load current for *C* and *L* filters.

C-Input Filters.   The ac component of a power supply can be effectively reduced by a C-input filter. A single capacitor is simply placed across the load resistor, as shown in Fig. 4-15. For alternation 1 of the circuit (Fig. 4-15(a)) the diode is forward biased. Current flows according to the arrows of the diagram. *C* charges very quickly to the peak voltage value of the first pulse. At the same point in time, current is also supplied to R_L. The initial surge of current through a diode is usually quite large. This current is used to charge *C* and supply R_L at the same time. A large capacitor, however, responds somewhat like low resistance when it is first being charged. Notice the amplitude of the I_d waveform during alternation 1.

When alternation 2 of the input occurs, the diode is reverse biased. Figure 4-15(b) shows how the circuit responds for this al-

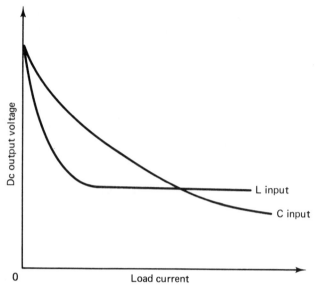

FIGURE 4-14 Load current versus output voltage comparisons for *L* and *C* filters.

ternation. Notice that there is no current flow from the source through the diode. The charge acquired by C during the first alternation now finds an easy discharge path through R_L. The resulting discharge current flow is indicated by the arrows between C and R_L. In effect, R_L is now being supplied current even when the diode is not conducting. The voltage across R_L is therefore maintained at a much higher value. See the V_{RL} waveform for alternation 2 in Fig. 4–15(c).

Discharge of C continues for the full time of alternation 2. Near the end of the alternation there is somewhat of a drop in the value of V_{RL}. This is due primarily to a depletion of capacitor charge current. At the end of this time, the next positive alternation occurs. The diode is again forward biased. The capacitor and R_L both receive current from the source at this time. With C still partially charged from the first alternation, less diode current is needed to recharge C. In Fig. 4-15(c), note the amplitude change in the second I_d pulse. The process from this point on is a repeat of alternations 1 and 2.

The effectiveness of a capacitor as a filter device is based on a number of factors. Three very important considerations are:

1. The size of the capacitor being used
2. The value of the load resistor R_L
3. The time duration of a given dc pulse

These three factors are related to one another by the formula

$$T = R \times C$$

where T is the time in seconds, R the resistance in ohms, and C the capacitance in farads. The product RC is an expression of the

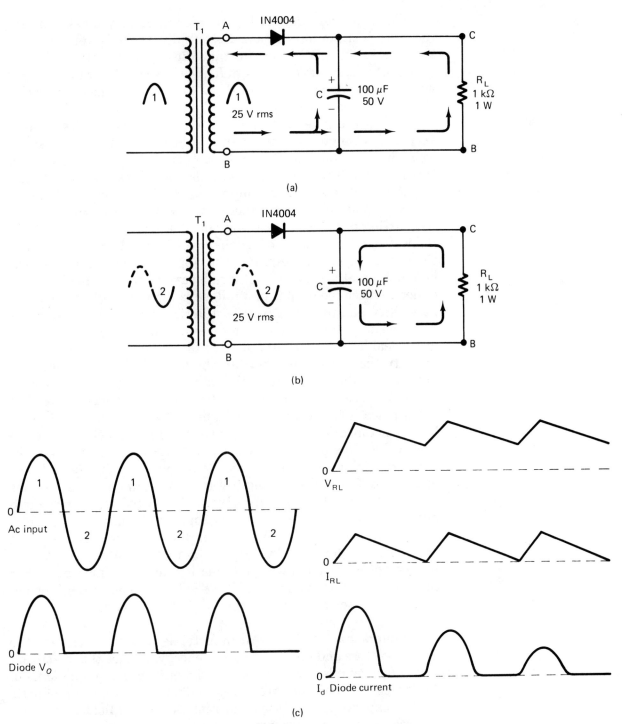

FIGURE 4-15 C-input filter action: (a) alternation 1; (b) alternation 2; (c) waveforms.

filter circuit's *time constant. RC* is a measure of how rapidly the voltage and current of the filter respond to changes in input voltage. A capacitor will charge to 63.2% of the applied voltage in one time constant. A discharging capacitor will have a 63.2% drop of its original value in one time constant. It takes five time constants to charge or discharge a capacitor fully.

The *filter capacitor* of Fig. 4-15 charges very quickly during the first positive alternation. Essentially, there is very little resistance for the RC time constant during this period. The discharge of C is, however, through R_L. If R_L is small, C will discharge very quickly. A large value of R_L will cause C to discharge rather slowly. For good filtering action, C must not discharge very rapidly during the time of one alternation. When this occurs, there is very little change in the value of V_{RL}. A C-input filter works very well when the value of R_L is relatively large. If the value of R_L is small, as in a heavy load, more ripple will appear in the output.

A rather interesting comparison of filtering occurs between half-wave and full-wave rectifier power supplies. The time between reoccurring peaks is twice as long in a half-wave circuit as it is for full-wave. The capacitor of a half-wave circuit therefore has more time to discharge through R_L. The ripple of a halfwave filter will be much greater than that of a full-wave circuit. In general, it is easier to filter the output of a full-wave rectifier. Figure 4-16 compares half- and full-wave rectifier filtering with a single capacitor.

The *dc output voltage* of a filtered power supply is usually a great deal higher than that of an unfiltered power supply. In Figure 4-16 the waveforms show an obvious difference between outputs. In the filtered outputs, the capacitor will charge to the peak value of the rectified output. The amount of discharge action that takes place is based on the resistance of R_L. For a light load or high resistance R_L, the filtered output will remain charged to the peak value. The peak value of the 10 V RMS is 14.14 V. For the unfiltered full-wave rectifier, the dc output is approximately 0.9 × RMS, or 9.0 V. Comparing 9 V to 14 V shows a rather decided difference in the two outputs. For the half-wave rectifier the output difference is even greater. The unfiltered half-wave output is approximately 45% of 10 V RMS, or 4.5 V. The filtered output is 14 V less 20% for the added ripple, or approximately 11.2 V. The filtered output of a half-wave rectifier is nearly 2.5 times more than that of the unfiltered output. These values are only rough generalizations of power supply output with a light load.

Inductance Filtering.   An inductor is a device that has the ability to store and release electrical energy. It does this by taking some of the applied current and changing it into a magnetic field. An increase in current through an inductor causes the magnetic field to expand. A decrease in current causes the field to collapse and release its stored energy.

The ability of an inductor to store and release energy can be used to achieve filtering. An increase in the current passing through an inductor causes a corresponding increase in its magnetic field. Voltage induced into the inductor due to a change in its field opposes a change in the current passing through it. A decrease in current flow causes a similar reaction. A drop in current causes its magnetic field to collapse. This action also induces voltage into the inductor. The induced voltage in this case causes

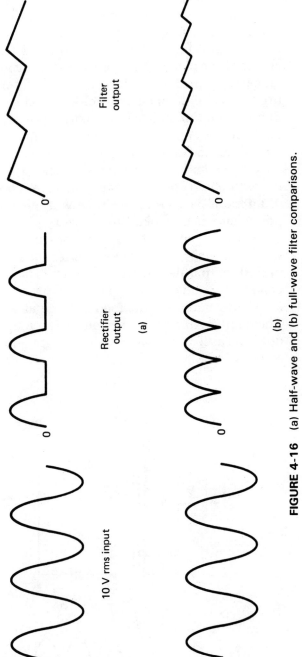

10 V rms input

Rectifier
output

(a)

Filter
output

(b)

FIGURE 4-16 (a) Half-wave and (b) full-wave filter comparisons.

a continuation of the current flow. As a result of action, the added current tends to bring up its decreasing value. An inductor opposes any change in its current flow. Inductive filtering is well suited for power supplies with a large load current.

An inductive input filter is shown in Fig. 4-17. The inductor is simply placed in series with the rectifier and the load. All the current being supplied to the load must pass through the inductor (L). The filtering action of L does not let a pronounced change in current take place. This prevents the output voltage from reaching an extreme peak or valley. Inductive filtering in general does not produce as high an output voltage as does capacitive filtering. Inductors tend to maintain the current at an average value. A larger load current can be drawn from an inductive filter without causing a decrease in its output voltage.

When an inductor is used as the primary filtering element, it is commonly called a *choke*. The term choke refers to the ability of an inductor to reduce ripple voltage and current. In electronic power supplies, choke filters are rarely used as a single filtering element. A combination inductor-capacitor or *LC* filter is more widely used. This filter has a series inductor with a capacitor connected in parallel with the load. The inductor controls large changes in load current. The capacitor, which follows the inductor, is used to maintain the load voltage at a constant voltage value. The combined filtering action of the inductor and capacitor produces a rather pure value of load voltage. Figure 4-18 shows a representative *LC* filter and its output.

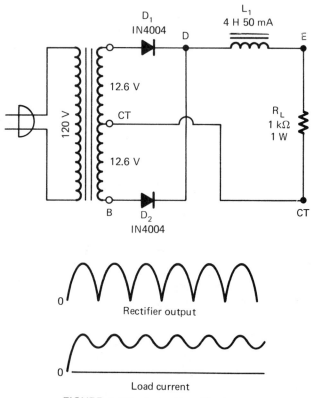

FIGURE 4-17 Inductive filtering.

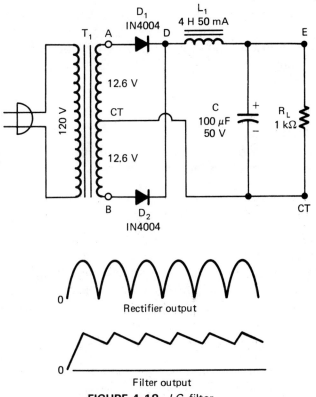

FIGURE 4-18   *LC* filter.

Pi Filters.   When a capacitor is placed in front of the inductor of an *LC* filter it is called a *pi filter*. In effect, this circuit becomes a *CLC* filter. The two capacitors are in parallel with R_L and the inductor is in series. Component placement of this circuit in a schematic diagram resembles the capital Greek letter pi (Π), which accounts for the name of the filter. See the schematic diagram of a pi filter in Fig. 4-19.

The operation of a pi filter can best be understood by considering L_1 and C_2 as an *LC* filter. This part of the circuit acts upon the output voltage developed by the capacitor-input filter C_1. C_1 charges to the peak value of the rectifier input. It has a ripple content that is very similar to that of C-input filter of Fig. 4-15. This voltage is then applied to C_2 through inductor L_1. C_1 charges C_2 through L_1. C_2 then holds its charge for a time interval deter-

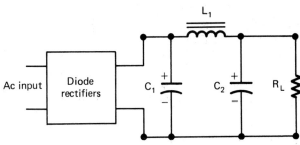

FIGURE 4-19   *LCL* (pi) filter.

mined by the time constant of C_2 and R_L. As a result of this action, there is additional filtering by L_1 and C_2. The ripple content of this filter is much lower than that of a single C-input filter. There is, however, a slight reduction in the dc supply voltage to R_L due to the voltage drop across L_1.

A pi filter has a very low ripple content when used with a light load. An increase in load, however, tends to lower its output voltage. This condition tends to limit the number of applications of pi filters. Today we find them used to supply radio circuits, in stereo amplifiers, and in TV receivers. The input to this filter is nearly always supplied by a full-wave rectifier.

RC Filters.   In applications where less filtering can be tolerated, an RC filter can be used in place of a pi filter. As shown in Figure 4-20, the inductor is replaced with a resistor. An inductor is rather expensive, quite large physically, and weighs a great deal more than a resistor. The performance of the RC filter is not quite as good as that of a pi filter. There is usually a reduction in dc output voltage and increased ripple.

In operating, C_1 charges to the peak value of the rectifier input. A drop in rectifier input voltage causes C_1 to discharge through R_1 and R_L. The voltage drop across R_1 lowers the dc output to some extent. C_2 charges to the peak value of the R_L voltage. The dc output of the filter is dependent on the load current. High values of load current cause more voltage drop across R_1. This, in turn, lowers the dc output. Low values of load current have less voltage drop across R_1. The output voltage therefore increases with a light load. In practice, RC filters are used in power supplies that have 100 mA or less of load current. The primary advantage of RC filters is reduced cost.

Voltage Regulation

The dc output of an unregulated power supply has a tendency to change value under normal operating conditions. Changes in ac input voltage and variations in the load are primarily responsible for these fluctuations. In some power-supply applications, voltage changes do not represent a serious problem. In many electronic circuits voltage changes may cause improper operation. When a stable dc voltage is required, power supplies must employ a volt-

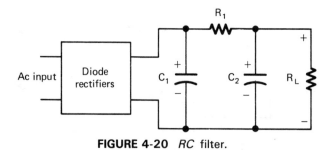

FIGURE 4-20 *RC* filter.

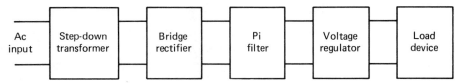

FIGURE 4-21   Block diagram of a power supply.

age regulator. The block diagram of a power supply in Fig. 4-21 shows where the *regulator* is located.

A number of voltage regulator circuits have been developed for use in power supplies. One very common method of regulation employs the *zener diode.* Figure 4-22 shows this type of regulator located between the filter and the load. The zener diode is connected in parallel or shunt with R_L. This regulator requires only a zener diode D_z and a series resistor R_s. Notice that D_z is placed across the filter circuit in the reverse bias direction. Connected in this way, the diode goes into conduction only when it reaches the zener breakdown voltage V_z. This voltage then remains constant for a large range of zener current I_z. Regulation is achieved by altering the conduction of I_z through the zener diode. The combined I_z and load current I_L must all pass through the series resistor. This current value then determines the amount of voltage drop across R_s. Variations in current through R_s are used to keep the output voltage at a constant value.

A schematic diagram of a 9-V *regulated power supply* is shown in Fig. 4-23. This circuit derives its input voltage from a 12.6-V transformer. *Rectification* is achieved by a self-contained bridge rectifier assembly. *Filtering* is accomplished by an *RC* filter. The series resistor of this circuit has two functions. It first couples capacitors C_1 and C_2 together in the filter circuit. Second, it serves as the series resistor for the regulator circuit. Diode D_z is a 9-V 1-W zener diode.

Operation of the regulated power supply is similar to that of the bridge circuit discussed earlier. Full-wave output from the rectifier is applied to C_1 of the filter circuit. C_1 then charges to the peak value of the RMS input less the voltage drop across two silicon diodes (1.2 V). This represents a value of $12.6 \times 1.414 - 1.2$,

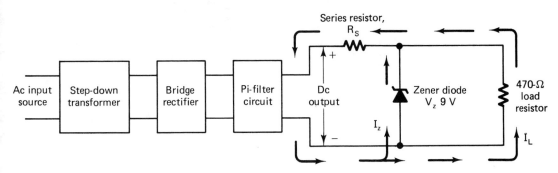

FIGURE 4-22   Zener diode regulator.

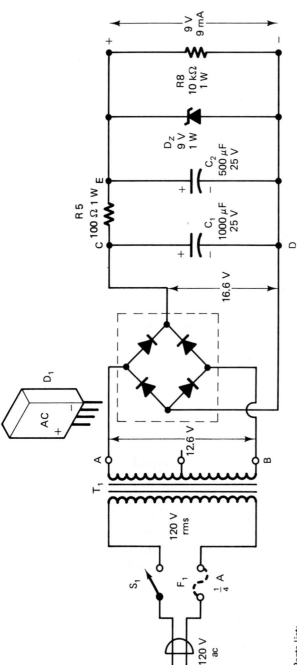

Parts list:

T_1 : 12.6-V 2-A Stancor P-8130, or Triad F-44X

D_1 : silicon bridge rectifier 2 A 50 V; General Instrument Co., KBF-005 or Motorola MDA-200

C_1 : 1000-μF 25-V electrolytic capacitor

C_2 : 500-μF 25-V electrolytic capacitor

R_5 : 100-Ω 1-W resistor

R_8 : 10-kΩ $\frac{1}{2}$-W resistor

D_z : 9-V 1-W zener diode IN5346A or HEP Z-2513

S_1 : SPST toggle switch

F_1 : Fuse, $\frac{1}{4}$ A 250 V

Misc. : metal chassis, PC board, line cord, fuse holder, grommets, solder, cabinet

FIGURE 4-23 9-V regulated power supply.

or 16.6 V dc input. R_s therefore has a voltage drop of 16.6 − 9 V, or 7.6 V. This represents a total current flow passing through R_5 of 7.6 V/100 Ω = 0.076 A or 76 mA. With the bleeder resistor R_8 serving as a fixed load, there is 9 V/10 kΩ = 0.9 mA, or 0.0009 A, of load current. The difference in I_{R5} and I_L is therefore 0.076 A − 0.0009 A = 0.0751 A, or 75.1 mA. This current must all pass through the zener diode when the circuit is in operation. Nine volts dc will then appear at the two output terminals of the power supply. This represents the *no-load (NL) condition* of operation.

When the power supply is connected to an external load, more current is required. Ideally, this current should be available with 9 V of output. Assume now that the power supply is connected to a 270-Ω external load. The total resistance that the power supply sees at its output is 10 kΩ in parallel with 270Ω. This represents a load resistance of 263Ω. With 9 V applied, the total load current is 9 V/263 = 0.0342, or 34.2 mA. The total output current available for the power supply is that which passes through R_s. This was calculated to be 0.076 A, or 76 mA. With the 270-Ω load, I_L is 34.2 mA with an I_Z of 41.8 mA. The output voltage therefore remains at 9 V with the increase in load.

With a slightly smaller external load, the power supply would go out of its regulation range. Should this occur, no I_Z will flow and the value of I_L alone will determine the current passing through R_5. An I_{R5} value in excess of 76 mA would cause a greater voltage drop across R_5. This, in effect, would cause the output voltage to be less than 9 V. All voltage regulator circuits of this type have a maximum load limitation. Exceeding this limit will cause the output voltage to go out of regulation.

If the resistance of the external load is increased in value, it causes a reduction in load current. This condition causes a reverse in the operation of the regulator. With a larger R_L there is less I_L. As a result of this, the zener diode must conduct more heavily. The maximum current-handling rating of D_z would determine this condition. An infinite load would, for example, demand no I_L. D_z would therefore conduct the full current passing through R_S. This value was calculated to be 76 mA for our circuit. For a 1-W, 9-V zener diode the maximum current rating is 1 W/9 V = 0.111A, or 111 mA. In this case, the diode is capable of handling the maximum possible I_L that will occur for an infinite load. In the actual circuit the bleeder resistor R_8 demands 0.9 mA of current when the load is infinite. The current through I_z will therefore not be in excess of 75.1 mA, which is well below its maximum rating. In effect, the regulating range of this power supply is from an infinite value to approximately 120 Ω. The output voltage will remain at 9 V over this entire range of load values.

A regulator must also be responsive to changes in input voltage. If the input voltage, for example, were to increase by 10%, it would cause the supply to see 13.86 V instead of 12.6 V. The peak value of 13.86 V would then be 19.6 V − 1.2 V, or 18.4 V instead of 16.6 V. The voltage drop across R_S would therefore increase from

7.6 V to 9.4 V. This, in turn, would cause an increase in total current through R_S of 94 mA. The current flow of the zener diode would increase to 94.0 – 0.9 = 93.1 mA with no external load. Since the zener is capable of handling up to 111 mA, it could respond to this change to maintain the output voltage at 9 V.

The response of a power supply to a decrease in input voltage must also be taken into account. A 10% decrease in input voltage would cause the supply to see only 11.34 V instead of 12.6 V. The peak value of 11.34 V is 16 V. Capacitor C_1 would then charge to 16.0 – 1.2 V, or 14.8 V. The 1.2 V is the voltage drop across two silicon diodes in the bridge. The voltage drop across R_5 would now be 5.8 V. With this reduced voltage, the total current would drop to 5.8 V/100Ω = 0.058 A, or 58 mA. This value can certainly be handled by the zener diode. The maximum low-resistance value of the load would increase somewhat. Under this condition the load may drop to only 160Ω before going out of regulation.

SELF-EXAMINATION

1. The function of a power supply is to _____.

2. A half-wave rectifier uses _____ diodes.

3. A full-wave rectifier uses _____ diodes.

4. A bridge rectifier uses _____ diodes.

5. A diode is forward biased when a _____ polarity is applied to its anode.

6. A diode is reverse biased when a _____ polarity is applied to its cathode.

7. The formula for finding dc voltage output of a bridge rectifier is: V_{dc} = _____.

8. The ac ripple frequency of a half-wave rectifier is _____ Hz.

9. The ac ripple frequency of a full-wave rectifier is _____ Hz.

10. The ac ripple frequency of a bridge rectifier is _____ Hz.

11. The purpose of a capacitor connected across the output of a rectifier circuit is _____.

12. The type of transformer used for full-wave rectification is a _____.

13. A power supply with a negative and a positive output is called a _____.

14. With 10 V RMS input to a half-wave rectifier, the output V_{dc} is _____.

15. With 30 V RMS input to a bridge rectifier, the output V_{dc} is _____.

16. The purpose of a zener diode in a power supply circuit is
    _____ _____.

17. Three types of voltage regulation circuits are
    _____, _____, and
    _____.

18. Three types of filter citcuits are _____,
    _____, and _____.

19. Two types of voltage doublers are _____
    and _____.

20. The purpose of a filter circuit is to increase _____
    and decrease _____.

ANSWERS

1. Convert ac to dc	2. One
3. Two	4. Four
5. Positive	6. Positive
7. 0.636 × peak ac input	8. 60
9. 120	10. 120
11. Filter	12. Center-tapped
13. Split power supply	14. 0.318 × (10 × 1.41) = 4.48 V
15. 0.636 × (30 × 1.41) = 26.9 V	16. Voltage regulation
17. Zener diodes; transistor regulators; three-terminal ICs.	18. Capacitor; *RC*; pi
19. Full-wave; half-wave	20. Dc output voltage; ac ripple voltage

EXPERIMENT 4-1
HALF-WAVE RECTIFIER

Most industrial applications that require direct-current power sources now use semiconductor rectifying devices.

The rectification method used in this experiment is only one method for converting alternating current into direct current. In the following experiments, you will investigate other methods of rectification. The semiconductor diode, single-phase, half-wave rectifier is the simplest method used for ac to dc conversion.

OBJECTIVE

To observe the characteristics of a single-phase, half-wave rectification circuit that uses a semiconductor diode.

EQUIPMENT

Variable ac power supply
Electronic multifunction meter
Resistor: 1000 Ω
Diode
Oscilloscope

PROCEDURE

1. Construct the single-phase, half-wave rectifier circuit shown in Fig. 4-1A.
2. With a voltmeter, measure the *ac values* across D_1 and R_L, and record them in Table 4-1A, using 10 V ac as the applied voltage.
3. Adjust the ac power source to 20 V ac and measure V_{D_1} and V_{R_L}. Record your values in Table 4-1A.
4. Adjust the ac power source to 30 V ac and measure V_{D_1} and V_{R_L}. Record your values in Table 4-1A.

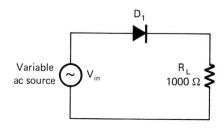

FIGURE 4-1A   Semiconductor diode, single-phase, half-wave rectifier circuit.

TABLE 4-1A.

Ac values

Applied Voltage	V_{in} (peak)	V_{D_1} (RMS)	V_{D_1} (peak)	V_{RL} (RMS)	V_{RL} (peak)
10 V ac					
20 V ac					
30 V ac					

5. Peak voltage is equal to 1.41 × RMS voltage. Calculate the peak values of V_{in}, V_{D_1}, and V_{RL} and record them in Table 4-1A.

6. Prepare the meter to measure *dc voltage* and complete the dc voltage values in Table 4-1B for 10, 20, and 30 V ac applied. (V_{in} is the dc voltage measurement across the ac source and should = 0).

7. Apply 10 V ac to the circuit. With an oscilloscope properly synchronized with the ac input, observe and record the necessary waveforms in Table 4-1C.

8. Also, record the V_{p-p} values in Table 4-1C.

ANALYSIS

1. What is the frequency of the ac ripple voltage across R_L?
   _____

2. What effect would a lower value of R_L have on the operation of this rectifier circuit? _____

3. What is the ratio of dc voltage output to peak ac voltage input for this rectifier circuit?
   $$\frac{V_{dc(out)}}{V_{peak(in)}} = \underline{\hspace{4cm}}.$$

4. What is the formula used for determining the value of dc voltage output for a semiconductor diode, single-phase, half-wave rectifier? _____

TABLE 4-1B.

Dc values

Applied Voltage	V_{in} (dc)	V_{D_1}	V_{RL} (V_{DC})
10 V ac			
20 V ac			
30 V ac			

TABLE 4-1C.
Oscilloscope waveforms

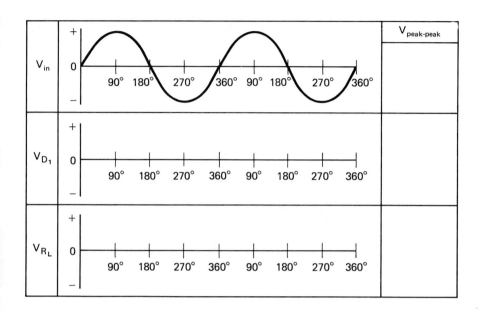

5. What effect does increasing the applied ac voltage have on the dc output of this rectifier circuit? _____

_____

6. What would happen if the polarity of D_1 was reversed?

_____

EXPERIMENT 4-2
FULL-WAVE RECTIFIER

Full-wave rectifiers have an advantage over half-wave rectifiers. They produce a higher dc voltage output. However, this method requires two diodes and a relatively expensive center-tapped transformer.

OBJECTIVE

To observe the characteristics of a single-phase, full-wave rectification circuit that uses semiconductor diodes and a center-tapped transformer. (A standard 6.3/12.6-V ac filament transformer will be used).

EQUIPMENT

Center-tapped 6.3/12.6-V ac transformer
Electronic multifunction meter
Resistor: 1000 Ω
Diodes: (2), or equivalent
Oscilloscope

PROCEDURE

1. Construct the circuit shown in Fig. 4-2A.
2. With a voltmeter, measure the *ac voltage* values across D_1, D_2, and R_L and record them in Table 4-2A.

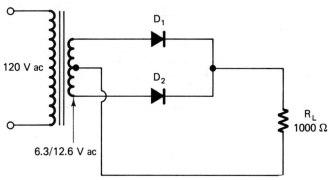

FIGURE 4-2A   Semiconductor diode, single-phase, full-wave rectifier circuit.

TABLE 4-2A.

Ac values

V_{in}	V_{in} (peak)	V_{D_1} (peak)	V_{D_2}	V_{D_2} (peak)	V_{RL}	V_{RL} (peak)
12.6 V ac						

3. Calculate the *peak* values of V_{in}, V_{D_1}, V_{D_2}, and V_{RL} and record in Table 4-2A.
4. Prepare the meter to measure *dc voltage* and complete the dc values in Table 4-2B.
5. With an oscilloscope properly synchronized with the ac input, observe and record the necessary waveforms in Table 4-2C.
6. Also, record $V_{p\text{-}p}$ values in Table 4-2C.

TABLE 4-2B.

Dc values

V_{in}	V_{in} (dc)	V_{D_1}	V_{D_2}	V_{RL} (V dc)
12.6 V ac				

TABLE 4-2C.

Oscilloscope waveforms

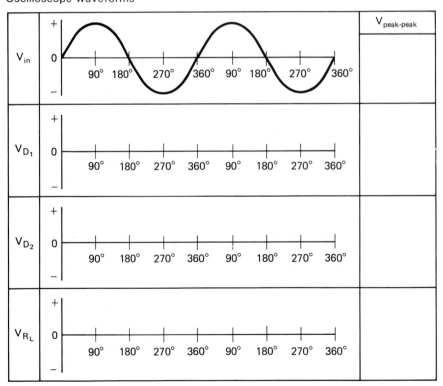

ANALYSIS

1. What is the frequency of the ac ripple voltage across R_L?

    _____

2. Would using a lower value resistance for R_L have the same effect as on the single-phase, half-wave rectifier? Why?

    _____

    _____

3. What is the ratio of dc voltage output to peak ac voltage input for the rectifier constructed in this experiment?

    $$\frac{V_{dc(out)}}{V_{peak(in)}} = \underline{\hspace{4cm}}$$

4. What is the formula for determining the value of dc voltage output for a single-phase, full-wave rectifier?

    _____

5. What would happen if the polarities of both D_1 and D_2 were reversed? _____

    _____

EXPERIMENT 4-3
BRIDGE RECTIFIER

Full-wave rectifier circuits are more desirable to use than half-wave circuits. It is possible to accomplish full-wave rectification by using a center-tapped transformer winding and two diode devices. However, the transformer is relatively expensive. Full-wave rectification is also accomplished by using four diodes connected in a bridge configuration. This circuit eliminates the need for a center-tapped transformer.

OBJECTIVE

To observe the operation of a single-phase bridge rectifier circuit, which is commonly used for converting alternating current to direct current.

EQUIPMENT

Variable ac power supply
Electronic multifunction meter
Resistor: 10,000 Ω
Diodes: (4), or equivalent
Oscilloscope

PROCEDURE

1. Construct the circuit shown in Fig. 4-3A. (All diodes are 1 Amp I_F or equivalent.)
2. With a voltmeter, measure the *ac voltage* values across each diode and R_L, using 10 V ac as the applied voltage. Record these values in Table 4-3A.

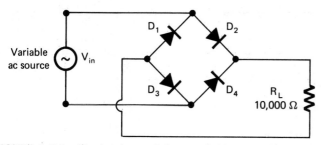

FIGURE 4-3A Single-phase, full-wave bridge rectifier circuit.

TABLE 4-3A.
Ac values

Applied Voltage	V_{in} (peak)	V_{D_1}	V_{D_2}	V_{D_3}	V_{D_4}	V_{RL} (RMS)	V_{RL} (peak)
10 V ac							
20 V ac							
30 V ac							

3. Adjust the ac power source to 20 V ac and measure the ac voltage across each diode and across R_L. Record your values in Table 4-3A.
4. Repeat the same measurements with 30 V ac applied and record in Table 4-3A.
5. Calculate the *peak* values of V_{in} and V_{RL} and record in Table 4-3A.
6. Prepare the meter to measure *dc voltage* and complete the dc voltage values in Table 4-3B for 10, 20, and 30 V ac applied.
7. Apply 10 V ac to the bridge rectifier circuit. With an oscilloscope properly synchronized with the ac input, observe and record the necessary waveforms in Table 4-3C.
8. Also, record the $V_{p\text{-}p}$ values in Table 4-3C.

ANALYSIS

1. What is the frequency of the ac ripple voltage for the bridge rectifier? _____
2. How do the dc output values with 10, 20, and 30 V ac applied to the bridge circuit compare to those values obtained when using a single-phase, half-wave rectifier circuit? _____

TABLE 4-3B.
Dc values

Applied Voltage	V_{D_1}	V_{D_2}	V_{D_3}	V_{D_4}	V_{RL} (V dc)
10 V ac					
20 V ac					
30 V ac					

TABLE 4-3C.
Oscilloscope waveforms

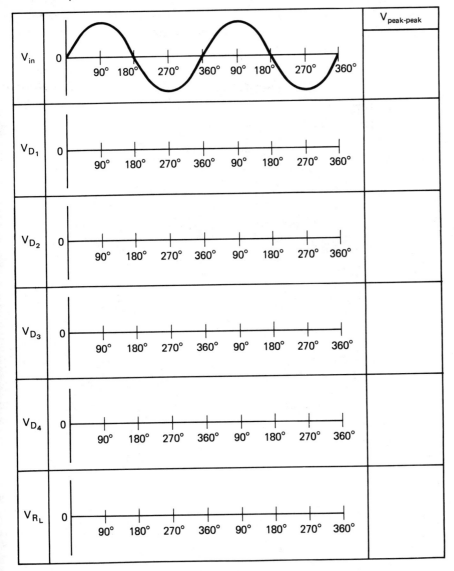

		$V_{peak\text{-}peak}$
V_{in}		
V_{D_1}		
V_{D_2}		
V_{D_3}		
V_{D_4}		
V_{R_L}		

3. What is the ratio of dc voltage output to the peak ac voltage input for this circuit?

$$\frac{V_{dc(out)}}{V_{peak(in)}} = \underline{\hspace{3cm}}$$

4. How does the ratio obtained in step 3 of the analysis compare to the full-wave rectifier that uses a center-tapped transformer? _____
_____

5. What would happen to the dc output if D_1 became short-circuited? _____
   _____

6. What would happen to the dc output if D_3 became an open-circuit path? _____
   _____

7. What would happen to the dc output if *all* diodes were reversed? _____
   _____

EXPERIMENT 4-4
CAPACITOR FILTERS

Rectification circuits are used as the part of a power supply that converts an ac input to a dc output. The output taken directly at the output of the rectifier circuit is a crude form of direct current that still contains a large amount of alternating current variation or ripple. Several different types of filtering circuits have been developed that can be added onto the output of a rectifier circuit. The purpose of filtering is to modify the *pulsating dc voltage* at the output of a rectifier circuit so that the resulting voltage will be a purer, or smoother, form of direct current.

OBJECTIVES

1. To see how a capacitor can be used to accomplish filtering.
2. To observe some of the important power-supply design considerations, such as ripple factor, percentage ripple, and voltage regulation.

EQUIPMENT

Electronic multifunction meter
Diodes: (4), or equivalent
Variable ac power supply
Oscilloscope
Resistors: 15,000 Ω, 10,000 Ω, 5000 Ω, 1000 Ω, and 500 Ω
Capacitor: 40-μF, 100-V dc

PROCEDURE

1. Construct the circuit shown in Fig. 4-4A.
2. Adjust the ac power supply to 10 V (RMS).
3. With a properly calibrated oscilloscope, measure and record the output waveform (across R_L) in Fig. 4-4 B.
4. Compute the following values for this rectification circuit:

   (a) $V_{peak} = 1.41 \times V_{RMS} =$ _____V.

   (b) $V_{dc} = 0.636 \times V_{peak} =$ _____V.

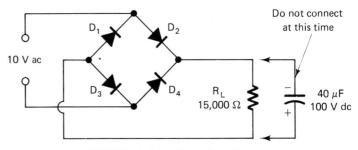

FIGURE 4-4A Capacitor filter.

(c) Ripple factor $= \dfrac{\text{ripple voltage (RMS)}}{V_{\text{dc}}}$

$= \underline{\hspace{2cm}}$

5. Using an electronic multifunction meter, measure V_{peak}, V_{dc}, and ripple voltage (RMS). Record the values in Table 4-4A.

6. Complete Table 4-4A for the indicated input voltage.

7. Turn off the power supply and connect a dc current meter in series with the load resistance. Adjust the power supply to 10 V ac and measure the load current.

Load current = $\underline{\hspace{5cm}}$ A dc.

V_{R_L} Waveform		$V_{\text{p-p}}$
+ 0 ┼──┼──┼──┼──┤ 90° 180° 270° 360° −		

FIGURE 4-4B Oscilloscope waveform of filtered output.

TABLE 4-4A.
Power-supply values

V_{in} (RMS)	V_{peak} Computed	V_{peak} Measured	V_{dc} Computed	V_{dc} Measured	Ripple Factor	%
10 V ac						
15 V ac						
20 V ac						
25 V ac						
30 V ac						

8. Turn off the power supply and connect a 40- μF, 100-V dc capacitor across R_L. Be sure to observe proper polarity.

9. Again, measure and record the load current with 10 V ac applied.

    Load current = _____ A dc.

10. With the oscilloscope, observe and record the waveform across R_L. Measure V_{dc} and ripple voltage (p-p) with the oscilloscope. Record in Fig. 4-4C.

11. Complete Table 4-4B for the indicated values of resistive loads with 10 V ac applied.

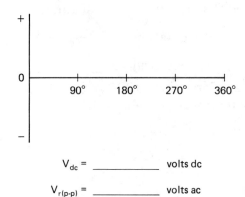

V_{dc} = _____ volts dc

$V_{r(p-p)}$ = _____ volts ac

FIGURE 4-4C   Output waveform.

TABLE 4-4B.
Power-supply resistive loads

R_L	I_{RL} Measured	V_R (RMS) Measured	V_{dc} Computed *	V_{dc} Measured	Ripple Factor
15,000 Ω					
10,000 Ω					
5000 Ω					
1000 Ω					
500 Ω					

* V_{dc} for a circuit with a capacitor filter is calculated as:

$$V_{dc} = V_{max} - \frac{V_{r\ (p-p)}}{2}$$

ANALYSIS

1. From Table 4-4A, when V_{in} was increased, what effect was observed on:
   a. V_{peak}? _____
   b. V_{dc}? _____
   c. Ripple factor? _____

2. What effect did the addition of a capacitor have on load current (see steps 7 and 9)? Why? _____

_____

3. What effect did the added capacitor have on:
   a. V_r (RMS)? _____
   b. V_{dc}? _____
   c. Ripple factor? _____

4. From Table 4-4B, what effect does increased load have on:
   a. V_r (RMS)? _____
   b. V_{dc}? _____
   c. Ripple factor? _____

5. Calculate the percentage voltage regulation from Table 4-4B values, using 15,000 Ω as the no-load resistance and 500 Ω as the full-load resistance:

$$\% \text{ regulation} = \frac{V_{NL} - V_{FL}}{V_{FL}}$$

$$\times \ 100 = \underline{\hspace{3cm}}\%.$$

6. Calculate the ideal value of filter capacitor for the 5000-Ω load of Table 4-4B.

$$C = \frac{2.4 \times I_{dc}}{V_{r(RMS)}} = \underline{\hspace{3cm}} \mu F.$$

EXPERIMENT 4-5
RC- AND PI-TYPE FILTERS

Two improvements over the simple capacitor filter for dc power supplies are the RC and pi (π) filter networks. An RC filter section, consisting of one resistor and two capacitors, may be connected across the output of a rectifier circuit. The (π)-type filter is similar, except an inductor is used rather than a resistor. These types of filters produce lower ripple voltage than the capacitor filter.

OBJECTIVES

1. To observe the operation of an RC filter section and then a π-type filter connected onto the output of a bridge rectifier.
2. To note the improved filtering action of these two types of filters compared to the capacitor filter.

EQUIPMENT

Ac power source
Electronic multifunction meter
Resistors: 1000 Ω, 470 Ω (2), and 270 Ω, 1.0 W
Diodes: (4), or equivalent
Capacitors: 2 -10 μF to 40 μF, with at least 50 V dc rating
Inductor: 8 H to 16 H

PROCEDURE

Section A: RC *Filter Circuit*

1. Construct the RC filter circuit shown in Fig. 4-5A.
2. Apply 6.3 V ac to the circuit. With a voltmeter, measure the ac voltage values across C_1, R_1, and R_L. Record these values in Table 4-5A.
3. Calculate the peak ac voltage values of V_{in}, V_{C_1}, V_{R_1}, and V_{RL}.
4. Repeat the same measurements and calculations with 12.6 V ac applied.
5. Prepare the meter to measure *dc voltage* and record the necessary dc values in Table 4-5B.

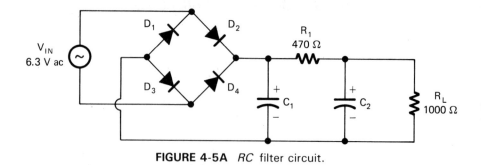

FIGURE 4-5A *RC* filter circuit.

Section B: π-Type Filter Circuit

1. Replace R_1 with an inductor with a value in the range of 8 H to 16 H. Refer to this device as L_1. This makes the circuit a π-type filter.
2. Again, prepare the meter to measure *ac voltage.* Measure and record in Table 4-5C the values of V_{C_1}, V_{L_1}, and V_{RL} with 6.3 V ac applied.
3. Calculate the *peak* values of V_{C_1}, V_{L_1}, and V_{RL} and record them in Table 4-5C.
4. Repeat the same measurements and calculations with 12.6 V ac applied. Record your values in Table 4-5C.

TABLE 4-5A.
Ac values (*RC* filter)

V_{in}	V_{in} *(peak)*	V_{C_1}	V_{C_1} *(peak)*	V_{R_1}	V_{R_1} *(peak)*	V_{RL}	V_{RL} *(peak)*
6.3 V ac							
12.6 V ac							

TABLE 4-5B.
Dc values (*RC* filter)

V_{in}	V_{C_1}	V_{R_1}	V_{RL}
6.3 V ac			
12.6 V ac			

TABLE 4-5C.
Ac values—pi (π)-type filter

V_{in}	V_{C_1}	V_{C_1} *(peak)*	V_{L_1}	V_{L_1} *(peak)*	V_{RL}	V_{RL} *(peak)*
6.3 V ac						
12. 6 V ac						

TABLE 4-5D.
Dc values—pi (π)-type filter

V_{in}	V_{C_1}	V_{L_1}	V_{R_L} (V dc)
6.3 V ac			
12.6 V ac			

5. Prepare the meter to measure *dc voltage*. Measure and record the necessary dc values in Table 4-5D.

6. With 12.6 V ac applied, substitute a 470-Ω and then a 270-Ω resistor for the load resistor R_L. Make the necessary measurements to complete Table 4-5E for the changes of load resistance.

7. Use the 1000-Ω resistor as the load (R_L) and maintain 12.6 V ac as the source voltage.

8. With an oscilloscope properly synchronized with the ac input, record the necessary waveforms in Table 4-5F.

9. Also, record $V_{p\text{-}p}$ values in Table 4-5F.

ANALYSIS

1. From the data of Table 4-5B, compare V_{C_1}, and V_{R_L} values. Why are these values different? _____
   _____
   _____

2. What characteristics of an *RC* filter does a comparison of the data of Tables 4-5A and B show? _____
   _____
   _____

3. For the *RC* filter, what is the ratio of dc voltage output to the peak ac voltage input?

   $$\frac{V_{dc(out)}}{V_{peak(out)}} = \text{_____}$$

4. What effect did replacing R_1 with an inductor have on the dc voltage output of the circuit? Compare Tables 4-5B and 4-5D. _____
   _____

TABLE 4-5E.
Load variations—dc values—pi (π)-type filter

R_L	V_{C_1}	V_{L_1}	V_{R_L} (V dc)	
470 Ω				
270 Ω				

TABLE 4-5F.
Oscilloscope waveforms—pi (π)-type filter

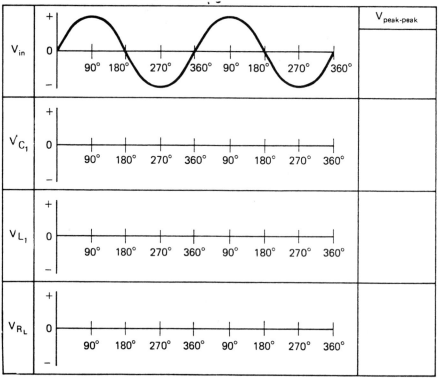

5. From Tables 4-5B and D, how do the values of V_{R_1} and V_{L1} compare? Why? _____

    _____

6. From the data of Table 4-5E, explain how a decrease in load resistance affects the dc output. _____

    _____

7. What effect would using a half-wave rectifier rather than a full-wave bridge rectifier have on the dc output from a π-filter network? _____

    _____

8. Which type of filter (RC or π-type) is most desirable to use? Why? _____

    _____

9. For the π-type filter, what is the ratio of V dc(out)/ Vpeak(in) with a 1000-Ω load? With a 470-Ω load? With a 270-Ω load? _____

    _____

EXPERIMENT 4-6
ZENER-DIODE VOLTAGE REGULATOR

An important characteristic of dc power sources is their ability to maintain a constant voltage output as load resistance values change. This characteristic is referred to as the *voltage regulation* of a power source. In order to improve voltage regulation, a zener diode may be connected across the output of a filtered dc power source. The reverse-bias operational characteristics of this device make it desirable to maintain a constant voltage in a circuit.

OBJECTIVE

To analyze a zener diode's operation as a dc power-supply voltage regulator.

EQUIPMENT

Ac power source
Diodes: (4), or equivalent
Electronic multifunction meter
Zener diode, or equivalent
Capacitors: 10 μF to 50 μF with 100 V dc rating
Inductor: 8 H to 16 H
Resistors: 270 Ω, 500 Ω, 1000 Ω, 5000 Ω, and 15,000 Ω

PROCEDURE

1. Construct the circuit shown in Fig. 4-6A.
2. Apply 20 V ac to the input of this zener-diode-regulated power supply. Measure voltages V_{C_1}, V_{L_1}, V_{C_2}, V_{R_1}, and

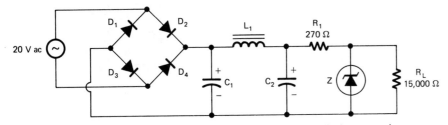

FIGURE 4-6A Circuit illustrating the use of a zener diode as a voltage regulating device.

TABLE 4-6A.
Regulated power-supply values

R_L	V_{C_1}	V_{L_1}	V_{C_2}	V_{R_1}	$V_{DC}\,(V_z)$	$I_{L_1} = \dfrac{V_{L_1}}{X_{L_1}}$	$I_{R_1} = \dfrac{V_{R_1}}{R_1}$	$I_{R_L} = \dfrac{V_{R_L}}{R_L}$	$I_z = I_{R_1} - I_{R_L}$
15,000 Ω									
5000 Ω									
500 Ω									

* $X_{L_1} = 2\,\pi \cdot f \cdot L$

V_Z. Use these values to calculate the dc current through L_1, R_1 Z, and R_L. Record the data in Table 4-6A.

3. Change the load resistance value to 5000 Ω and then to 500 Ω and complete Table 4-6A for each of these load resistance values.

ANALYSIS

1. Draw a simplified graph to show the forward- and reverse-bias characteristics of a zener diode in Fig. 4-6B.
2. How does the zener diode differ from a conventional PN junction diode? _____
_____
_____
3. Why is the reverse-bias characteristic of a zener diode useful for voltage regulation _____
_____
_____
4. How do the forward-bias characteristics of a zener diode compare with a PN junction diode? _____
_____
_____
5. From the data of Table 4-6A, discuss the effect of a lower load resistance on:

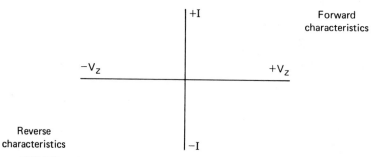

FIGURE 4-6B Zener diode characteristics.

a. V_{C1} _____

b. V_{C2} _____

c. V_{R1} _____

d. V_Z _____

e. V_{RL} _____

f. I_{R1} _____

g. I_Z _____

h. I_{RL} _____

6. Calculate the voltage regulation of this circuit using 15,000 Ω as no load and 500 Ω as full load.

$$\% \text{ regulation} = \frac{V_{NL} - V_{FL}}{V_{FL}}$$

$$= \underline{\hspace{3in}}\%$$

7. What is the ratio of $V_{dc(out)}$ and $V_{peak(in)}$ for this circuit? (Use 15,000-Ω values.)

$$\frac{V_{dc(out)}}{V_{peak(in)}} = \underline{\hspace{2.5in}}$$

EXPERIMENT 4-7
HALF-WAVE VOLTAGE DOUBLER

A half-wave voltage doubler can be used to increase dc voltage levels from a fixed value ac source without using a transformer. In many cases, particularly where high voltages at low current levels are required, voltage-multiplier circuits are beneficial to use.

OBJECTIVE

To observe the operation of a half-wave voltage doubler.

EQUIPMENT

Variable ac power supply
Electronic multifunction meter
Resistor: 10,000 Ω
Diodes: (2), or equivalent
Capacitors: (2), -10 μF to 40 μF, with at least 50-V dc rating

PROCEDURE

1. Construct the circuit shown in Fig. 4-7A.
2. Apply 6.3 V ac to the circuit. With a voltmeter, measure the *ac voltage* values across the diodes, capacitors, and R_L. Record these values in Table 4-7A.
3. Calculate the *peak* values of V_{in}, V_{C1}, V_{C2}, and V_{RL}. Record these values in Table 4-7A.
4. Repeat the same measurements and calculations with 12.6 V ac applied.

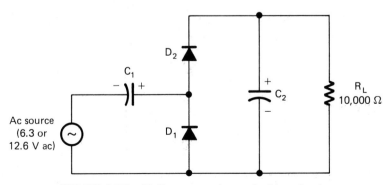

FIGURE 4-7A   Half-wave, voltage-doubler circuit.

TABLE 4-7A.
Ac values

V_{in}	V_{in} (peak)	V_{D_1}	V_{D_2}	V_{C_1}	V_{C_1} (peak)	V_{C_2}	V_{C_2} (peak)	V_{R_L}	V_{R_L} (peak)
6.3 V ac									
12.6 V ac									

TABLE 4-7B.
Dc values

V_{in}	V_{D_1}	V_{C_1}	V_{R_L} (V dc)
6.3 V ac			
12. 6 V ac			

TABLE 4-7C.
Oscilloscope waveforms

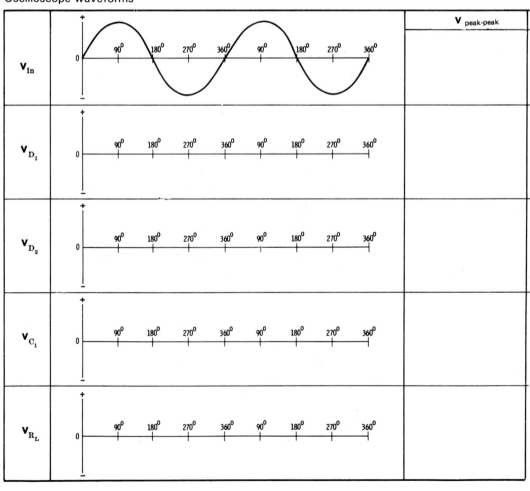

5. Prepare the meter to measure *dc voltage* and complete the necessary dc values in Table 4-7B.
6. With an oscilloscope properly synchronized with the ac input, record the necessary waveforms in Table 4-7C. Use 12.6 V ac as the applied voltage.
7. Also, record $V_{p\text{-}p}$ values in Table 4-7C.

ANALYSIS

1. How is voltage multiplication accomplished? _____

2. How do the dc values of V_{C_1} and V_{R_L} compare in this circuit? Why? _____

3. What effect would a lower value resistor for R_L have on circuit operation? _____

4. What would happen if D_2 was reversed in this circuit?

5. What is the ratio of dc voltage output to peak ac voltage input for this circuit?

$$\frac{V_{dc(out)}}{V_{peak(in)}} = \underline{\hspace{3cm}}$$

EXPERIMENT 4-8
FULL-WAVE VOLTAGE DOUBLER

The full-wave doubler circuit is similar in many respects to the half-wave voltage doubler. You should be able to observe some key differences between these two types of voltage doublers by comparing the data of this experiment with that of the previous experiment.

OBJECTIVE

To observe the operation of a full-wave voltage doubler.

EQUIPMENT

Variable ac power supply
Electronic multifunction meter
Resistor: 10,000 Ω
Diodes: (2), or equivalent
Capacitors: (2), 10 μF to 40 μF, with at least 50 V dc rating

PROCEDURE

1. Construct the circuit shown in Fig. 4-8A.
2. Apply 6.3 V ac to the circuit. With a voltmeter, measure the ac voltage values across the diodes, capacitors, and R_L. Record these values in Table 4-8A.
3. Calculate the peak values of V_{in} and V_{RL}. Record these values in Table 4-8A.

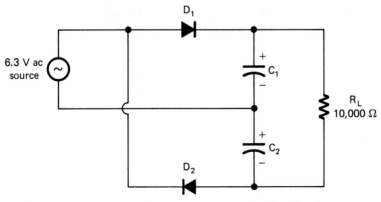

FIGURE 4-8A   Full-wave, voltage-doubler circuit.

TABLE 4-8A.
Ac values

V_{in}	V_{in} (peak)	V_{D_1}	V_{D_2}	V_{C_1}	V_{C_2}	V_{RL}	V_{R_L} (peak)
6.3 V ac							
12.6 V ac							

4. Repeat the same measurements and calculations with 12.6 V ac applied.
5. Prepare the meter to measure *dc voltage* and complete the necessary dc values in Table 4-8B.
6. With an oscilloscope properly synchronized with the ac input, record the necessary waveforms in Table 4-8C. Use 12.6 V ac as the applied voltage.
7. Also, record $V_{p\text{-}p}$ values in Table 4-8C.

ANALYSIS

1. What is the difference between the half-wave and full-wave voltage doubler? _____
   _____
   _____

2. How do the dc values of V_{C_1} and V_{RL} compare in this circuit? Why? _____
   _____
   _____

3. What effect would a lower value resistor for R_L have on circuit operation? _____
   _____

4. What would happen if D_2 was reversed in this circuit?
   _____
   _____

5. What is the ratio of dc voltage output to peak ac voltage input for this circuit?

$$\frac{V_{dc(out)}}{V_{peak(in)}} = \underline{\hspace{4cm}}$$

TABLE 4-8B.
Dc values

V_{in}	V_{D_1}	V_{D_2}	V_{C_1}	V_{C_2}	V_{RL} (V dc)
6.3 V ac					
12.6 V ac					

TABLE 4-8C.
Oscilloscope waveforms

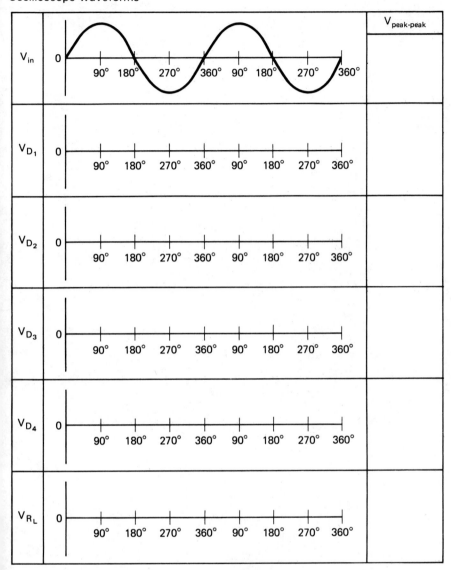

6. Compare the oscilloscope waveforms of Table 4-8C of this activity to Table 4-7C of the half-wave rectifier circuit.

_____

_____

_____

UNIT 4 EXAMINATION
POWER-SUPPLY CIRCUITS

Instructions: For each of the following, circle the answer that most correctly completes the statement:

1. The circuit shown in Fig. E-4-1 is a:
    a. Half-wave voltage doubler
    b. Full-wave voltage doubler
    c. Half-wave voltage tripler
    d. Full-wave voltage tripler

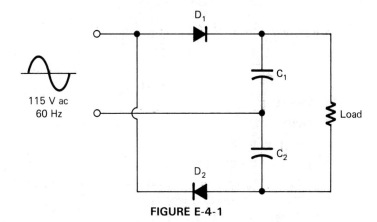

FIGURE E-4-1

2. The minimum number of components needed to build a half-wave voltage doubler is:
    a. One capacitor and two diodes
    b. Two capacitors and two diodes
    c. Two capacitors and three diodes
    d. Three capacitors and three diodes

3. The circuit shown in Fig. E-4-3 is a:
    a. Half-wave voltage doubler
    b. Full-wave voltage doubler
    c. Half-wave voltage tripler
    d. Full-wave voltage tripler

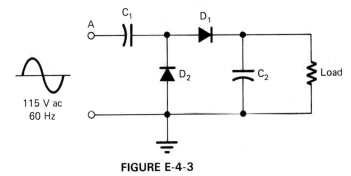

FIGURE E-4-3

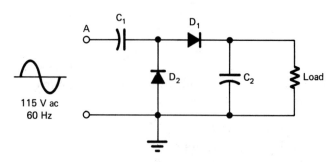

FIGURE E-4-4

4. In Fig. E-4-4, D_1 conducts when point A is:
   a. Positive
   b. Negative
   c. Positive or negative
   d. Used as an output

5. The dc voltage output across the output capacitor of a half-wave voltage doubler is:
   a. 2 × RMS ac
   b. 2 × peak ac
   c. 2 × effective ac
   d. 2 × average ac

6. The ripple frequency of the circuit of Fig. E-4-6 is:
   a. 30 Hz
   b. 60 Hz
   c. 120 Hz
   d. 240 Hz

7. The ripple frequency of the circuit shown in Fig. E-4-6 is:
   a. 30 Hz
   b. 60 Hz
   c. 120 Hz
   d. 240 Hz

8. The approximate output of a half-wave voltage doubler with 115 V RMS input is:
   a. 115 V dc
   b. 162.5 V dc
   c. 230 V dc
   d. 325 V dc

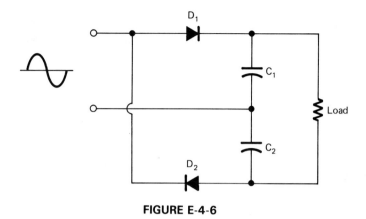

FIGURE E-4-6

9. The approximate output of a full-wave voltage doubler with 115 V ac RMS input is:
   a. 115 V dc
   b. 162.5 V dc
   c. 230 V dc
   d. 325 V dc

10. The ripple frequency of a full-wave rectifier circuit with a 60 Hz ac input is:
    a. 30 Hz
    b. 60 Hz
    c. 120 Hz
    d. 240 Hz

11. The simplest type of voltage regulator circuit is the:
    a. Shunt regulator
    b. Series regulator
    c. Emitter follower regulator
    d. Zener diode regulator

12. The effect of increasing the value of a capacitor in an RC filter network on the output of a power supply is to:
    a. Increase voltage and increase ripple
    b. Decrease voltage and decrease ripple
    c. Increase voltage and decrease ripple
    d. Decrease voltage and increase ripple

13. The circuit listed below which is the most effective filter is the:
    a. RC filter
    b. LC (pi) filter
    c. Capacitor filter
    d. Inductor filter

14. The ripple frequency of a half-wave rectifier circuit with a 60-Hz ac input is:

a. 30 Hz

b. 60 Hz

c. 120 Hz

d. 240 Hz

15. The circuit listed below that is not a type of voltage regulator is:

a. Series

b. Shunt

c. Zoner diode

d. Constant current

16. The formula used to calculate % regulation is:

a. $\% \text{ regulation} = \dfrac{V_{\text{no load}} + V_{\text{full load}}}{E_{\text{full load}}} \times 100$

b. $\% \text{ regulation} = \dfrac{V_{\text{no load}} + V_{\text{full load}}}{V_{\text{no load}}} \times 100$

c. $\% \text{ regulation} = \dfrac{V_{\text{no load}} - V_{\text{full load}}}{V_{\text{no load}}} \times 100$

d. $\% \text{ regulation} = \dfrac{V_{\text{no load}} - V_{\text{full load}}}{V_{\text{full load}}} \times 100$

17. A power supply delivers 12 V with no load and 10 V with a full load. The % regulation is:

a. 2%

b. 20%

c. 80%

d. 200%

18. The type of regulator shown in the circuit of Fig. E-4-18 is:

a. Series

b. Shunt

c. Zener diode

d. Constant current

19. In the circuit shown in Fig. E-4-19, the component that functions as a series regulator is:

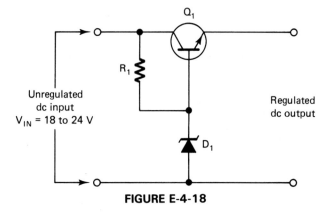

FIGURE E-4-18

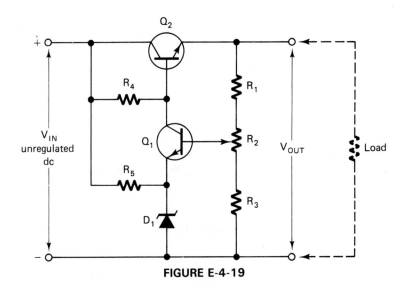

FIGURE E-4-19

    a. Q_1

    b. Q_2

    c. R_5 and D_1

    d. R_4 and R_5

20. The type of filter used in Fig. E-4-20 is a(n):

    a. *RC*

    b. *LC*

    c. Capacitor

    d. Inductor

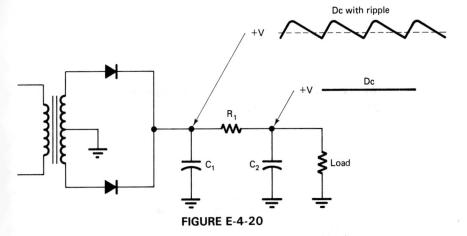

FIGURE E-4-20

21. The type of filter used in Fig. E-4-21 is a(n):

    a. RC

    b. LC

    c. Capicator

    d. Inductor

22. The circuit shown in Fig. E-4-22 is a:

    a. Half-wave rectifier with a positive output

    b. Half-wave rectifier with a negative output

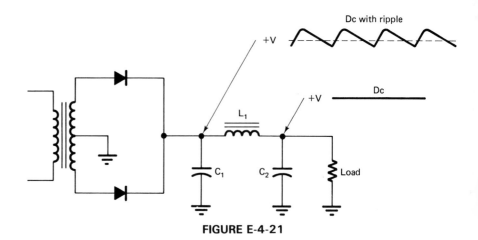

FIGURE E-4-21

   c. Full-wave rectifier with a positive output

   d. Full-wave rectifier with a negative output

23. In the circuit shown in Fig. E-4-22, the diodes which conduct when point A is negative and point B is positive are:

   a. D_1 and D_2

   b. D_3 and D_4

   c. D_2 and D_3

   d. D_1 and D_4

24. In the circuit of Fig. E-4-22, the diodes which conduct when point A is positive and point B is negative are:

   a. D_1 and D_2

   b. D_3 and D_4

   c. D_2 and D_3

   d. D_1 and D_4

25. The minimum number of diodes needed to construct a half-wave voltage doubler is:

   a. One

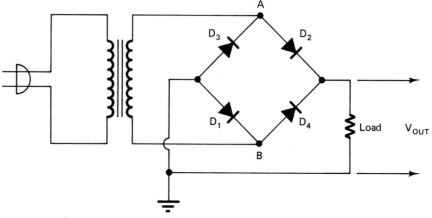

FIGURE E-4-22

    b. Two

    c. Three

    d. Four

26. The number of diodes needed to construct a bridge rectifier circuit is:

    a. One

    b. Two

    c. Three

    d. Four

27. The minimum number of diodes needed to construct a full-wave rectifier is:

    a. One

    b. Two

    c. Three

    d. Four

28. The circuit shown in Fig. E-4-28 is a:

    a. Half-wave rectifier with a positive output

    b. Half-wave rectifier with a negative output

    c. Full-wave rectifier with a positive output

    d. Full-wave rectifier with a negative output

29. If a half-wave rectifier has 115 V ac (RMS) input, the maximum voltage across the filter capacitor (without a load) is:

    a. 115 V dc

    b. 150 V dc

    c. 162 V dc

    d. 324 V dc

30. The ripple frequency of a bridge rectifier with a 60 Hz ac input is:

    a. 30 Hz

    b. 60 Hz

    c. 120 Hz

    d. 240 Hz

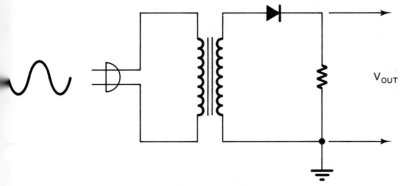

FIGURE E-4-28

UNIT FIVE

Oscillator Circuits

UNIT INTRODUCTION

Many electronic systems employ circuits that convert the dc energy of a power supply into a useful form of ac. Oscillators, generators, and electronic clocks are typical circuits. In a radio receiver for example, dc is converted into high-frequency ac to achieve signal tuning. Television receivers also have oscillators in their tuners. Oscillators are also used to produce horizontal and vertical sweep signals in TVs. These signals control the electron beam of the picture tube. In the same way, calculators and computers employ an electronic clock circuit. Timing pulses are produced by this circuit. Every RF transmitter employs an oscillator. This part of the system generates the signal that is radiated into space. Oscillators are extremely important in electronics.

An oscillator generally uses an amplifying device to aid in the generation of the ac output signal. Transistors and ICs can be used for this function. In this unit, we study solid-state oscillator circuits.

UNIT OBJECTIVES

Upon completion of this unit, you will be able to:

1. Calculate the frequency of oscillators using their *RC* component values.
2. Explain the operation of a blocking oscillator.
3. Construct an oscillator circuit.
4. Explain the sine wave output and damped oscillation of an *LC* tank circuit.
5. Measure output frequency of an oscillator circuit with an oscilloscope.
6. Compare measured and calculated frequency of an *LC* tank circuit.
7. Construct Colpitts and Hartley oscillators and observe variation in frequency with different values of *L* and *C*.
8. Describe the operation of astable, monostable, and bistable multivibrator circuits.
9. Explain the operation of a Schmitt trigger circuit.
10. Investigate the 555 timer integrated circuit and its applications.
11. List the classes of feedback oscillators.
12. Calculate the frequency of common *LC* oscillators.
13. Identify and explain the operation of oscillator circuits.
14. Recognize crystal oscillator circuits and explain their frequency selective characteristics.

IMPORTANT TERMS

Astable multivibrator. A free-running generator that develops a continuous square-wave output.

Blocking oscillator. An oscillator circuit that drives an active device into cutoff or blocks its conduction for a certain period of time.

Capacitor divider. A network of series-connected capacitors, each having a voltage drop of ac.

Comparator. An amplifier that has a reference signal input and a test signal input. When the test signal differs in value from the reference signal, a state change in the output occurs.

Continuous wave (CW). Uninterrupted sine waves, usually of the RF type, that are radiated into space by a transmitter.

Damped wave. A wave in which successive oscillations decrease in amplitude.

Differentiator network. A resistor-capacitor combination where the output appears across the resistor. An applied square wave produces a positive and negative spiked output.

Electromagnetic field. The space around an inductor that changes due to current flow through a coil. The field expands and collapses with applied ac.

Electrostatic field. The space or area around a charged body where there is some reaction to the developed charge.

Feedback. Transferring voltage from the output of a circuit back to its input.

Ferrite core. An inductor core made of molded iron particles.

Flip-flop. A multivibrator circuit having two stable states that can be switched back and forth.

Free running. An oscillator circuit that develops a continuous waveform that is not stabilized, such as an astable multivibrator.

Monostable. A multivibrator with one stable state. It changes to the other state momentarily and then returns to its stable state.

Nonsinusodial. Waveforms that are not sine waves, such as square, sawtooth, and pulsing waves.

Radio-frequency choke (RFC). An inductor or coil that offers high impedance to RF ac.

Regenerative feedback. Feedback from the output to the input that is in phase so that it is additive.

Relaxation oscillator. A nonsinusodial oscillator that has a resting or nonconductive period during its operation.

Resonant frequency. The frequency at which a tuned circuit oscillates.

Saturation. An active device operational region where the output current levels off to a constant value.

Stability. The ability of an oscillator to stay at a given frequency or given condition of operation without variation.

Symmetrical. A condition of balance where parts, shapes, and sizes of two or more items are the same.

Synchronized. A condition of operation that occurs at the same time. The vertical and horizontal sweep of an oscilloscope are often synchronized or placed in sync.

Tank circuit. A parallel resonant *LC* circuit.

Threshold. The beginning or entering point of an operating condition. A terminal connection of the LM555 IC timer.

Time constant (RC). The period required for voltage of a capacitor to increase 63.2% of the maximum value or decrease to 36.7% of the maximum value.

Toggle. A switching condition that changes back and forth.

Triggered. A control technique that causes a device to change its operational state, such as a triggered monostable multivibrator.

Vertical blocking oscillator. A TV circuit that generates the vertical sweep signal for deflection of the cathode-ray tube or picture tube.

OSCILLATOR TYPES

There are several possible ways of describing an oscillator. This can be done according to generated frequency, operational stability, power output, signal waveforms, and components. In this chapter it is convenient to divide oscillators according to their method of operation. This includes feedback oscillator and relaxation oscillators. Each group has a number of distinguishing features.

In *feedback oscillators* a portion of the output power is returned to the input circuit. This type of oscillator usually employs a tuned *LC* circuit. The operating frequency of the oscillator is established by this circuit. Most sine-wave oscillators are of this type. The frequency range is from a few hertz to millions of hertz. Applications of the feedback oscillator are widely used in radio and television receiver tuning circuits and in transmitters. Feedback oscillators respond very well to RF generator applications.

Relaxation oscillators respond to an electronic device that goes into conduction for a certain time, then turns off for a period of time. This condition of operation repeats itself on a continuous basis. This oscillator usually responds to the charge and discharge of an *RC* or an *RL* network. Oscillators of this type usually generate square- or triangular-shaped waves. The active device of the oscillator is "triggered" into conduction by a change in voltage. Applications include the vertical and horizontal sweep generators of a television receiver and computer clock circuits. Relaxation oscillators respond extremely well to low-frequency applications.

FEEDBACK OSCILLATORS

Feedback is a process by which a portion of the output signal of a circuit is returned to the input. This function is an important characteristic of any oscillator circuit. Feedback may be accomplished by inductance, capacitance, or resistance coupling. A variety of different circuit techniques have been developed today. As a general rule, each technique has certain characteristics that distinguishes it from others. This accounts for the large number of different circuit variations.

If you have ever been to a public gathering where a sound system was being used, you have probably had an opportunity to experience feedback. If a microphone is placed too close to a speaker, feedback can occur. Sound from the speaker is picked up by the microphone and returned to the amplifier. It then comes out of the speaker and is returned again to the microphone. The process continues until a hum or high-pitched howl is produced. Figure 5-1 shows an example of sound system feedback. This condition is considered to be mechanical feedback. In sound-system operation, feedback is an undesirable feature. In an oscillator, circuit operational feedback is a necessity.

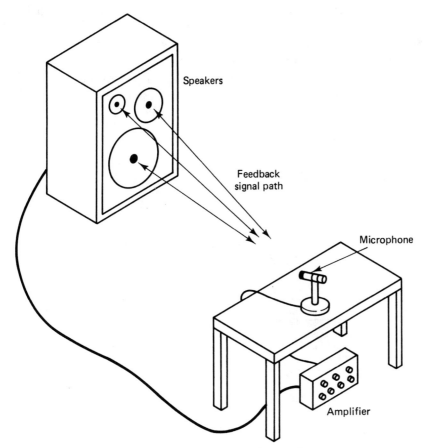

FIGURE 5-1   Sound-system feedback.

Oscillator Fundamentals

An *oscillator* is an electronic circuit that generates a continuously repetitive output signal. The output signal may be alternating current or some form of varying dc. The unique feature of an oscillator is its generation function. It does not receive an external input signal. It develops the resulting output signal from the dc operating power provided by the power supply. An oscillator is, in a sense, an electronic power converter. It changes dc power into ac or some useful form of varying dc. A block diagram of a feedback oscillator is shown in Fig. 5-2.

An oscillator has an amplifying device, a feedback network, frequency-determining components, and a dc power source. The *amplifier* is used primarily to increase the output signal to a usable level. Any device that has signal gain capabilities can be used for this function. The *feedback network* can be inductive, capacitive, or resistive. It is responsible for returning a portion of the amplifier's output back to the input. The feedback signal must be of the correct phase and value in order to cause oscillations to occur. Inphase or *regenerative* feedback is essential in an oscillator. An inductor-capacitor (*LC*) network determines the frequency of the oscillator. The charge and discharge action of this network establishes the oscillating voltage. This signal is then applied to the input of the amplifier. In a sense, the *LC network* is energized by the feedback signal. This energy is needed to overcome the internal resistance of the LC network. With suitable feedback, a continuous ac signal can be generated. The output of a good oscillator must be uniform. It should not vary in frequency of amplitude.

LC Circuit Operation

The operating frequency of a feedback oscillator is usually determined by an inductance-capacitance network. An *LC* network is

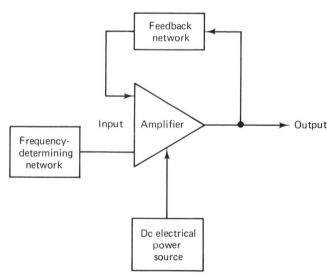

FIGURE 5-2 Fundamental oscillator parts.

sometimes called a *tank circuit*. The tank, in this case, has a storage capability. It stores an ac voltage that occurs at its *resonant frequency*. This voltage can be stored in the tank for a short period of time. In an oscillator, the tank circuit is responsible for the frequency of the ac voltage that is being generated.

In order to see how ac is produced from dc, let us take a look at an *LC* tank circuit. Figure 5-3(a), shows a simple tank circuit connected to a battery. In Fig. 5-3(b), the switch is closed momentarily. This action causes the capacitor to charge to the value of the battery voltage. Note the direction of the charging current. After a short time the switch is opened. Figure 5-3(c) shows the accumulated charge voltage on the capacitor.

We will now see how a tank circuit develops a sine-wave volt-

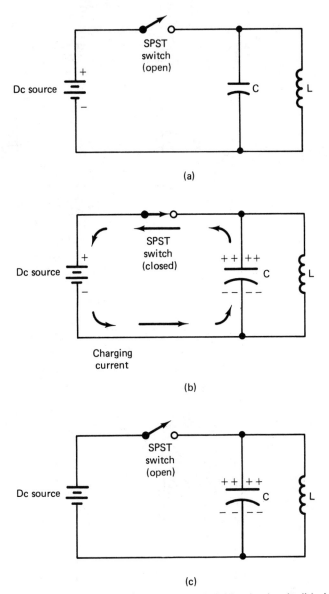

FIGURE 5-3 *LC* tank circuit being charged: (a) basic circuit; (b) charging action; (c) charged capacitor.

age. Assume now that the capacitor of Fig. 5-4(a) has been charged to a desired voltage value by momentarily connecting it to a battery. Figure 5-4(b) shows the capacitor discharging through the inductor. Discharge current flowing through L causes an electromagnetic field to expand around the inductor. Figure 5-4(c) shows

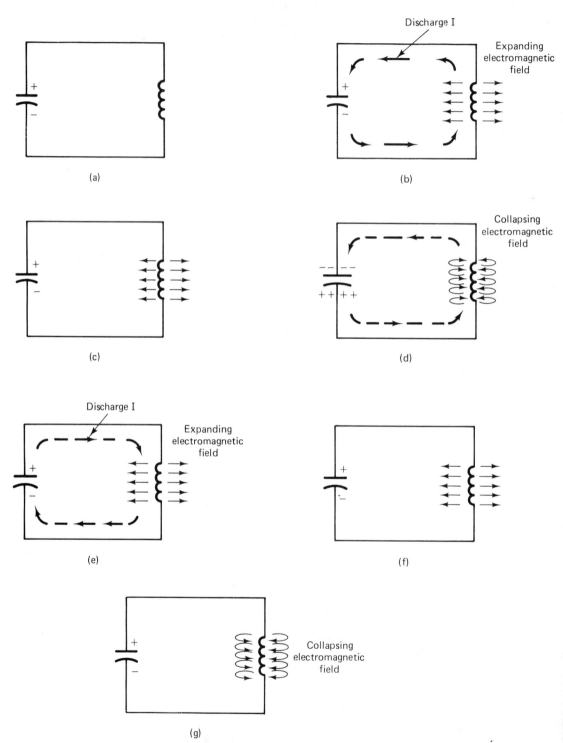

FIGURE 5-4 Charge-discharge action of an LC circuit.

the capacitor charge depleted. The field around the inductor then collapses. This causes a continuation of current flow for a short time. Figure 5-4(d) shows the capacitor being charged by the induced voltage of the collapsing field. The capacitor once again begins to discharge through L. In Fig. 5-4(e), note that the direction of the discharge current is reversed. The electromagnetic field again expands around L. Its polarity is reversed. Figure 5-4(f) shows the capacitor with its charge depleted. The field around L collapses at this time. This causes a continuation of current flow. Figure 5-4(g) shows C being charged by the induced voltage of the collapsing field. This causes C to be charged to the same polarity as the beginning step. The process is then repeated. The charge and discharge of C through L causes an ac voltage to occur in the circuit.

The frequency of the ac voltage generated by a tank circuit is based on the values of L and C. This is called the *resonant frequency* of a tank circuit. The formula for resonant frequency is

$$\text{Frequency } (f_r) = \frac{1}{2 \pi \sqrt{LC}}$$

where f_r is in hertz, L is the inductance in henries, and C is the capacitance in farads. This formula is extremely important because it applies to all LC sine-wave oscillators. Resonance occurs when the inductive reactance (X_L) equals the capacitive reactance (X_C). A tank circuit will oscillate at this frequency for a rather long period of time without a great deal of circuit opposition.

At resonant frequency an LC tank always has some circuit resistance. This resistance tends to oppose the ac that circulates through the circuit. The ac voltage will therefore decrease in amplitude after a few cycles of operation. Figure 5-5(a) shows the resulting wave of a tank circuit. Note the gradual decrease in signal amplitude. This is called a *damped sine wave*. A tank circuit produces a wave of this type.

A tank circuit, as we have seen, generates an ac signal when it is energized with dc. The generated wave gradually diminishes in amplitude. After a few cycles of operation, the lost energy is transformed into heat. When the lost energy is great enough, the signal stops oscillating.

The oscillations of a tank circuit could be made continuous if energy were periodically added to the circuit. The added energy would restore that lost by the circuit through heat. The added energy must be in step with the circulating tank energy. If the two are additive, the circulating signal will have a continuous amplitude. Figure 5-5(b) shows the *continuous wave* (CW) of an energized tank circuit.

Our original tank circuit was energized by momentarily connecting a dc source to the capacitor. This action caused the capacitor to become charged by manual switching action. To keep a tank circuit oscillating by manual switch operation is physically

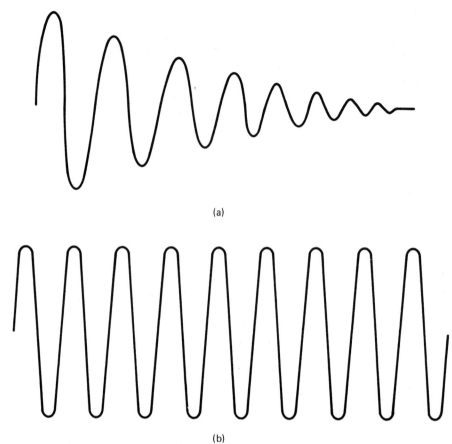

(a)

(b)

FIGURE 5-5   Wave types: (a) damped oscillatory wave; (b) continuous wave.

impossible. Switching of this type is normally achieved by an electronic device. Transistors are commonly used to perform this operation in an oscillator circuit. The output of the transistor is applied to the tank circuit in proper step with the circulating energy. When this is achieved the oscillator generates a CW signal of a fixed frequency.

The inductance (coil) of a tank circuit varies a great deal, depending on the frequency being generated. Most applications of LC oscillators are in the RF range. Some representative *RF oscillator coils* are shown in Fig. 5-6. Coil inductance is usually a variable. Inductance is changed by moving the position of a *ferrite core* inside the coil. Adjustment of the inductance permits some change in the frequency of the tank circuit.

Armstrong Oscillator

The Armstrong oscillator (often called a "tickler coil") of Fig. 5-7 is used to show the basic operation of an LC oscillator. The characteristic curve of the transistor and its ac load line are also used to explain its operation. Note that conventional bias voltages are applied to the transistor. The emitter-base junction is forward bi-

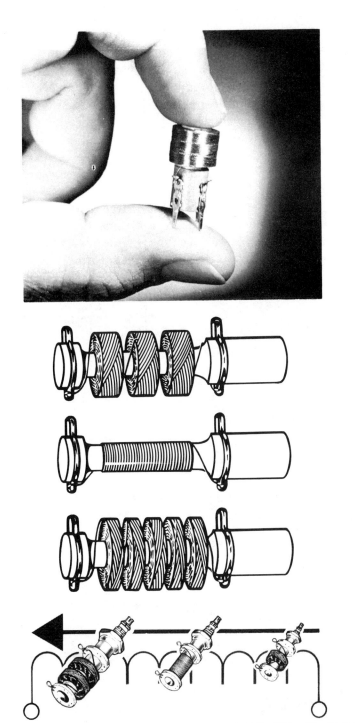

FIGURE 5-6 RF oscillator coils. (Courtesy of J.W. Miller Div./Bell Industries.)

ased and the collector is reverse biased. Emitter biasing is achieved by R_3. R_1 and R_2 serve as a divider network for the base voltage.

When power is first applied to the transistor, R_1 and R_2 establish the operating point (Q) near the center of the load line. The output of the transistor, at the collector, should ideally be O V ac. Due to the initial surge of current at turn-on time, some noise will

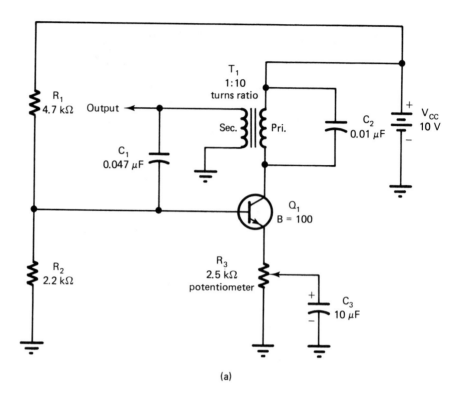

(a)

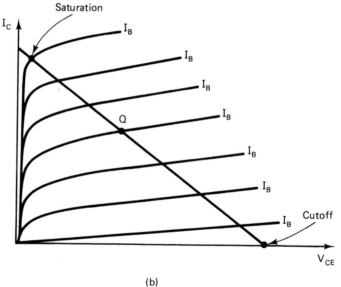

(b)

FIGURE 5-7 Armstrong oscillator: (a) circuit; (b) characteristic curves.

immediately appear at the collector. This is usually of a very small value. For our discussion, let us assume that a -1-mV signal appears at the collector. The transformer inverts this voltage and steps it down by a factor of 10. T_1 of the circuit has a 10:1 turns ratio. A $+0.1$-mV signal is therefore applied to C_1 in the base circuit.

It should be noted at this time that the transistor has a beta of 100. With $+0.1$ V applied to the base, Q_1 will develop a

−10-mV output signal at the collector. The polarity change from + to − in the output signal is due to the inverting characteristic of a common-emitter amplifier. The output signal voltage is again stepped down by the transformer and applied to the base of Q_1. A collector signal of −10 mV will now cause a +1.0-mV base voltage. With transistor gain, the collector voltage will immediately rise to −100 mV. The process will continue, thus producing collector voltages of −1.0 V and finally −10 V. At this point, transistor operation will be driven up the load line until saturation is reached. Note this point on the load line. The collector voltage of Q_1 will not change beyond this point.

With no change in V_C across the primary winding of T_1, the secondary voltage drops immediately to zero. The base voltage returns immediately to the Q point. This decrease of negative-going base voltage (from saturation to point Q) causes V_C to be positive-going. Through transformer action this appears as a negative-going base voltage. This action continues to drive the transistor past point Q. The process continues until the *cutoff* point is reached. The transformer then stops supplying input voltage to the base. Transistor operation swings immediately in the opposite direction. R_1 and R_2 cause the base voltage to again rise to point Q. The process is then repeated. Q_1 goes to saturation, to point Q, to cutoff, and returns again to point Q. Operation is continuous. An ac voltage appears across the secondary of T_1.

The frequency of the Armstrong oscillator is based on the value of C_1 and S. S is the inductance of the transformer secondary winding. The frequency of the generated wave is determined by the LC resonant frequency formula. C_1 and S form a tank circuit with the emitter-base junction of Q_1 and R_3 included.

The output of the Armstrong oscillator of Fig. 5-7 can be changed by altering the value of R_3. Gain is highest when the variable arm of R_3 is at the top of its range. This adjustment may cause a great deal of signal distortion. In some cases, the output voltage may appear as a square wave instead of a sine wave. When the position of the variable arm is moved downward, the output wave becomes a sine wave. Adjusting R_3 to the bottom of its range may cause oscillations to stop. The gain of Q_1 may not be capable of developing enough feedback for regeneration at this point. Normally 20 to 30% of output must be returned to the base to assure continuous oscillation.

Hartley Oscillator

The Hartley oscillator of Fig. 5-8 is used rather extensively in AM and FM radio receivers. The resonant frequency of the circuit is determined by the values of T_1 and C_1. Capacitor C_2 is used to couple ac to the base of Q_1. Biasing for Q_1 is provided by R_2 and R_1. Capacitor C_4 couples ac variations in collector voltage to the lower side of T_1. The *RF choke coil* (L_1) prevents ac from going into the power supply. L_1 also serves as the load for the circuit.

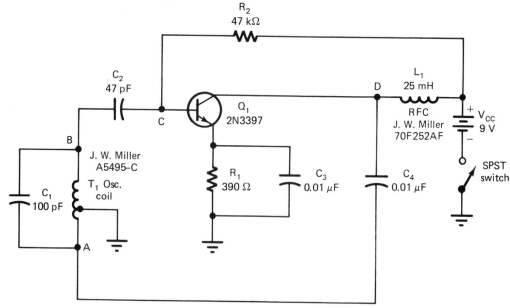

FIGURE 5-8 Hartley oscillator.

Q_1 is an NPN transistor connected in a common-emitter circuit configuration.

When dc power is applied to the circuit, current flows from the negative side of the source through R_1 to the emitter. The collector and base are both connected to the positive side of V_{CC}. This forward biases the emitter-base junction and reverse biases the collector. I_E, I_B, and I_C flow initially through Q_1. With I_C flowing through L_1, the collector voltage drops in value. This negative-going voltage is applied to the lower side of T_1 by capacitor C_4. This causes current flow in the lower coil. An electromagnetic field expands around the coil. This, in turn, cuts across the upper part of the coil. By transformer action, the top of the upper coil swings positive. Capacitor C_1 is charged by this voltage. This same voltage is also added to the forward bias voltage of Q_1 through C_2. Q_1 is eventually driven into saturation. When saturation occurs, there is no change in the value of V_C. The field around the lower part of T_1 collapses immediately. This causes a change in the polarity of the voltage at the top of T_1. The top plate of C_1 now becomes negative and the bottom becomes positive.

The accumulated charge on C_1 will begin to discharge immediately through T_1 by normal tank circuit action. This negative voltage at the top of C_1 causes the base of Q_1 to swing negative. Q_1 is driven to cutoff. This in turn causes the V_C of Q_1 to rise very quickly. A positive-going voltage is then transferred to the lower part of T_1 by C_4, providing feedback. This adds to the voltage of C_1. C_1 continues to discharge. The value change in V_C eventually stops, and no voltage is fed back through C_4. C_1 is fully discharged by this time. The field around the lower side of L_1 then collapses. C_1 charges again, with the bottom side being positive and the top negative. Q_1 again becomes conductive. The process is repeated on a continuous basis. The tank circuit produces a continuous wave. The circuit losses of the tank are restored by regenerative feedback.

A distinguishing feature of the Hartley oscillator is its *tapped coil.* A number of circuit variations are possible. The coil may be placed in series with the collector. Collector current flows through the coil in normal operation. A variation of this type is called a *series-fed Hartley oscillator.* The circuit of Fig. 5-8 is a *shunt-fed Hartley oscillator.* I_C does not flow through T_1. The coil is connected in parallel or shunt with the dc voltage source. Only ac flows through the lower part of T_1. Shunt-fed Hartley oscillators tend to produce more stable output.

Colpitts Oscillator

A *Colpitts oscillator* is very similar to the shunt-fed Hartley oscillator. The primary difference is in the tank circuit structure. A Colpitts oscillator uses two capacitors instead of a divided coil. Feedback is developed by an *electrostatic field* across the capacitor divider network. Frequency is determined by two capacitors in series and the inductor.

Figure 5-9 shows a schematic of the Colpitts oscillator. Bias voltage for the base is provided by resistors R_1 and R_2. The emitter is biased by R_4. The collector is reverse biased by connection to the positive side of V_{CC} through R_3. This resistor also serves as the collector load. The transistor is connected in a common-emitter circuit configuration.

When dc power is applied to the circuit, current flows from

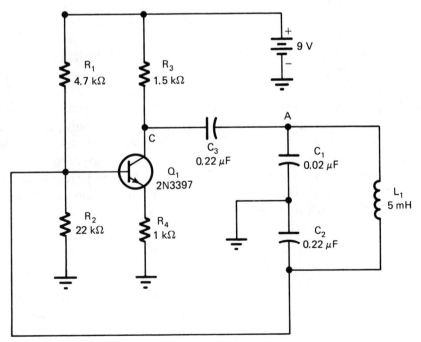

FIGURE 5-9   Colpitts oscillator.

the negative side of V_{CC} through R_4, Q_1, and R_3. I_C flowing through R_3 causes a drop in the positive value of V_C. This negative-going voltage is applied to the top plate of C_1 through C_3. With C_1 and C_2 connected in series, the bottom plate of C_2 takes on a positive charge. This adds to the positive voltage of the base, which in turn causes an increase in I_B. The conduction of Q_1 increases. The process continues until Q_1 is saturated.

When Q_1 becomes saturated there is no further increase in I_C. The changing value of V_C also stops. There is no feedback to the top side of C_1. The composite charge across C_1 and C_2 discharges through L_1. Discharge current flow through L_1 is from the top plate of C_1 to the bottom of C_2. The positive charge on C_2 soon diminishes. The electromagnetic field around L_1 collapses. The discharge current continues for a short time. The bottom plate of C_2 now becomes negatively charged, and the top plate of C_1 becomes positive. The negative charge voltage of C_2 reduces the forward bias voltage of Q_1. I_C decreases in value. V_C begins to increase in value. This change in voltage is again fed back to the top plate of C_1 through C_3. C_1 becomes more positively charged, and the bottom of C_2 becomes more negative. The process continues until Q_1 is driven to cutoff.

When Q_1 reaches its cutoff point, no I_C flows. There is no feedback voltage supplied to C_1. The composite charge across C_1 and C_2 discharges through L_1. Discharge current flows from the bottom of C_2 to the top of C_1. The negative charge on C_2 soon diminishes. The electromagnetic field around L_1 collapses. Current flow continues. The bottom plate of C_2 now becomes positive and the top plate of C_1 becomes negative. The positive charge voltage of C_2 pulls Q_1 out of the cutoff region. I_C begins to flow again. The process then repeats itself from this point. Feedback energy is added to the tank circuit momentarily during each alternation. As a general rule, the Colpitts circuit is a very reliable oscillator.

The amount of feedback developed by the Colpitts oscillator is based on the *capacitance ratio* of C_1 and C_2. The value of C_1 in our circuit is much smaller than C_2. The capacitive reactance (X_C) of C_1 is, therefore, much greater for C_1 than for C_2. The voltage across C_1 is much greater than that across C_2. By making the value of C_2 smaller, the feedback voltage can be increased. As a general rule, large amounts of feedback may cause the generated wave to be distorted. Small values of feedback do not permit the circuit to oscillate. In practice, 10 to 50% of the collector voltage is returned to the tank circuit as feedback energy.

Crystal Oscillator

When extremely high frequency stability is desired, crystal oscillators are used. The crystal of an oscillator is a finely ground wafer of quartz or Rochelle salt. It has the property to change electrical energy into vibrations of mechanical energy. It can also change mechanical vibrations into electrical energy. These two conditions of operation are called the *piezoelectric effect.*

The crystal of an oscillator is placed between two metal plates. Contact is made to each surface of the crystal by these plates. The entire assembly is then mounted in a holding case. Connection to each plate is made through terminal posts. When a crystal is placed in a circuit, it is plugged into a socket.

A crystal by itself behaves like a series resonant circuit. It has inductance (L), capacitance (C), and resistance (R). Figure 5-10(a) shows an equivalent circuit. The L function is determined by the mass of the crystal. Capacitance deals with its ability to change mechanically. Resistance corresponds to the electrical equivalent of mechanical friction. The series resonant equivalent circuit changes a great deal when the crystal is placed in a holder.

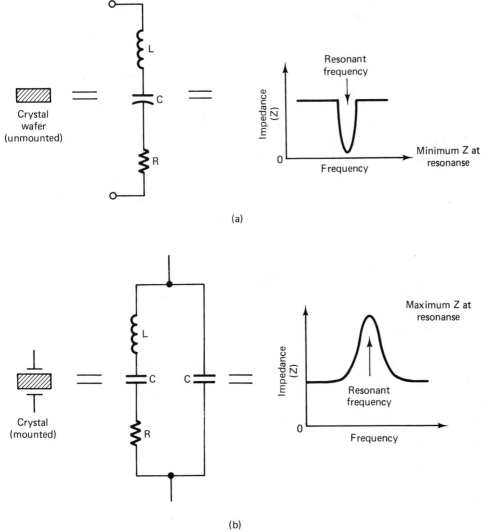

(a)

(b)

FIGURE 5-10   Crystal equivalent circuits: (a) series resonant; (b) parallel resonant.

Capacitance due to the metal support plates is added in parallel with the crystal equivalent circuit. Figure 5-10(b) shows the equivalent circuit of a crystal placed in a holder. In a sense, a crystal can have either a series resonant or a parallel resonant characteristic.

In an oscillator, the placement of a crystal in the circuit has a great deal to do with how it responds. If placed in the tank circuit, it responds as a parallel resonant device. In some applications the crystal serves as a tank circuit. When a crystal is placed in the feedback path, it responds as a series resonant device. It essentially responds as a sharp filter. It permits feedback only of the desired frequency. Hartley and Colpitts oscillators can be modified to accommodate a crystal of this type. The operating stability of this type of oscillator is much better than the circuit without a crystal. Figure 5-11 shows crystal-controlled versions of the Hartley and Colpitts oscillators. Operation is essentially the same.

Pierce Oscillator

The *Pierce oscillator* of Fig. 5-12 employs a crystal as its tank circuit. In this circuit the crystal responds as a parallel resonant circuit. In a sense, the Pierce oscillator is a modification of the basic Colpitts oscillator. The crystal is used in place of the tank circuit inductor. A specific crystal is selected for the desired frequency to be generated. The parallel resonant frequency of the crystal is slightly higher than its equivalent series resonant frequency.

Operation of the Pierce oscillator is based on feedback from the collector to the base through C_1 and C_2. These two capacitors provide a combined 180° phase shift. The output of the common-emitter amplifier is therefore inverted to achieve in-phase or regenerative feedback. The value ratio of C_1 and C_2 determines the level of feedback voltage. From 10 to 50% of the output must be fed back to energize the crystal. When it is properly energized, the resonant frequency response of the crystal is extremely sharp. It will vibrate only over a narrow range of frequency. The output at this frequency is very stable. The output of a Pierce oscillator is usually quite small. A crystal can be damaged by excessive mechanical strains and the heat caused by excessive power.

RELAXATION OSCILLATORS

Relaxation oscillators are primarily responsible for the generation of *nonsinusodial* waveforms. *Sawtooth, rectangular,* and a variety of irregular-shaped waves are included in this classification. These oscillators generally depend on the charge and discharge of a capacitor-resistor network for their operation. Voltage changes from the network are used to alter the conduction of an electronic device. Transistors, unijunction transistors (UJTs), and ICs can be used to perform the control function of this oscillator.

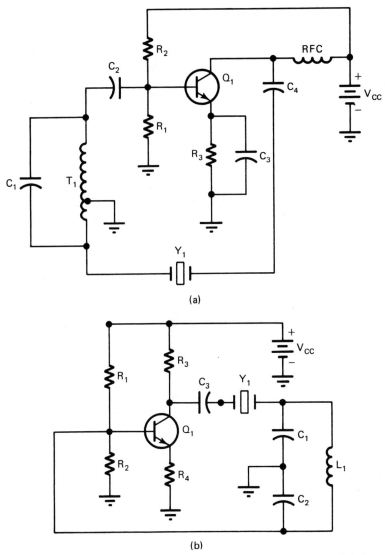

FIGURE 5-11 Crystal-controlled oscillators: (a) Hartley; (b) Colpitts.

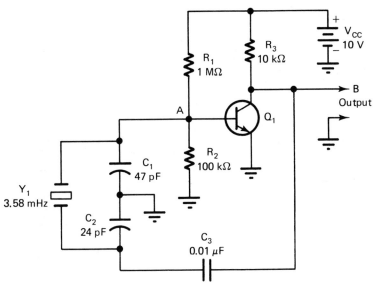

FIGURE 5-12 Pierce oscillator.

RC Circuit

When a source of dc is connected in series with a resistor and a capacitor, we have an *RC* circuit. (See Fig. 5-13(a).) The SPDT switch is used to charge and discharge the capacitor. In the charge position, the voltage source is applied to the *RC* circuit. *C* charges through *R*. In the discharge position, the source is removed from the circuit. *C* discharges through *R*.

Individual *time constant curves* are shown in Fig. 5-13(b) for each component of the *RC* circuit. V_C shows the capacitor voltage with respect to time. V_R is the resistor voltage. Circuit current

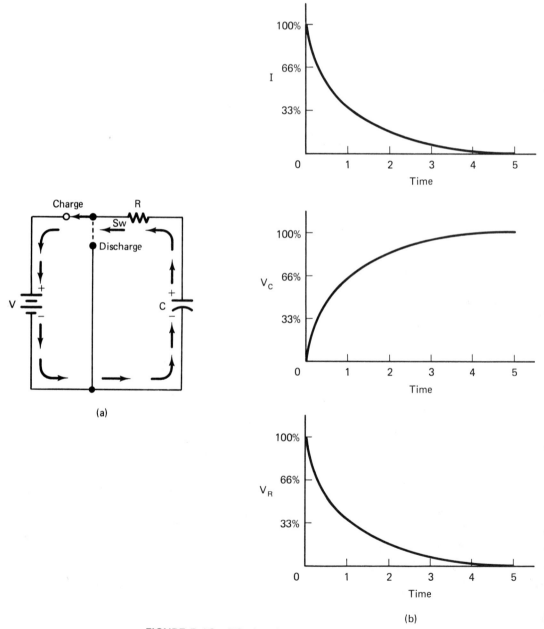

(a)

(b)

FIGURE 5-13 *RC* charging action: (a) circuit; (b) circuit values.

flow is displayed by the *I* curve. Note how the value of these curves change with respect to time.

The circuit switch is initially placed in the charge position. Note the direction of charge current indicated by the circuit arrows. In one time constant the V_C curve rises to 63% of the source voltage. Beyond this point there is a smaller change in the value of V_C. After five time constants the capacitor is considered to be fully charged. Note also how the circuit current and the resistor voltage change with respect to time. Initially, *I* and V_R rise to maximum values. After one time constant, only 37% of *I* flows. V_R, which is current dependent, follows the change in *I*. After five time constants there is zero current flow. This shows that *C* has been fully charged. With *C* charged there is no circuit current flow, and V_R is zero. The circuit remains in this state as long as the switch remains in the charge position.

Assume now that the capacitor of the circuit of Fig. 5-14(a) has been fully charged. The switch is now placed in the discharge position. In this situation, the dc source is removed from the circuit. The resistor is now connected across the capacitor. *C* will discharge through *R*. Current flows from the lower plate of *C* through *R* to the upper plate of *C*. Note the direction of the discharge current path indicated by circuit arrows.

The initial discharge current is of a maximum value. See the individual time constant curves for V_C, *I*, and V_R in Fig. 5-14(b). The discharge current shows *I* to be a maximum value initially. After one time constant it drops to 37% of the maximum value. After five time constants *I* drops to zero. V_C and V_R follow *I* in the same manner. After five time constants, *C* is discharged. There is no circuit *I*. V_C and V_R are both zero. The circuit remains in this state as long as the switch remains in the discharge position.

UJT Oscillator

The charge and discharge of a capacitor through a resistor can be used to generate a sawtooth waveform. The charge-discharge switch of Figs. 5-13 and 5-14 can be replaced with an active device. Transistors or ICs can be used to accomplish the switching action. Conduction and nonconduction of the active device regulates the charge and discharge of the RC network. Circuits connected in this manner are classified as *relaxation oscillators*. When the circuit device is conductive, it is active. When it is not conductive, it is relaxed. The active device of a relaxation oscillator switches states between conduction and nonconduction. This regulates the charge and discharge rate of the capacitor. A sawtooth waveform appears across the capacitor of this type of oscillator.

A UJT is used in the relaxation oscillator of Fig. 5-15. The RC network is composed of R_1 and C_1. The junction of the network is connected to the emitter (*E*) of the UJT. The UJT will not go into conduction until a certain voltage value is reached. When conduction occurs, the emitter-base 1 junction becomes low resistant.

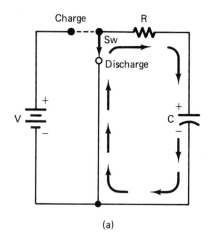

(a)

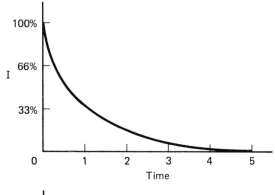

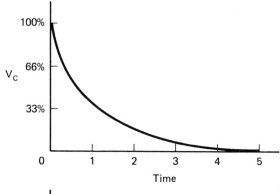

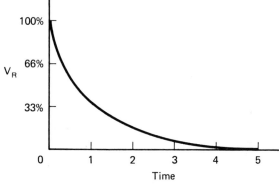

(b)

FIGURE 5-14 *RC* discharging action: (a) circuit; (b) circuit values.

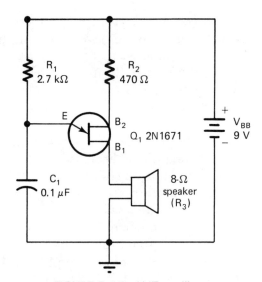

FIGURE 5-15 UJT oscillator.

This provides a low-resistant discharge path for C. Current flows through R_3 only when the UJT is conducting. R_3 refers to the resistance of the speaker in this circuit.

Assume now that power is applied to the UJT oscillator circuit. The UJT is in a nonconductive state initially. The emitter-B_1 junction is reverse biased. C_1 begins to accumulate charge voltage after a short period of time. Time equals $R \times C$ in this case. The capacitor eventually charges to a voltage value that will cause the E-B_1 junction to become conductive. When this point is reached, the E-B_1 junction becomes low resistant. C_1 discharges immediately through the low-resistant E-B_1 junction. This action removes the forward bias voltage from the emitter. The UJT immediately becomes nonconductive. C_1 begins to charge again through R_1. The process is then repeated on a continuous basis.

UJT oscillators are found in applications that require a signal with a slow rise time and a rapid fall time. The E-B_1 junction of the UJT has this type of output. Between B_1 and circuit ground, the UJT produces a *spiked pulse*. This type of output is frequently used in timing circuits and counting applications. The time that it takes for the waveform to repeat itself is called the *pulse repetition rate (PRR)*. This term is very similar to the hertz designation for frequency. As a general rule, UJT oscillators are very stable and are quite accurate when used with time constants of one or less.

Astable Multivibrator

Multivibrators are a very important classification of relaxation oscillator. This type of circuit employs an RC network in its physical makeup. A rectangular-shaped wave is developed by the output. Astable multivibrators are commonly used in television receivers to control the electron beam deflection of the picture tube. Computers use this type of generator to develop timing pulses.

Multivibrators are considered to be either *triggered* devices or *free running*. A triggered multivibrator requires an input signal or timing pulse to be made operational. The output of this multivibrator is controlled, or *synchronized*, by an input signal. The electron beam deflection oscillators of a television receiver are triggered into operation. When the receiver is tuned to an operationing channel, its oscillators are synchronized by the incoming signal. When it is tuned to an unused channel the oscillators are free running. Free-running oscillators are self-starting. They operate continuously as long as electrical power is supplied. The shape and frequency of the waveform is determined by component selection. An astable multivibrator is a free-running oscillator.

A multivibrator is composed of two amplifiers that are *cross-coupled*. The output of amplifier 1 goes to the input of amplifier 2. The output of amplifier 2 is coupled back to the input of amplifier 1. Since each amplifier inverts the polarity of the input signal, the combined effect is a positive feedback signal. With positive feedback an oscillator is regenerative and produces continuous output.

Figure 5-16 shows a representative multivibrator using bipolar transistors. These amplifiers are connected in a common-emitter circuit configuration. R_2 and R_3 provide forward bias voltage for the base of each transistor. Capacitor C_1 couples the collector of transistors Q_1 to the base of Q_2. Capacitor C_2 couples the collector of Q_2 to the base of Q_1. Because of the cross-coupling, one transistor will be conductive and one will be cut off. After a short period of time, the two transistors change states. The conducting transistor is cut off and the off transistor becomes conductive. The circuit changes back and forth between these two

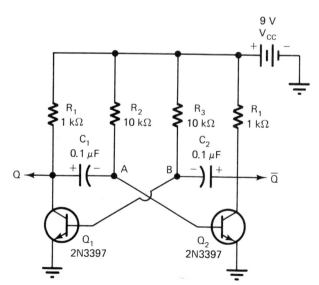

FIGURE 5-16   Astable multivibrator.

states. The output of the circuit is a rectangular-shaped wave. An output signal can be obtained from the collector of either transistor. As a rule the output is labeled Q or $\overline{Q}$. This denotes that outputs are of an opposite polarity.

When power is first applied to the multivibrator of Fig. 5-16, one transistor goes into conduction first. A slight difference in component tolerances usually accounts for this condition. For explanation purposes, assume that Q_1 conducts first. When Q_1 becomes conductive, there is a voltage drop across R_1. V_C becomes less than V_{CC}. This causes a negative-going voltage to be applied to C_1. The positive base voltage of Q_1 is reduced by this voltage. The conduction of Q_2 decreases. The collector voltage of Q_2 begins to rise to the value of V_{CC}. A positive-going voltage is applied to C_2. This voltage is added to the base voltage of Q_1. Q_1 becomes more conductive. The process continues until Q_1 becomes saturated and Q_2 is cut off.

When the output voltages of each transistor becomes stabilized, there is no feedback voltage. Q_2 is again forward biased by R_2. Conduction of Q_2 causes a drop in V_C. This negative-going voltage is coupled to the base of Q_1 through C_2. Q_1 becomes less conductive. The V_C of Q_1 begins to rise toward the value of V_{CC}. This is coupled to the base of Q_2 by C_1. The process continues until Q_2 is saturated and Q_1 is cut off. The output voltages then become stabilized. The process is then repeated.

The oscillation frequency of a multivibrator is determined by the time constants of R_2 and C_1 and R_3 and C_2. The values of R_2 and R_3 are usually selected to cause each transistor to reach saturation. C_1 and C_2 are then chosen to develop the desired operating frequency. If C_1 equals C_2 and R_2 equals R_3, the output is symmetrical. This means that each transistor is on and off for an equal amount of time. The output frequency of a *symmetrical multivibrator* is determined by the formula

$$\text{Frequency } (f) = \frac{1}{1.4RC}$$

If the resistor and capacitor values are unequal, the output is not symmetrical. One transistor could be on for a long period with the alternate transistor being on for only a short period. The output of a nonsymmetrical multivibration is described as a rectangular wave.

Monostable Multivibrator

A *monostable multivibrator* has one stable state of operation. It is often called a *one-shot* multivibrator. One trigger pulse causes the oscillator to change its operational state. After a short period of time, however, the oscillator returns to its original starting state. The RC time constant of this circuit determines the time period of the state change. A monostable multivibrator always returns to its original state. No operational change will occur until a trigger pulse is applied. A monostable multivibrator is considered to be a *triggered* oscillator.

Figure 5-17 shows a schematic of a monostable multivibrator. This circuit has two operational states. Its stable state is based on conduction of Q_2 with Q_1 cut off. The stable state occurs when Q_1 is conductive and Q_2 is cut off. The circuit relaxes in its stable state when no trigger pulse is supplied. The unstable state is initiated by a trigger pulse. When a trigger pulse arrives at the input, the circuit changes from its stable state to the unstable state. After the time of $0.7 \times R_2 C_1$, the circuit goes back to its stable state. No circuit change occurs until another trigger pulse is applied to the input.

Consider now the operation of a monostable multivibrator when power is first applied. No trigger pulse is applied initially. Q_2 is forward biased by a divider network consisting of R_2, D_1, and R_5. The value of R_2 is selected to cause Q_2 to reach saturation. Resistors R_1 and R_3 reverse-bias each collector. With the base of Q_2 forward biased, it is driven into saturation immediately. The collector voltage of Q_2 drops to a very small value. This voltage coupled through R_4 is applied to the base of Q_1. V_B is not great enough to cause conduction of Q_1. The circuit therefore remains in this conduction state as long as power is applied. This represents the stable state of the circuit.

To initiate a state change in a monostable multivibrator a triggered pulse must be applied to its input. Figure 5-18 shows a representative trigger pulse, a wave-shaped pulse, and the resulting output of the multivibrator. C_2 and R_5 of the input circuit form

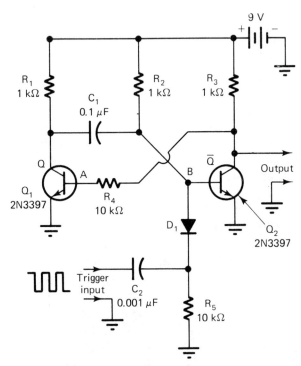

FIGURE 5-17 Monostable multivibrator.

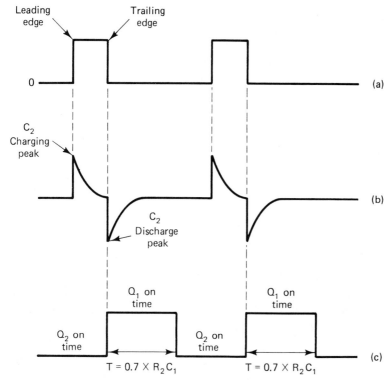

FIGURE 5-18 Monostable multivibrator waveforms: (a) trigger input waveform; (b) differentiator output waveform; (c) monostable multivibrator output waveform.

a *differentiator* network. The *leading edge* of the applied trigger pulse causes a large current flow through R_5. After C_2 begins to charge, the current through R_5 begins to drop. When the *trailing edge* of the pulse arrives, the voltage applied to C_2 drops to zero. With no source voltage applied to C_2, the capacitor will discharge through R_5. An opposite-polarity pulse therefore occurs at the trailing edge of the input pulse. The input pulse is thus changed into a positive and a negative spike that appear across R_5. D_1 conducts only during the time of the negative spike. This will feed a negative spike to the base of Q_2. With a square trigger pulse applied to the input, a single negative spike pulse is applied to the base of Q_2. This initiates a state change in the multivibrator.

When the base of Q_2 receives a negative spike, it is driven into cutoff. This causes the collector voltage of Q_2 to rise very quickly to the value of $+V_{CC}$. This, in turn, causes the base of Q_1 to become positive. Q_1 therefore becomes conductive and Q_2 is driven into cutoff. When Q_1 conducts, the emitter-collector junction becomes very low resistant. Charging current flows through Q_1, C_1, and R_2. The bottom of R_2 immediately becomes negative due to the charging current of C_1. This drives the base of Q_2 negative. Q_2 remains in its cutoff state. The process continues until C_1 becomes charged. The charging current through R_2 then begins to slow down, and the top of R_2 eventually becomes positive. Q_2 immediately goes into conduction. This in turn drives Q_1 to cutoff.

The circuit has therefore changed back to its stable state. It will remain in this state until the next trigger pulse arrives at the input.

Bistable Multivibrator

A bistable multivibrator has two stable states of operation. A trigger pulse applied to the input will cause the circuit to assume one stable state. A second pulse will cause it to switch to the alternate stable state. This type of multivibrator will change states only when a trigger pulse is applied. It is often called a *flip-flop*. It flips to one state when triggered and flops back to the other state when triggered. The circuit becomes stable in either state. It will not change states, or *toggle*, until commanded to do so by a trigger pulse. Figure 5-19 shows a schematic of a bistable multivibrator using bipolar transistors.

When electrical power is first applied to a multivibrator, it assumes one of its stable states. One transistor goes into conduction faster than the other. In the circuit of Fig. 5-19, let us assume that Q_1 goes into conduction faster than Q_2. The collector voltage of Q_1 therefore begins to drop very quickly. *Direct coupling* between the collector and base causes a corresponding drop in the voltage of Q_2. A reduction in Q_2 voltage causes a decrease in I_B and I_C.

The V_C of Q_2 rises to the value of $+V_{CC}$. This positive-going voltage is coupled back to the base of Q_1 by R_3. This increases the conduction of Q_1, which in turn decreases conduction of Q_2. The process continues until Q_1 is *saturated* and Q_2 is cut off. The circuit remains in the stable state.

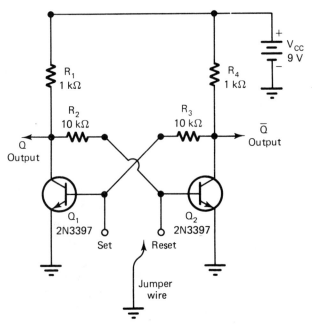

FIGURE 5-19 Bistable multivibrator.

To initiate a state change, a trigger pulse must be applied. A negative pulse applied to the base of Q_1 causes it to go into the cutoff region. A positive pulse applied to the base of Q_2 causes it to go into conduction. This polarity applies to NPN transistors. The pulse polarity is reversed for PNP transistors.

In our circuit assume that a negative pulse is applied to the base of Q_1. When this occurs the I_B and I_C of Q_1 are reduced immediately. The V_C rises toward the value of $+V_{CC}$. This positive-going voltage is coupled back to the base of Q_2. The I_B and I_C of Q_2 rise very quickly. This causes a corresponding drop in the V_C of Q_2. Direct coupling of V_C through R_3 causes a decrease in the I_B and I_C of Q_1. The process continues until Q_1 is cut off and Q_2 reaches saturation. This represents the second stabilized state of operation. The circuit will remain in this state until commanded to change or the power is removed.

IC Waveform Generators

The *NE/SE 555* IC is a multifunction device that is widely used today. It can be modified to respond as an astable multivibrator. This particular circuit can be achieved with a minimum of components and a power source. Circuit design is easy to accomplish and operation is very reliable. This specific chip is available through a number of manufacturers. As a rule, the number 555 usually appears in the manufacturer's part identification number. SN72555, MC14555, SE555, LM555, XR555, and CA555 are some of the common part numbers for this chip.

The internal circuitry of a 555 IC is generally viewed in functional blocks. In this regard, the chip has two *comparators*, a bistable *flip-flop*, a resistive *divider*, a *discharge transistor*, and an *output* stage. Figure 5-20 shows the functional blocks of a 555 IC.

The *voltage divider* of the IC consists of three 5-kΩ resistors. The network is connected internally across the $+V_{CC}$ and ground-supply source. Voltage developed by the lower resistor is one-third of V_{CC}. The middle divider point is two-thirds of the value of V_{CC}. This connection is terminated at pin 5. Pin number 5 is designated as the control voltage.

The two *comparators* of the 555 respond as an amplifying switch circuit. A reference voltage is applied to one input of each comparator. A voltage value applied to the other input initiates a change in output when it differs with the reference value. The comparator is referenced at two-thirds of V_{CC} at its negative input. This is where pin 5 is connected to the middle divider resistor. The other input is terminated at pin 6. This pin is called the *threshold* terminal. When the voltage at pin 6 rises above two-thirds of V_{CC}, the output of the comparator swings positive. This is then applied to the *reset* input of the flip-flop.

Comparator 2 is referenced to one-third of V_{CC}. The positive input of comparator 2 is connected to the lower divider network resistor. External pin connection 2 is applied to the negative input

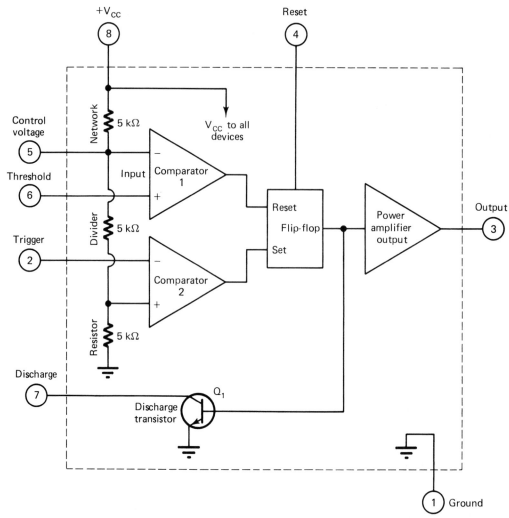

FIGURE 5-20   Internal circuits of an LM555 IC.

of comparator 2. This is called the *trigger* input. If the voltage of the trigger drops below one-third of V_{CC} the comparator output will swing positive. This is applied to the *set* input of the flip-flop.

The flip-flop of the 555 IC is a *bistable multivibrator*. It has reset and set inputs and one output. When the reset input is positive, the output goes positive. A positive voltage to the set input causes the output to go negative. The output of the flip-flop is dependent on the status of the two comparator inputs.

The output of the flip-flop is applied to both the *output* stage and the *discharge* transistor. The output stage is terminated at pin 3. The discharge transistor is connected to terminal 7. The output stage is a power amplifier and a signal inverter. A load device connected to terminal 3 will see either $+V_{CC}$ or ground, depending on the state of the input signal. The output terminal switches between these two values. Load current values of up to 200 mA can be controlled by the output terminal. A load device connected to $+V_{CC}$ is energized when pin 3 goes to ground. When the output goes to $+V_{CC}$ the output is off. A load device connected to ground turns on when the output goes to $+V_{CC}$. It is off when the output goes to ground. The output switches back and forth between these two states.

Transistor Q_1 is called a discharge transistor. The output of the flip-flop is applied to the base of Q_1. When the flip-flop is reset (positive), it forward-biases Q_1. Pin 7 connects to ground through Q_1. This causes pin 7 to be grounded. When the flip-flop is set (negative), it reverse-biases Q_1. This causes pin 7 to be infinite or open with respect to ground. Pin 7 therefore has two states, shorted to ground or open.

We will now see how the internal circuitry of the 555 responds as a multivibrator.

IC Astable Multivibrator

When used as an astable multivibrator, the 555 is an *RC oscillator*. The shape of the waveform and its frequency are determined primarily by an *RC* network. The astable multivibrator circuit is self-starting and operates continuously for long periods of time. Figure 5-21(a) shows the LM 555 connected as an astable multivibrator. A common application of this circuit is the *time base generator* for clock circuits and computers.

Connection of the 555 IC as an astable multivibrator requires two resistors, a capacitor, and a power source. The output of the circuit is connected to pin 3. Pin 8 is $+V_{CC}$ and pin 1 is ground. The supply voltage can be from 5 to 15 V dc. Resistor R_A is connected between $+V_{CC}$ and the discharge terminal (pin 7). Resistor R_B is connected between pin 7 and the threshold terminal (pin 6). A capacitor is connected between the threshold and ground. The trigger (pin 2) and threshold (pin 6) are connected together.

When power is first applied, the capacitor will charge through R_A and R_B. When the voltage at pin 6 (threshold) rises slightly

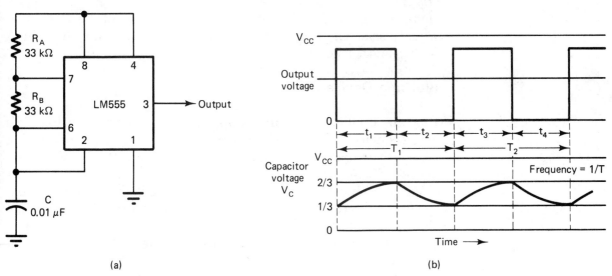

(a)   (b)

FIGURE 5-21 Astable multivibrator: (a) circuit; (b) waveforms.

above two-thirds of V_{CC}, it changes the state of comparator 1. This resets the flip-flop and causes its output to go positive. The output (pin 3) goes to ground and the base of Q_1 is forward biased. Q_1 discharges C through R_B to ground.

When the charge voltage of C drops slightly below one-third of V_{CC}, it energizes comparator 2. The trigger (pin 2) and pin 6 are still connected together. Comparator 2 causes a positive voltage to go to the set input of the flip-flop. This sets the flip-flop, which causes its output to go negative. The output (pin 3) swings to $+V_{CC}$. The base of Q_1 is reverse biased. This opens the discharge (pin 7). C begins to charge again to V_{CC} through R_A and R_B. The process is repeated from this point. The charge value of C varies between one-third and two-thirds of V_{CC}. See the resulting waveforms of Fig. 5-21(b).

The *output frequency* of the astable multivibrator is represented as $f = 1/T$. This represents the total time needed to charge and discharge C. The charge time is represented by spaces t_1 and t_3. In seconds t_1 is $0.693(R_A + R_B)C$. The *discharge time* is shown as t_2 and t_4. In seconds t_2 is $0.693R_BC$. The combined time for one operational cycle is, therefore, $T = t_1 + t_2$ or $t_3 + t_4$. Expressed as frequency, this is

$$\text{Frequency} = \frac{1}{\text{time}}, \quad \text{or} \quad f = \frac{1}{T}$$

Combining t_1 and t_2 or t_3 and t_4 makes the frequency formula

$$f = \frac{1}{T}, \quad \text{or} \quad = \frac{1.44}{(R_A + 2R_B)C}$$

The resistance ratio of R_A and R_B is quite critical in the operation of an astable multivibrator. If R_B is more than half the value of R_A, the circuit does not oscillate. Essentially, this prevents the trigger from dropping in value from two-thirds of V_{CC} to one-third of V_{CC}. This means that the IC is not capable of retriggering itself. It is therefore unprepared for the next operational cycle. Most IC manufacturers provide data charts that would assist the user in selecting the correct R_A and R_B values with respect to C.

Blocking Oscillators

Blocking oscillators are an example of the relaxation principle. The active device, which is ordinarily a transistor, is cut off during most of the operational cycle. It turns on for a short operational period of time to discharge an RC network. In appearance this circuit closely resembles the Armstrong oscillator. Feedback from the output to the input is needed to achieve the blocking function. The output of the oscillator is used to change the shape of a waveform. This circuit was commonly used as the vertical oscillator of a television receiver. The *vertical blocking oscillator* (VBO) transformer provides regenerative feedback from the output to the input. An iron core transformer is used in this circuit because the generated frequency is 60 Hz.

A transistor vertical blocking oscillator is shown in Fig. 5-22. A sawtooth-forming capacitor (C_1) charges when the source voltage is initially applied. It discharges through the transistor when it becomes conductive. The output across the capacitor is a sawtooth waveform.

When the supply voltage is initially applied, the charge of C_1 is zero. This is represented by T_1 on the output waveform. The transistor is cut off by the reverse bias voltage developed by emitter resistor R_7. C_1 begins to charge to $-V_{CC}$, as indicated by the solid arrows. As C_1 charges, voltage is applied to the base of Q_1 through R_1 and the secondary of T_1. This voltage forward-biases Q_1. The time of the wave from T_1 to T_2 represents the charging action of C_1. At T_2, the base voltage overcomes the reverse-bias emitter voltage. Q_1 goes into conduction and current flows through the primary winding of T_1. Feedback from primary to secondary

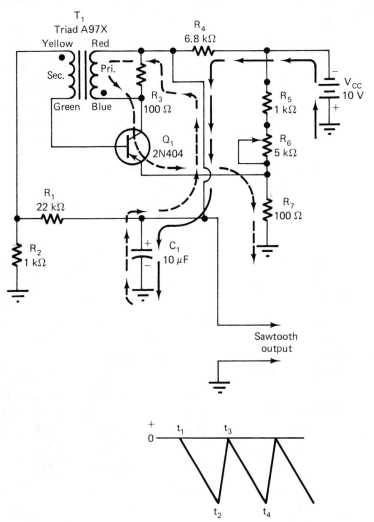

FIGURE 5-22 Transistor vertical blocking oscillator. Solid arrows denote C_1 charge path; dashed arrows denote C_1 discharge path.

drives the base more negative. The process continues until Q_1 saturates.

When the transistor is conductive, it provides a low-resistance discharge path for C_1 (see the dashed-line arrows of the discharge path). C_1 discharges very quickly because of the low-impedance path. This is represented by the space between T_2 and T_3 of the waveform. When C_1 is fully discharged, Q_1 is cut off again by R_7. The process then repeats itself with C_1 charging again.

The frequency of the sawtooth output is dependent on the reverse bias voltage of Q_1. An increase in negative voltage from R_6 increases the cutoff time for Q_1. This increases the charge time that C_1 needs to develop voltage to bring Q_1 into conduction. Value changes in R_6 alter the sawtooth frequency of the oscillator.

SELF-EXAMINATION

1. The four parts of an oscillator circuit are
   _____, _____, _____, and _____.
2. The frequency of an *LC* circuit with a 0.5-uF capacitor and a 100-mH inductor is _____ Hz.
3. Four types of oscillators are _____, _____, _____, and _____.
4. Three types of multivibrators are _____, _____, and _____.
5. An IC oscillator may be constructed using a _____.
6. A process by which a portion of a circuit's output is returned to its input is called _____.
7. A circuit that generates a continuously repetitive output signal is called an _____.
8. An *LC* network is sometimes called a _____ circuit.
9. The frequency of ac voltage generated by an *LC* tank circuit is called _____ frequency.
10. The waveform produced by a tank circuit (without feedback) is called a _____ sine wave.
11. The frequency of an Armstrong oscillator is based on the values of _____ and _____.
12. A distinguishing feature of a Hartley oscillator is its _____ _____.
13. A Copitts oscillator uses _____ instead of a tapped coil.
14. _____ oscillators are used for high-frequency stability.

15. A _____ oscillator uses a crystal as its tank circuit.

16. _____ oscillators are used to generate nonsinusoidal waveforms.

17. A _____ transistor may be used as a relaxation oscillator.

18. Multivibrators are classified as either _____ or _____.

19. A _____ multivibrator is a free-running oscillator.

20. A monostable multivibrator is also called a _____ multivibrator.

21. A bistable mutivibrator is also called a _____.

22. An IC used as a multivibrator circuit is the _____.

ANSWERS

1. Amplifier; feedback network; frequency-determining network; power source

2. $$\frac{1}{2\pi\sqrt{LC}}$$
   $$= \frac{1}{6.28\sqrt{100 \times 10^{-3} \times 0.5 \times 10^{-6}}}$$
   $$= 712 \text{ Hz}$$

3. Armstrong; Hartley; Colpitts; crystal

4. Astable; monostable; bistable

5. 555 timer

6. Feedback

7. Oscillator

8. Tank

9. Resonant

10. Damped

11. L; C

12. Tapped coil

13. Two capacitors

14. Crystal

15. Pierce

16. Relaxation

17. Unijunction

18. Triggered; free-running

19. Astable

20. One-shot

21. Flip-flop

22. 555

EXPERIMENT 5-1
HARTLEY OSCILLATORS

Hartley oscillators are commonly used in radio receivers and transmitter circuits. This type of oscillator has unusual stability and can be used to produce a wide range of frequencies. The basic oscillator includes an LC tank circuit, a feedback path and an amplifying device. A transistor is used to achieve amplification. Both variable-frequency and fixed-frequency oscillators can be built with the Hartley circuit.

OBJECTIVES

1. To construct a Hartley oscillator circuit.
2. To examine the operation of a Hartley oscillator circuit.
3. To use an oscilloscope to observe wave forms at key test points of the oscillator circuit.

EQUIPMENT

Multimeter
Oscilloscope
Variable power supply, 0–15 V
Variable capacitor: 365 pF
SPST switch
Capacitor: 100 pF, 47 pF, 0.01 μF
NPN transistor
Hartley oscillator coil
Resistors: 470 Ω, 56 kΩ
Inductor: 25 mH

PROCEDURE

1. Connect the Hartley oscillator circuit of Fig. 5-1A.
2. Measure the resistance of the oscillator coil with a multimeter and determine the tapped connection. Connect it into the circuit according to the indicated resistance value.
3. Adjust the variable dc power supply to 10 V dc before turning on the circuit switch.
4. Prepare the meter to measure voltage with negative po-

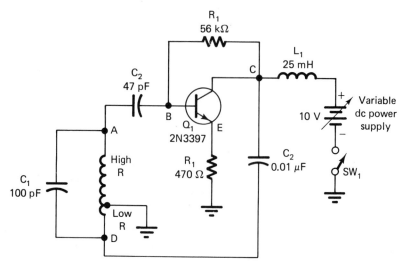

FIGURE 5-1A Hartley oscillator circuit.

larity. Connect it across the output at terminal A and ground. If the oscillator is working, there should be a negative voltage at the output: output = _____ V.

5. Turn on the oscilloscope and prepare it for operation with the sweep-selector switch set at the 1-μs or 1.0-MHz range. Set the sync selector or trigger to the internal sync position.

6. Connect the oscilloscope ground to the circuit ground and the vertical probe to test point *A*. If the oscillator is working, a sine wave will appear at this test point.

7. Make a sketch of the observed wave form in the space of Fig. 5-1B. Indicate the peak-to-peak value of the wave form.

8. Using the same procedure, connect the vertical probe to test points *B* and *C*. Make sketches of the observed waveforms.

9. Turn off the circuit switch and disconnect the 100-pF capacitor from points *A* and *D*. In its place connect a 365-pF variable capacitor.

10. Turn on the circuit switch and connect the vertical probe of the oscilloscope to test point *A*. Connect the common probe to ground.

11. Adjust the variable capacitor to where the plates are fully meshed. The capacitor is at its _____ (smallest or largest) capacitance value at this position. Adjust the capacitor to where the movable plates are completely extended out of the stationary plates. The capacitor is at its _____ (smallest or largest) capacitance value at this position.

12. Observing the oscilloscope shows that frequency is _____ (high or low) when the ca-

FIGURE 5-1B

pacitor plates are meshed and _____
(high or low) when plates are extended.

13. If an AM radio receiver is available, turn it on and place it near the oscillator circuit. Tune the receiver to a location near center of the dial that is not occupied by a station.

14. Adjust the variable capacitor of the oscillator circuit through its range while listening for a whistling sound on the AM radio receiver. When the approximate frequency has been located, try tuning the receiver to a precise spot where whistling occurs. The frequency of the receiver is approximately _____ KHz.

ANALYSIS

1. What components determine the frequency of the oscillator in this experiment? _____

2. What achieves amplification in this experiment?

3. How does feedback from the output of the transistor occur? _____

4. What does feedback actually achieve in the operation of an oscillator? _____

EXPERIMENT 5-2
COLPITTS OSCILLATOR

A Colpitts oscillator is very similar in operation to the Hartley oscillator of Experiment 5-1. The Colpitts oscillator, however, develops feedback by means of an electrostatic field across two capacitors. In a strict sense, a Colpitts oscillator has split capacitors and a Hartley oscillator has a tapped inductor. In a Colpitts oscillator, two capacitors of the tank circuit form a voltage divider circuit for the ac feedback signal.

OBJECTIVES

1. To construct a Colpitts oscillator circuit.
2. To use an oscilloscope to trace the feedback path of the oscillator circuit.
3. To measure circuit voltages and observe wave forms at key test points.

EQUIPMENT

Multimeter
Oscilloscope
Variable power supply, 0–15 V
NPN transistor
Variable capacitor: 365/125 pF, two-ganged
Capacitors: 47 pF, 100 pF, 0.001 μF, 0.01 μF
Oscillator coil
Inductor: 25 mH
Resistors: 470Ω, 10 kΩ, 22 Ω
SPST switch

PROCEDURE

1. Connect the Colpitts oscillator circuit of Fig. 5-2A.
2. Turn on the variable dc power supply and adjust it to 5 V dc before turning on the circuit switch.
3. Prepare the oscilloscope for operation by connecting it for external sync as indicated in Fig. 5-2A. Connect the vertical probe of the oscilloscope to the oscillator output at test point A and observe the waveform.

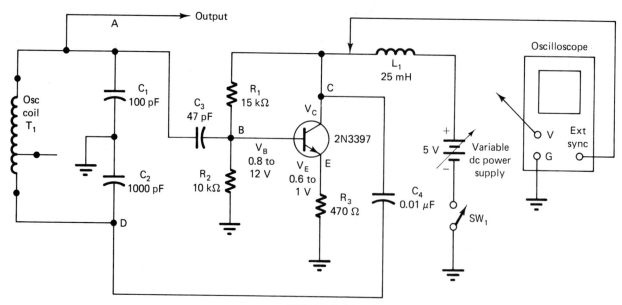

FIGURE 5-2A Colpitts oscillator circuit.

4. Make a sketch of the waveform in the space of Fig. 5-2B. Indicate the phase and peak-to-peak voltage.

5. Using the same procedure as in step 3, connect the vertical probe of the oscilloscope to test points *B*, *C*, and *E*. Observe the waveforms and indicate the phase and peak-to-peak voltage.

6. Using a multimeter, measure the operating voltages at the emitter, base and collector. Measured values should fall within voltage ranges indicated at *B-C-E* of Fig. 5-2A. Note that the negative lead of the power source should be connected to the ground of the circuit. With the meter ground connected to the circuit ground, all voltage readings will be positive values.

7. Turn off the circuit switch and disconnect capacitors C_1 and C_2 from the circuit. Connect a two-ganged 365-pF/125-pF variable capacitor in place of C_1 and C_2. The ca-

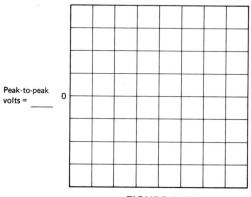

FIGURE 5-2B

pacitor section with the largest plates (365 pF) should be connected in place of C_2, with the smaller capacitor (125 pF) in place of C_1.

8. Turn on the circuit switch and test the output signal. What is the most obvious difference in the modified output (with variable capacitor) and original circuit output?

_____

_____

9. Adjust the variable capacitor through its range while observing the oscilloscope. The highest frequency output occurs when the variable capacitor is _____.

ANALYSIS

1. What is the major difference between a Hartley and a Colpitts oscillator? _____

_____

2. How does the Colpitts oscillator of this experiment achieve feedback? _____

_____

3. What are the frequency-determining components of the Colpitts oscillator? _____

_____

EXPERIMENT 5-3
CRYSTAL OSCILLATORS

Crystal oscillators are widely used in radio frequency circuits that require a precise frequency. Oscillators of this type are designed to generate very stable frequencies for communication circuits. Radio broadcast stations, amateur radio transmitters and citizen band transmitters usually are controlled by crystal oscillators.

OBJECTIVES

1. To construct a crystal oscillator circuit.
2. To observe waveforms at key test points with an oscilloscope.
3. To measure circuit voltages with a multimeter.

EQUIPMENT

Multimeter
Variable power supply, 0–15 V
Oscilloscope
Crystal
NPN transistor
Capacitors: 0.001 μF, 0.01 μF, 0.1 μF
Resistor: 470 k Ω
Inductor: 25 mH
SPST switch
AM radio receiver (optional)

PROCEDURE

1. Connect the crystal oscillator circuit of Fig. 5-3A.
2. Turn on variable dc power supply and adjust it to 10 V before turning on the circuit switch.
3. Prepare the oscilloscope for external sync operation with the sweep frequency at 100 kHz or a horizontal time in the 5-ms/cm range. Connect the external sync input of the oscilloscope to test point C.
4. Connect the ground lead of the oscilloscope to the ground of the circuit. Connect the vertical probe to test points A, B, and C.

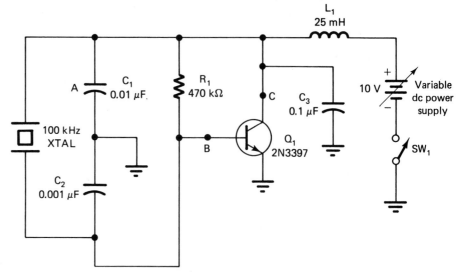

FIGURE 5-3A Crystal oscillator circuit.

5. Make a sketch of the observed waveforms at test points *A*, *B*, and *C* in the space of Fig. 5-3B. Indicate the peak-to-peak voltages.

6. Return the vertical probe of the oscilloscope to test point *A* and change the voltage of the dc source from 5 to 15 V while observing the output. Did you notice a change in oscillator frequency? _____Why?

_____

7. Prepare the multimeter to measure dc voltage. Measure voltage at the base and collector of the transistor with respect to common ground: V_B = _____ V; V_C = _____V.

8. If an AM radio receiver is available, turn it on and place it near the oscillator circuit. Tune the receiver through its entire range. How many spots on the receiver do you hear an oscillator? _____

Test point B _____
(output)

Test point A _____
(input)

Test point C _____

B _____ $V_{p\text{-}p}$; _____ Phase

A _____ $V_{p\text{-}p}$, _____ Phase

C _____ $V_{p\text{-}p}$; _____ Phase

FIGURE 5-3B

9. Turn off the circuit switch, power supply, and oscilloscope. Disconnect the circuit and return all components to the storage cabinet.

ANALYSIS

1. How is the piezoelectric effect used in a crystal oscillator?
   _____
   _____

2. Why are crystals used in oscillators for communication circuits? _____
   _____
   _____

3. Why does the frequency of this oscillator produce a number of different oscillations on a radio receiver? _____
   _____
   _____

EXPERIMENT 5-4
ASTABLE MULTIVIBRATORS

An astable multivibrator is a rather unusual oscillator circuit that repeats itself by switching on and off during an operational cycle. In practice, this type of circuit is often used to generate square waves. It frequently serves as a clock pulse generator for digital circuits or as vertical or horizontal oscillators of television circuits.

OBJECTIVES

1. To construct an astable multivibrator circuit.
2. To observe waveforms at key test points of the circuit.
3. To change the switching speed of a multivibrator by changing the value of selected components.

EQUIPMENT

Multimeter
Variable power supply, 0–15 V
Oscilloscope
NPN transistors (2)
Lamps with sockets (2)
Capacitors: 0.01 μF (2), 100 μF (2)
Resistors: 1 k Ω (2), 4.7 k Ω
SPST switch

PROCEDURE

1. Connect the astable multivibrator of Fig. 5-4A.
2. Turn on the power supply and adjust it to 5 V dc before turning on the circuit switch.
3. Describe the switching action of the multivibrator output across the two lamps. _____
   _____
4. Turn off the switch and remove C_1 and C_2 from the circuit. Install 0.01-μF capacitors in place of the 100-μF capacitors.
5. Turn on the circuit switch. Describe the switching action of the two lamps. _____
   _____

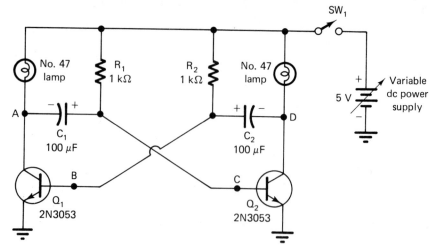

FIGURE 5-4A   Astable multivibrator circuit.

6. Prepare the oscilloscope for external sync operation. Connect the external sync input to test point *D*.

7. Connect the vertical probe of the oscilloscope to test points *A*, *B*, *C*, and *D* shown in Fig. 5-4A. Make a sketch of the wave form appearing at each test point in the spaces of Fig. 5-4B.

8. Turn off the circuit switch and remove R_1 from the circuit. Replace R_1 with a 4.7-kΩ resistor.

9. Turn on the circuit switch and observe the waveforms at test points *A*, *B*, *C*, and *D*. Describe the primary differences in appearance of the waveforms. _____

_____

_____

ANALYSIS

1. Which component values in the circuit determine the shape of the waveform of an astable multivibrator?

_____

Point A 0 _____   Point D 0 _____
Q_1 output Q_2 output

Point B 0 _____   Point C 0 _____
Q_1 input Q_2 input

FIGURE 5-4B   Waveform diagrams.

2. How would the lamps used in this astable multivibrator circuit respond if two 500-μF capacitors were used in place of the two 100-μF capacitors? _____

3. If the capacitors of an astable multivibrator were made smaller than 0.01 μF, how would it respond? _____

EXPERIMENT 5-5
TRIGGERED MULTIVIBRATORS

When an external voltage is used to cause a state change in a multivibrator, it is considered to be triggered. Triggering is used when it is desirable to control the operation of an oscillator circuit by an outside signal source. Most multivibrator circuits can be triggered. In fact, the bistable and monostable multivibrators studied in this experiment must be triggered in order to function. A *monostable multivibrator* has one stable state, in which it remains when not being triggered. When it is triggered, it will change states, then return to its original stable condition. A *bistable multivibrator* is often called a *flip-flop* because it has two stable states of operation. A trigger pulse of the correct polarity applied to either transistor initiates a state change. After triggering occurs, the circuit remains in a stable state until the next trigger pulse arrives.

OBJECTIVES

1. To construct bistable and monostable multivibrator circuits.
2. To determine the stable state and conduction time of a transistor by voltage measurements.
3. To trigger multivibrators into a state change by applying proper voltage polarities.

EQUIPMENT

Multimeter
Variable power supply, 0–15 V
NPN transistors (2)
Resistors: 1 kΩ (2), 4.7 kΩ, 10 kΩ, 22 kΩ 56 kΩ
Capacitor: 100 μF
SPST switch

PROCEDURE

1. Construct the monostable multivibrator of Fig. 5-5A.
2. Turn on the variable power supply and adjust it to 5 V dc before turning on the circuit switch.
3. With a multimeter measure dc voltage at test points *A*

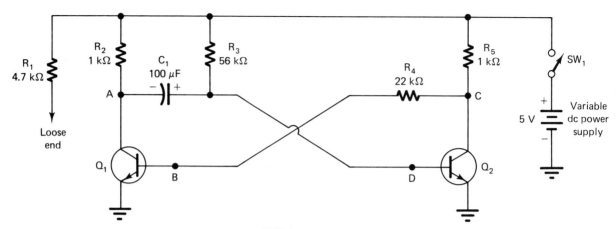

FIGURE 5-5A Monostable multivibrator circuit.

and *C*. A conductive transistor will indicate a low voltage at its collector. A nonconductive transistor will indicate source voltage at its collector.

4. According to your voltage measurements: Q_1 is _____, and Q_2 is _____.

5. With the multimeter connected to test point *C*, momentarily connect the loose end of R_5 to test point *B*. Note the voltage value after a few seconds. Trigger the multivibrator several times to verify its operation. Describe your findings. _____

6. Measure the voltage at test point *A*. Trigger Q_2 by connecting test point *D* to ground while observing the voltage at test point *A*. Describe your findings. _____

7. Turn off the circuit switch and alter the circuit to form the bistable circuit of Fig. 5-5B.

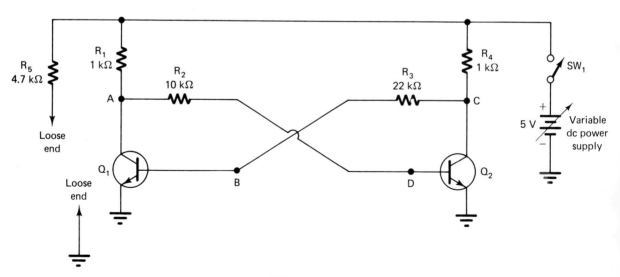

FIGURE 5-5B Bistable multivibrator circuit.

8. Turn on the variable power supply and adjust it to 5 V dc before turning on the circuit switch.

9. With a multimeter, measure dc voltage at test points A and C. If a transistor is conducting, there will be a very low voltage at its collector. A nonconductive transistor will have source voltage at its collector: Q_1 is _____, and Q_2 is _____.

10. Momentarily touch the loose end of the ground lead to the base of the conductive transistor at either test point B or D.

11. Measure collector voltage at test pionts A and C. Q_1 is _____, and Q_2 is _____.

12. Trigger the bistable multivibrator several times, noting procedure. To trigger it into a state change, momentarily touch the ground lead to _____ transistor.

13. With the loose end of 4.7-kΩ resistor, change states of the multivibrator by triggering the base. A positive voltage triggers _____ transistor into a state change.

ANALYSIS

1. Describe what it takes to trigger a multivibrator into a state change. _____
   _____
   _____

2. What are the differences between a bistable and a monostable multivibrator? _____
   _____
   _____

3. Why does a monostable multivibrator have a delay in state changes? _____
   _____
   _____

EXPERIMENT 5-6
FUNCTION GENERATORS

A number of ICs are on the market today that can be used to produce sine, square, or triangular waveforms from a single chip. These devices are commonly called *function generators*. This type of generator serves as a very useful signal source for a number of applications.

OBJECTIVES

1. To connect the 555 timer IC to produce sawtooth and square-wave outputs.
2. To observe an astable multivibrator that generates a non-symmetrical square-wave output.

EQUIPMENT

Integrated circuit: 555 timer
Resistors: 10 kΩ (2), 22 kΩ
Capacitors: 0.01 μF, 0.01 μF, 10 μF
Oscilloscope
Power supply, 0–15 V
SPST switch

PROCEDURE

1. Construct the 555 timer function generator circuit of Fig. 5-6A.
2. Prepare the oscilloscope for operation and connect it to the square-wave output. Make a sketch of the observed waveform in Fig. 5-6B. Indicate the peak-to-peak output or volts per division.
3. If your oscilloscope has a calibrated horizontal time base, determine the frequency of the waveform with the formula $f = 1/T$: $f = $ _____ Hz.
4. The operational time of this circuit is based upon the time it takes to charge C through resistor R_A and to discharge C through R_B. The frequency is found by using the formula $f = 1.44/(R_A + 2R_B)C$.
5. Calculate the frequency of the square-wave output: $f = $ _____Hz. How closely does your

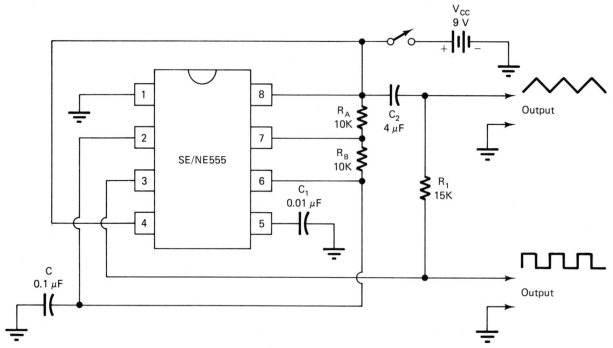

FIGURE 5-6A SE/NE555 function generator circuit.

calculated value correspond to the oscilloscope value?

_____

6. Connect the oscilloscope across the sawtooth output. Make a sketch of the observed waveform in Fig. 5-6C. Indicate the peak-to-peak output or volts per division.

7. With the oscilloscope determine the frequency of the waveform: f = _____ Hz. How does this compare with that of the square-wave output?

_____

How do you account for this change? _____

_____

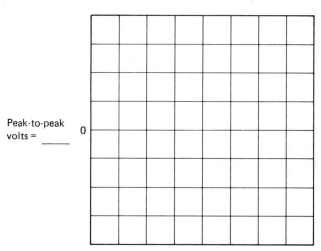

FIGURE 5-6B Observed square-wave output waveform.

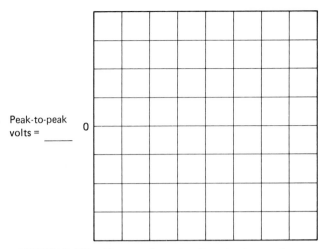

Peak-to-peak
volts = _____

0

FIGURE 5-6C   Observed sawtooth output waveform.

ANALYSIS

1. How could the frequency of the circuit output be made variable? _____
   _____
   _____

2. How does the triangular wave compare with the square wave? _____
   _____
   _____

UNIT 5 EXAMINATION
OSCILLATOR CIRCUITS

Instructions: For each of the following, circle the answer that most correctly completes the statement.

1. The oscillator that has the best frequency stability is the:
   - a. Crystal
   - b. Hartley
   - c. *LC*
   - d. Colpitts

2. In Fig. E-5-2, the 555 timer is set up as a:
   - a. Monostable multivibrator
   - b. Bistable multivibrator
   - c. Tristable multivibrator
   - d. Astable multivibrator

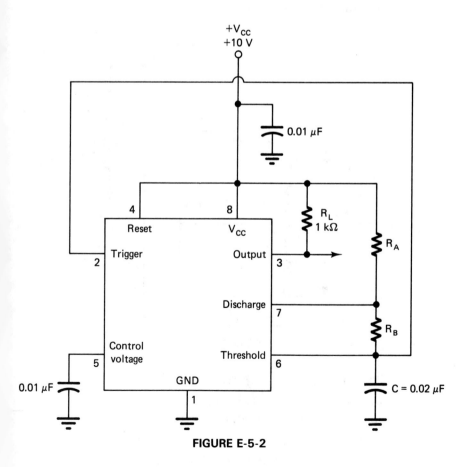

FIGURE E-5-2

3. Figure E-5-3 shows the output of a 555 timer. The output frequency is:

   a. 138.6 Hz   **b.** 7.215 KHz

   c. 10.823 KHz   **d.** 21.645 KHz

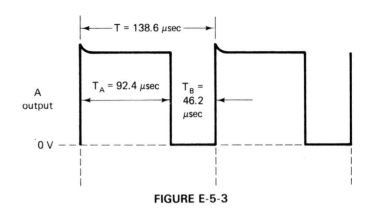

FIGURE E-5-3

4. The circuit shown in Fig. E-5-4 is a:

   a. Clapp oscillator

   b. Series-fed Hartley oscillator

   c. Shunt-fed Hartley oscillator

   d. Basic Colpitts oscillator

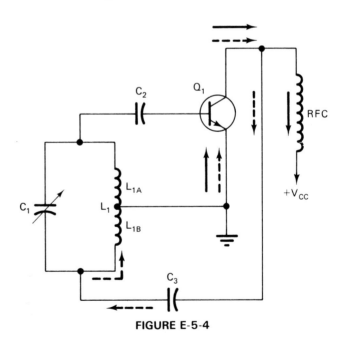

FIGURE E-5-4

5. The circuit shown in Fig. E-5-5 is a:

   a. Armstrong crystal oscillator

   b. Pierce crystal oscillator

   c. Crystal-controlled Colpitts oscillator

   d. Crystal-controlled Hartley oscillator

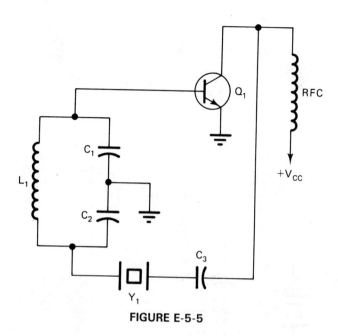

FIGURE E-5-5

6. The circuit shown in Fig. E-5-6 is a:
   a. Armstrong crystal oscillator
   b. Pierce crystal oscillator
   c. Crystal controlled Colpitts oscillator
   d. Crystal controlled Hartley oscillator

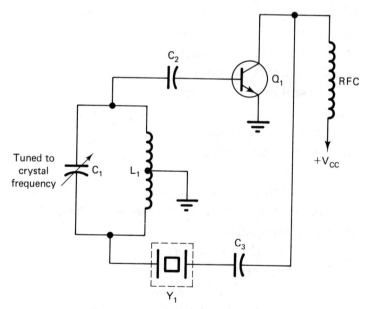

FIGURE E-5-6

7. The circuit shown in Fig. E-5-7 is a:
   a. Armstrong crystal oscillator
   b. Pierce crystal oscillator
   c. Crystal controlled Colpitts oscillator
   d. Crystal controlled Hartley oscillator

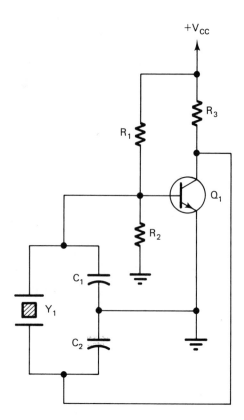

FIGURE E-5-7

8. An identifying feature of the Hartley oscillator is its:
   a. Center-tapped inductor
   b. Center-tapped capacitor
   c. Two input transistors
   d. Crystal in the feedback path
9. An identifying feature of the Colpitts oscillator is its:
   a. Center-tapped inductor
   b. Center-tapped capacitors
   c. Two input transistors
   d. Crystal in the feedback path
10. An identifying feature of the Pierce oscillator is its:
    a. Center-tapped inductor
    b. Center-tapped capacitors
    c. Two input transistors
    d. Crystal in the feedback path
11. The feedback amplitude and phase of an Armstrong oscillator is controlled by:
    a. A crystal
    b. An *RC* circuit
    c. Transistor leakage
    d. A tickler coil

12. The method that is *not* used for feedback is:
    a. *LC*   **b.** *RC*
    c. Crystal   **d.** Blocking

13. The components that determine the feedback ratio of the circuit shown in Fig. E-5-13 are:
    a. R_1 and R_2 **b.** R_1 and R_3
    c. R_2 and R_3 **d.** C_1 and C_2

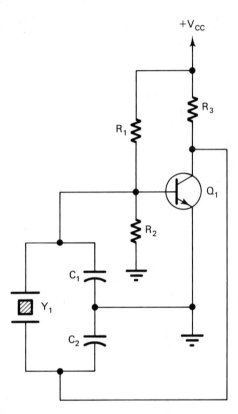

$+V_{CC}$

FIGURE E-5-13

14. In Fig. E-5-14, the components that determine the feedback ratio are:
    a. R_1 and R_2 **b.** C_1 and C_2
    c. C_1 and L_1 **d.** L_{1A} and L_{1B}

15. In Fig. E-5-15, the components that determine the feedback ratio are:
    a. R_1 and R_2 **b.** C_1 and C_2
    c. L_1 and C_1 **d.** L_1 and L_2

16. An oscillator must have:
    a. Positive feedback
    b. Negative feedback
    c. Dampened feedback
    d. Neutralized feedback

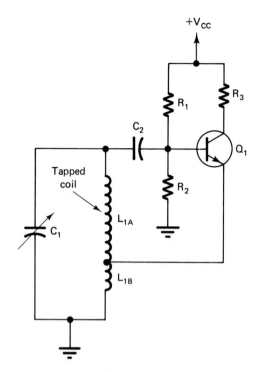

FIGURE E-5-14

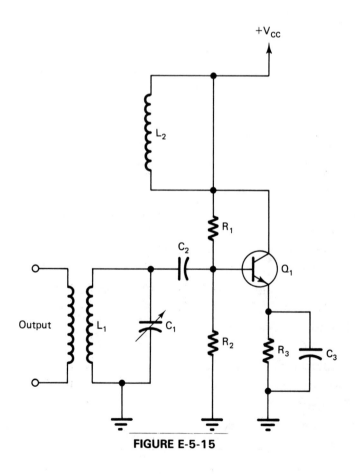

FIGURE E-5-15

17. The correct formula for calculating the frequency of an *LC* oscillator is:

    a. $F = \dfrac{1}{2\pi \sqrt{GRC}}$

    b. $F = \dfrac{1}{2\pi \sqrt{LC}}$

    c. $F = \dfrac{1}{2\pi \sqrt{R_1 R_2 C_1 C_2}}$

    d. $F = \dfrac{1}{2\pi \sqrt{RC}}$

18. The type of oscillator that would most likely be used as a reference for a clock is the:
    a. Clapp   b. Hartley
    c. Pierce   d. Wein-bridge

19. The number of stable states of an astable multivibrator is:
    a. One   b. Two
    c. Three   d. None

20. Another name for an astable multivibrator is:
    a. Monostable   b. Bistable
    c. Tristable   d. Free-running

21. The number of stable states of a monostable multivibrator is:
    a. One   b. Two
    c. Three   d. None

22. The number of possible inputs of an astable multivibrator is:
    a. One   b. Two
    c. Three   d. None

23. The output of an astable multivibrator is a:
    a. Sine wave   b. Triangular wave
    c. Square wave   d. Saw-tooth wave

24. In the circuit of Fig. E-5-24, the 555 timer is set up as a(n):
    a. Monostable multivibrator
    b. Bistable multivibrator
    c. Tristable multivibrator
    d. Astable multivibrator

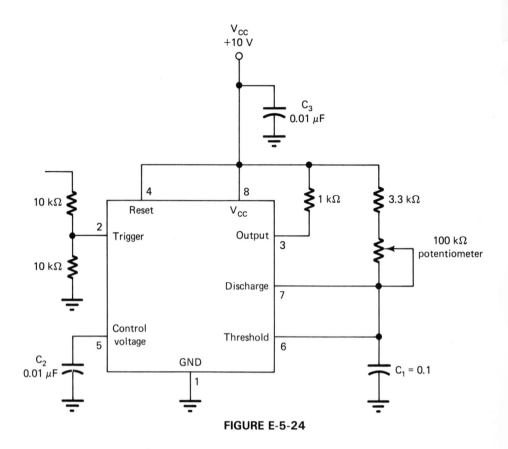

FIGURE E-5-24

UNIT SIX

COMMUNICATIONS CIRCUITS

UNIT INTRODUCTION

High-level sound amplification makes it possible to communicate over long distances. A public-address amplifier system, for example, permits an announcer to communicate with a large number of people in a stadium or an arena. Even the most sophisticated sound system, however, has some limitations. Sound waves moving away from the source have a tendency to become somewhat weaker the farther they travel. An increase in signal strength does not, therefore, necessarily solve this problem. People near large speakers usually become very uncomfortable when the sound is increased to a high level. It is possible, however, to communicate with people over long distances without increasing sound levels. *Electromagnetic waves* make this type of communication possible. Radio, television, and long-distance telephone communication are achieved by this process.

Electromagnetic wave communication systems play a very important role in our daily lives. We listen to radio receivers, watch television, talk to friends over long distances on the telephone, and even communicate with astronauts by electromagnetic waves. This type of communication is widely used today. It is important that we have some basic understanding of this type of communication.

In this unit we investigate how sound is transmitted through the air by electromagnetic waves. This type of communication uses high-frequency ac or RF energy for its operation. These signals travel through the air at 186,000 mi/s. Electromagnetic communication permits sound to travel long distances instantaneously.

UNIT OBJECTIVES

Upon completion of this unit you will be able to:

1. Explain the advantages, disadvantages, and characteristics of amplitude modulation and frequency modulation.
2. Explain the operation of AM and FM detectors.
3. Draw block diagrams of basic AM and FM transmitters and receivers.
4. Discuss the effects of heterodyning.
5. Discuss how intelligence is added to and removed from an RF carrier.
6. Calculate percentage of modulation.
7. Explain the operation of a cathode-ray tube (CRT).
8. Describe how a picture is produced on the face of a picture tube.
9. Draw block diagrams of a black-and-white and a color TV set, and explain the function of each block.
10. Explain the functions of sweep signals and how they are synchronized.
11. Explain how high voltage is produced in a TV receiver.
12. Describe how video amplifiers obtain wide bandwidth and list ways of increasing bandwidth.

IMPORTANT TERMS

Beat frequency. A resulting frequency that develops when two frequencies are combined in a nonlinear device.

Beat-frequency oscillator. An oscillator of a CW receiver. Its output beats with the incoming CW signal to produce an audio signal.

Blanking pulse. A part of the TV signal where the electron beam is turned off during the retrace period. There is both vertical and horizontal blanking.

Buffer amplifier. An RF amplifier that follows the oscillator of a transmitter. It isolates the oscillator from the load.

Carrier wave. An RF wave to which modulation is applied.

Center frequency. The carrier wave of an FM system without modulation applied.

Chroma. Short for *chrominance*. Refers to color in general.

Compatible. A TV system characteristic in which broadcasts in color may be received in black and white on sets not adapted for color.

Deflection. Electron beam movement of a TV system that scans the camera tube or picture tube.

Deflection yoke. A coil fixture that moves an electron beam vertically and horizontally.

Diplexer. A special TV transmitter coupling device that isolates the audio carrier and the picture carrier signals from each other.

Field, even lined. The even-numbered scanning lines of one TV picture or frame.

Field, odd lined. The odd-numbered scanning lines of one TV picture or frame.

Frame. A complete electronically produced TV picture of 525 horizontally scanned lines.

Ganged. Two or more components connected together by a common shaft—a three-ganged variable capacitor.

Heterodyning. The process of combining signals of independent frequencies to obtain a different frequency.

High-level modulation. A situation in which the modulating component is added to an RF carrier in the final power output of the transmitter.

Hue. A color, such as red, green, or blue.

Interlace scanning. An electronic picture production process in which the odd lines, then the even lines, are all scanned to make a complete 525-line picture.

Intermediate frequency (IF). A single frequency that is developed by heterodyning two input signals together in a superheterodyne receiver.

I-signal (I). A color signal of a TV system that is in phase with the 3.58-MHz color subcarrier.

Line-of-sight transmission. An RF signal transmission that radiates out in straight lines because of its short wavelength.

Modulating component. A specific signal or energy used to change the characteristic of an RF carrier. Audio and picture modulation are very common.

Modulation. The process of changing some characteristic of an RF carrier so that intelligence can be transmitted.

Monochrome. Black-and-white television.

Negative picture phase. A video signal characteristic where the darkest part of a picture causes the greatest change in signal amplitude.

Phasor. A line used to denote value by its length and phase by its position in a vector diagram.

Plumbicon. A television camera tube that operates on the photoconduction principle.

Q signal. A color signal of a TV system that is out of phase with the 3.58-MHz color subcarrier.

Retrace. The process of returning the scanning beam of a camera or picture tube to its starting position.

Saturation. The strength or intensity of a color used in a TV system.

Scanning. In a TV system, the process of moving an electron beam vertically and horizontally.

Selectivity. A receiver function of picking out a desired RF signal. Tuning achieves selectivity.

Sidebands. The frequencies above and below the carrier frequency that are developed because of modulation.

Skip. An RF transmission signal pattern that is the result of signals being reflected from the ionosphere or the earth.

Sky-wave. An RF signal radiated from an antenna into the ionosphere.

Sync. An abbreviation for synchronization.

Synchronization. A control process that keeps electronic signals in step. The sweep of a TV receiver is synchronized by the transmitted picture signal.

Telegraphy. The process of conveying messages by coded telegraph signals.

Trace time. A period of the scanning process where picture information is reproduced or developed.

Vestigial sideband. A transmission procedure where part of one sideband is removed to reduce bandwidth.

Videcon. A TV camera tube that operates on the photoconductive principle.

Wavelength. The distance between two corresponding points that represents one complete wave.

Y signal. The brightness or luminance signal of a TV system.

Zero beating. The resulting difference in frequency that occurs when two signals of the same frequency are heterodyned.

COMMUNICATIONS SYSTEMS

A communications system is very similar to any other electronic system. It has an energy source, transmission path, control, a load device, and one or more indicators. These individual parts are all essential to the operation of the system. The physical layout of the communication system is one of its most distinguishing features.

Figure 6-1 shows a block diagram of an RF communication system. The signal source of the system is an RF *transmitter*. The transmitter is the center or focal point of the system. The *RF signal* is sent to the remaining parts of the system through space. Air is the transmission path of the system. RF finds air to be an excellent signal path. The control function of the system is directly related to the signal path. The distance that the RF signal must travel has a great deal to do with its strength. The load of the system is an infinite number of radio *receivers*. Each receiver picks the signal out of the air and uses it to do work. Any number of receivers can be used without directly influencing the output of the transmitter. System indicators may be found at a number of locations. Meters, lamps, and waveform monitoring oscilloscopes are typical indicators. These basic functions apply in general to all RF communication systems.

A number of different RF communication systems are in op-

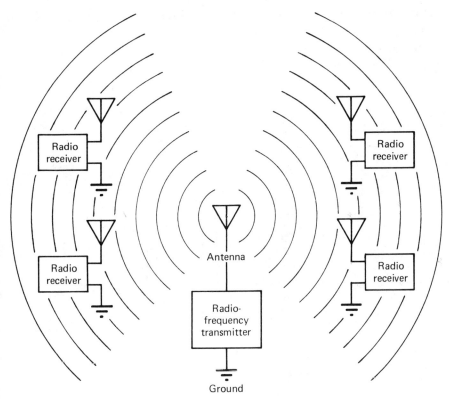

FIGURE 6-1   RF communication system.

eration today. As a general rule, these systems are classified according to the method by which signal information or intelligence is applied to the transmitted signal. Three very common communication systems are *continuous wave (CW), amplitude modulation (AM),* and *frequency modulation (FM).* Each system has a number of unique features that distinguishes it from the others. The transmitter and receiver of the system are quite different. They are all similar to the extent that they use RF signal energy sources as sources of operation.

A more realistic way to look at an RF communication system is to divide the transmitter and receiver into separate systems. The transmitter then has all its system parts in one discrete unit. Its primary function is to generate an RF signal that contains intelligence and to radiate this signal into space. The receiver can also be viewed as an independent system. It has a complete set of system parts. Its primary function is to pick up the RF transmitted signal, extract the intelligence, and develop it into a usable output. A one-way communication system has one transmitter and an infinite number of receivers. Commercial AM, FM, and TV communication is achieved by this method. *Two-way mobile communication systems* have a transmitter and a receiver at each location. This type of system permits direct communication between each location. *Citizen's band (CB) radio* is considered to be two-way communication.

Figure 6-2 shows a simplification of the transmitter and receiver as independent systems. The transmitter has an RF *oscillator, amplifiers,* and a *power source.* The antenna-ground at the output circuit serves as the load device for the system. Intelligence is applied according to the design of the system. The system may be CW, AM, or FM.

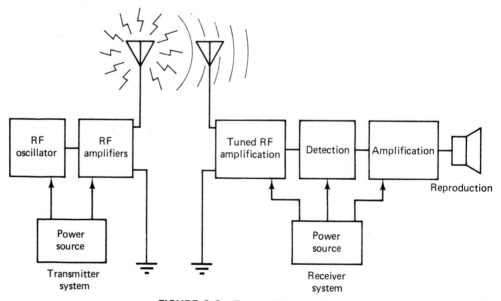

FIGURE 6-2 Transmitter-receiver systems.

The RF receiver function is represented as an independent system. It will not be operational unless a signal is sent out by the transmitter. The receiver has tuned RF amplification, detection, reproduction, and a power supply. The detection function picks out the intelligence from the received signal. The reproduction unit, which is usually a speaker, serves as the load device. The receiver responds only to the RF signal sent out by the transmitter.

ELECTROMAGNETIC WAVES

Electromagnetic communication systems rely on the radiation of high-frequency energy from an antenna-ground network. This particular principle was discovered by Heinrich Hertz around 1885. He found that high-frequency waves produced by an electrical spark caused electrical energy to be induced in a coil of wire some distance away. He also discovered that current passing through a coil of wire produces a strong electromagnetic field. The fundamental unit of frequency is the hertz, in recognition of his electromagnetic discoveries.

When direct current is applied to a coil of wire, it causes the field to remain stationary as long as current is flowing. When dc is first turned on the field expands. When it is turned off, the field collapses and cuts across the coil. As a general rule, dc electromagnetic field development is of no significant value in radio communication.

When low-frequency ac is applied to an inductor it produces an electromagnetic field. Since ac is in a constant state of change, the resulting field is also in a state of change. It changes polarity twice during each operational cycle.

Figure 6-3 shows how the electromagnetic field of a coil changes when ac is applied. In step (a), the field is expanding during the first half of the positive alternation. At the peak of the alternation, coil current begins to decrease in value. Step (b) shows the field *collapsing*. The current value decreases to the zero baseline at this time. This completes the positive alternation, or first 180°, of the sine wave. Step (c) shows the field *expanding* for the negative alternation. The polarity of this field is now reversed. At the peak of the negative alternation the field begins to collapse. Step (d) shows the field change for the last part of the alternation. The cycle is complete at this point. The sequence is repeated for each new sine wave.

When high-frequency ac is applied to an inductor, it also produces an electromagnetic field. The directional change of high frequency ac, however, occurs very quickly. The change is so rapid that there is not enough time for the collapsing field to return to the coil. As a result, a portion of the field becomes detached and is forced away from the coil. In a sense, the field radiates out from the coil. The word *radio* was developed from these two terms.

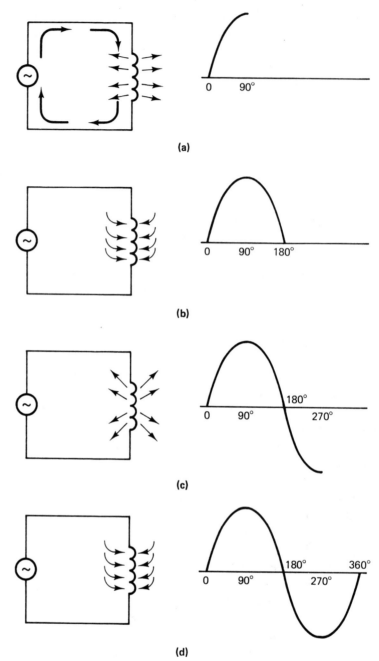

FIGURE 6-3 Electromagnetic field changes.

The electromagnetic wave of an RF communication system radiates out from the antenna of the transmitter. This wave is the end result of an interaction between electric and magnetic fields. The resulting wave is invisible, cannot be heard, and travels at the speed of light. These waves become weaker after leaving the antenna. Stronger electromagnetic waves can be effectively used in a radio communication system. These waves do not effect human beings like high-powered sound waves. RF waves are also easier to work with in the receiver circuit.

Electromagnetic wave radiation is directly related to its frequency. *Low-frequency (LF)* waves of 30 to 300 kHz and *medium-frequency (MF)* waves of 300 to 3000 kHz tend to follow the curvature of the earth. These waves are commonly called *ground waves.* They are usable during the day or night for distances up to 100 miles. Daytime reception of commercial AM radio is largely through ground waves.

A different part of the electromagnetic wave radiated from an antenna is called a *sky wave.* Depending on its frequency, sky waves are reflected back to the earth by a layer of ionized particles called the *ionosphere.* These waves permit signal reception well beyond the range of the ground wave. Sky-wave signal patterns change according to ion density and the position of the ionized layer. During the daylight hours the ionosphere is very dense and near the surface of the earth. Signal reflection or *skip* of this type is not very suitable for long-distance communication. In the evening hours the ionosphere is less dense and moves to higher altitudes. Sky-wave reflection patterns have larger angles and travel greater distances. Figure 6-4 shows some sky-wave patterns and the ground wave.

Very high frequency (VHF) signals of 30 to 300 MHz tend to move in a straight line. The height of the antenna and its radiation angle has a great deal to do with the direction of the signal path. *Line-of-sight transmission* describes this type of radiation pattern. FM, TV, and satellite communication are achieved by VHF signal radiation. The physical size of the wave, or its *wavelength,* decreases with an increase in frequency. High-frequency signals have a short wavelength. These signals pass very readily through the ionosphere without being reflected or distorted. Communication between an FM or TV transmitter and a receiver on earth is limited to a few hundred miles. Communication between an earth station and a satellite can involve thousands of miles.

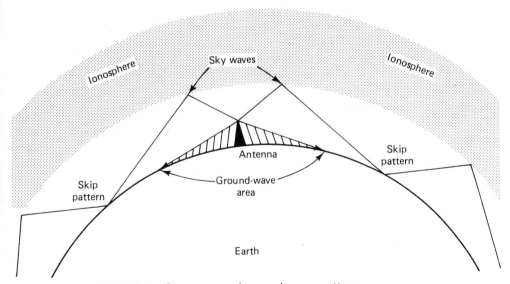

FIGURE 6-4 Sky-wave and ground-wave patterns.

CONTINUOUS-WAVE COMMUNICATION

Continuous-wave (CW) communication is the simplest of all communication systems. It consists of a transmitter and a receiver. The *transmitter* employs an oscillator for the generation of an RF signal. The oscillator generates a continuous sine wave. In most systems the CW signal is amplified to a desired power level by an RF power amplifier. The output of the final power amplifier is then connected to the antenna-ground network. The antenna radiates the signal into space. Information or intelligence is applied to the CW signal by turning it on and off. This causes the signal to be broken into a series of pulses, or short bursts of RF energy. The pulses conform to an intelligible code. The international *Morse code* is in common usage today. Figure 6-5 shows a simplification of the Morse code. A short burst of RF represents a dot. A burst three times longer represents a dash. Signals of this type are called keyed or coded continuous waves. The *Federal Communications Commission (FCC)* classified keyed CW signals as type A1. This is considered to be *radio telegraphy* with on-off keying. Figure 6-6 shows a comparison of a CW signal and a keyed CW signal.

The receiver function of a CW communication system is somewhat more complex than the transmitter. It has an *antenna-ground network*, a *tuning* circuit, an *RF amplifier*, a *heterodyne detector*, a *beat-frequency oscillator*, an *audio amplifier*, and a *speaker*. The receiver is designed to pick up a CW signal and convert it into a sound signal that drives the speaker. Figure 6-7 shows a block diagram of the CW communication system.

A •—	K —•—	U ••—
B —•••	L •—••	V •••—
C —•—•	M ——	W •——
D —••	N —•	X —••—
E •	O ———	Y —•——
F ••—•	P •——•	Z ——••
G ——•	Q ——•—	
H ••••	R •—•	PERIOD •—•—•—
I ••	S •••	COMMA ——••——
J •———	T —	QUESTION ••——••

1 •————	5 •••••	8 ———••
2 ••———	6 —••••	9 ————•
3 •••——	7 ——•••	0 —————
4 ••••—		

FIGURE 6-5 Morse code.

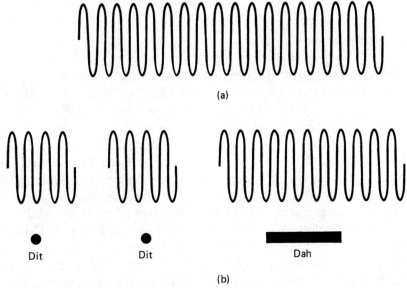

(a)

Dit   Dit   Dah

(b)

FIGURE 6-6 Comparison of CW signals: (a) CW signal; (b) coded or keyed continuous wave, letter D.

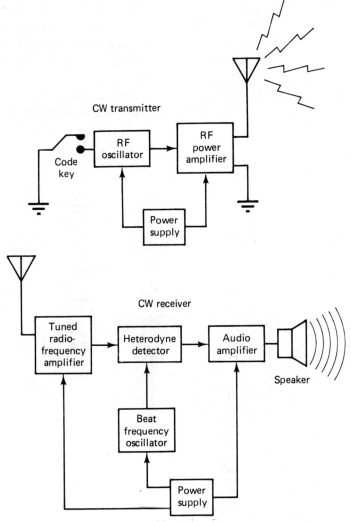

FIGURE 6-7 Block diagram of a CW communication system.

CW Transmitter

A CW transmitter in its simplest form has an oscillator, a power supply, an antenna-ground network, and a keying circuit. The power output developed by the circuit is usually quite small. In this type of system, the active device serves jointly as an oscillator and a power amplifier. This is done to reduce the number of circuit stages in front of the antenna. An oscillator used in this manner generally has reduced frequency stability. As a rule, frequency stability is a very important operational consideration. Single-stage transmitters are rarely used today because of their instability.

A single-stage CW transmitter is shown in Fig. 6-8. This particular circuit is a tuned-based *Hartley oscillator*. The output frequency is adjustable up to 3.5 MHz. The power output is approximately 0.5 W. When the output of the oscillator is connected to a load, it usually causes a shift in frequency. The load should, therefore, be of a constant impedance value in order for this type of transmitter to be effective.

Before a CW transmitter can be placed into operation, it is necessary to close the code key. This operation supplies forward bias voltage to the base through the divider network of R_1 and R_2. With the transistor properly biased, a feedback signal is supplied to the T_1-C_1 tank circuit. This charges C_1 and causes the circuit to oscillate. Opening the key causes the oscillator to stop functioning. Intelligence can be injected into the CW signal in the form

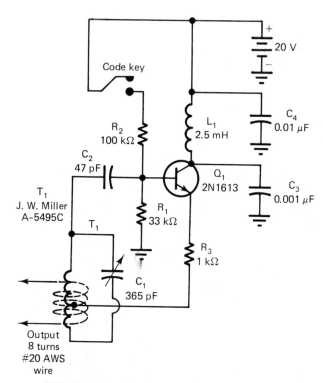

FIGURE 6-8   Single-stage CW transmitter.

of code. The code key simply responds as a fast-acting on-off switch.

The output of the transmitter is developed across a coil wound around T_1. Inductive coupling of this type is very common in RF transmitter circuits. A low-impedance output is needed to match the antenna-ground impedance. When this is accomplished there is a maximum transfer of energy from the oscillator to the antenna.

A single-stage CW transmitter has a number of shortcomings. The power output is usually held to a low level. The frequency stability is rather poor. These two problems can be overcome to some extent by adding a power amplifier after the oscillator. Systems of this type are called *master oscillator power amplifier (MOPA) transmitters.*

Figure 6-9 shows the circuitry of a low-power MOPA transmitter system. Q_1 is a modified *Pierce oscillator.* The crystal (Y_1) provides feedback from the collector to the base. Transformer T_1 is used to match the collector impedance of Q_1 to the base of transistor Q_2. Q_2 is the power amplifier. It operates as a class C amplifier. The oscillator is therefore isolated from the antenna through Q_2. This provides a fixed load for the oscillator. Operational stability is improved and power output is increased with this modification.

CW Receiver

The receiver of a CW communication system is responsible for intercepting an RF signal and recovering the information it con-

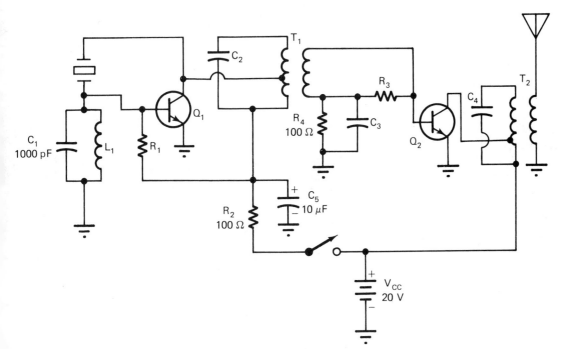

FIGURE 6-9 MOPA transmitter.

tains. Information of this type is in the form of a *radiotelegraph code*. The signal is an RF carrier wave that is interrupted by a coded message. A CW receiver must perform a number of functions in order to achieve this operation. This includes signal reception, selection, RF amplification, detection, AF amplification, and sound reproduction. These functions are also used in the reception of other RF signals. In fact, CW reception is usually only one of a number of operations performed by a communications receiver. This function is generally achieved by placing the receiver in its CW mode of operation. A switch is usually needed to perform this operation.

Figure 6-10 shows the functions of a CW communication receiver in a block diagram. Note that electrical power is needed to energize the active components of the system. Not all blocks of the diagram are supplied operating power. This means that some of the functions can be achieved without solid state components. The antenna, tuning circuit, and detector do not require power supply energy for operation. These functions are achieved by signal energy. A radio receiver circuit has a signal energy source and operational energy source. The power supply provides operational energy directly to the circuit. The signal source for RF energy is the transmitter. This signal must be intercepted from the air and processed by the receiver during its operation.

Antennas. The antenna of a CW receiver is responsible for the reception function. It intercepts a small portion of the RF signal that is sent out by the transmitter. Ordinarily, the receiving antenna only develops a few microwatts of power. In a strict sense, the antenna is a *transducer*. It is designed to convert electromagnetic wave energy into RF signal voltage. The signal voltage then causes a corresponding current flow in the antenna-ground net-

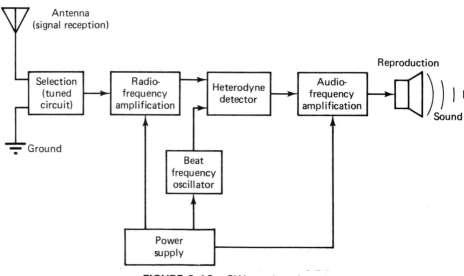

FIGURE 6-10 CW receiver functions.

work. RF antenna power is the product of signal voltage and current.

Receiving antennas come in a variety of styles and types. In two-way communication systems, the transmitter and receiving antenna are of the same unit. The antenna is switched back and forth between the transmitter and receiver. In one-way communication systems antenna construction is not particularly critical. A long piece of wire can respond as a receiving antenna. In weak signal areas, antennas may be tuned to resonate at the particular frequency being received. Portable radio receivers use small *antenna coils* that are attached directly to the circuit.

When the electromagnetic wave of a transmitter passes over the receiving antenna, ac voltage is induced into it. This voltage causes current to flow, as shown in Fig. 6-11. Starting at point *A* of the waveform, the electrons are at a standstill. The induced signal then causes the voltage to rise to its positive peak. The resulting current flow is from the antenna to the ground. It rises to a peak, then drops to zero at point *C* on the curve. The polarity of the induced voltage changes at this point. Between points *C* and *E* it rises to the peak of the negative alternation and returns again to zero. The resulting current flow is from the ground into the antenna. This induced signal is repeated for each succeeding alternation.

With radio frequency induced into the antenna, a corresponding RF is brought into the receiver by the antenna coil. The antenna coil serves as an *impedance-matching* transformer. It matches the impedance of the antenna to the impedance of the receiver input circuit. When outside antennas are used, the primary winding is an extension of the antenna. The loop antenna coil of a portable receiver is attached directly to the input circuit.

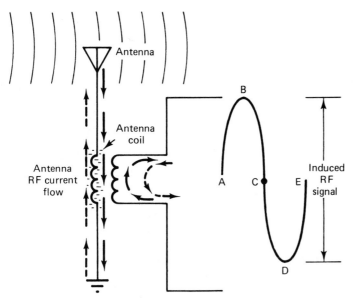

FIGURE 6-11 Antenna signal voltage and current.

Signal Selection. The signal-selection function of a radio receiver refers to its ability to pick out a desired RF signal. In receiver circuits, the antenna is generally designed to intercept a band or range of different frequencies. The receiver must then select the desired signal from all those intercepted by the antenna. Signal selection is primarily achieved by an *LC resonant circuit*. This circuit provides a low-impedance path for its resonant frequency. Nonresonant frequencies see a very high impedance. In effect, the resonant frequency signal is permitted to pass through the tuner without opposition.

Figure 6-12 shows the input *tuner* of a CW receiver. In this circuit L_2 is the secondary winding of the antenna transformer. C_1 is a variable capacitor. By changing the value of C_1, the resonant frequency of L_2C_1 is tuned to the desired frequency. The selected frequency is then permitted to pass into the remainder of the receiver. Signal selection for AM, FM, and TV receivers all respond to a similar tuning circuit.

In communication receiving circuits a number of tuning stages are needed for good signal selection. Each stage of tuning permits the receiver to be more selective of the desired frequency. The tuning *response curve* of one tuned stage is shown in Fig.

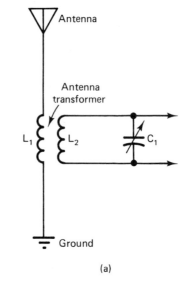

(a)

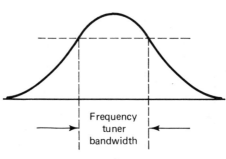

(b)

FIGURE 6-12 (a) Input tuning circuit; (b) tuner frequency response curve.

6-12. In a crowded RF band, several stations operating near the selected frequency might be received at the same time. Additional tuning improves the selection function by narrowing the response curve. Figure 6-13 shows some representative loop antenna coils.

RF Amplification. Most CW receivers employ at least one stage of RF amplification. This specific function is designed to amplify the weak RF signal intercepted by the antenna. Some degree of amplification is generally needed to boost the signal to a level where its intelligence can be recovered. *RF amplification* can be achieved by a variety of active devices. Vacuum tubes were used for a number of years. Solid-state circuits may now employ bipolar transistors and MOSFETs in their design. Several manufacturers have now developed IC RF amplifiers. Any of these active devices can be used to achieve RF signal amplification.

A representative RF amplifier circuit is shown in Fig. 6-14. The primary function of the circuit is to determine the level of

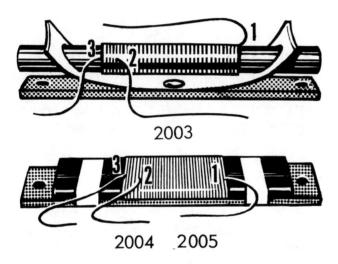

2003

2004 2005

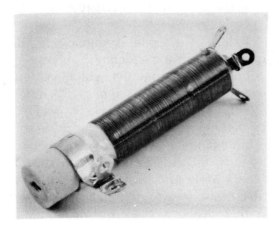

FIGURE 6-13   Antenna coils. (Courtesy of J.W. Miller Div./Bell Industries.

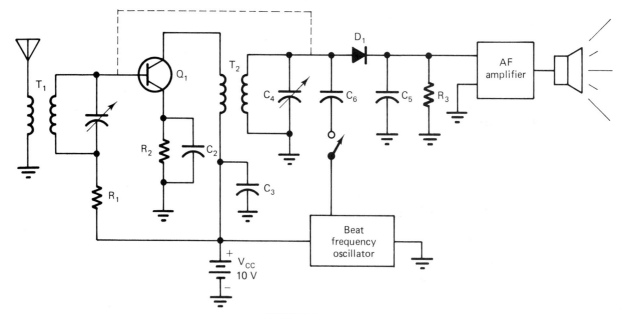

FIGURE 6-14   Tuned radio frequency receiver.

amplification. Note that the amplifier has both input and output tuning circuits. This circuit is called a *tuned radio-frequency (TRF) amplifier*. The broken line between the two variable capacitors indicates that they are *ganged* together. When one capacitor is tuned, the second capacitor is adjusted to a corresponding value.

Heterodyne Detection.   Heterodyne detection is an essential function of the CW receiver. *Heterodyning* is simply a process of mixing two ac signals in a nonlinear device. In the CW receiver, the nonlinear device is a diode. The applied signals are called the fundamental frequencies. The resulting output of this circuit has four frequencies. Two of these are the original fundamental frequencies. The other two are *beat frequencies*. Beat frequencies are the addition and the difference in the fundamental frequencies. Heterodyne detectors used in other receiver circuits are commonly called *mixers*.

Figure 6-15 shows the signal frequencies applied to a heterodyne detector. Fundamental frequency F_1 is representative of the incoming RF signal. This signal has been keyed on and off with a telegraphic code. Fundamental frequency F_2 comes from the *beat frequency oscillator (BFO)* of the receiver. It is a CW signal of 1.001 MHz. Both RF signals are applied to the receiver's diode.

The resulting output of a heterodyne detector is F_1, F_2, $F_1 + F_2$, and $F_2 - F_1$. F_1 is 1.0 MHz and F_2 is 1.001 MHz. Beat frequency B_1 is 2.001 MHz and beat frequency B_2 is 1000 Hz. The output of the detector contains all four frequencies. The difference beat frequency of 1000 Hz is the range of human hearing. F_1, F_2, and F_3 are RF signals.

When two RF signals are applied to a diode, the resulting

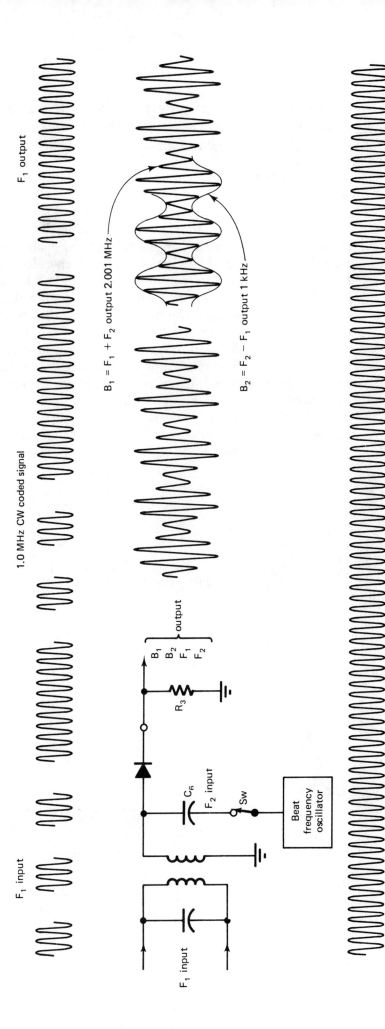

F₁ input

F₁ output

1.0 MHz CW coded signal

$B_1 = F_1 + F_2$ output 2.001 MHz

$B_2 = F_2 - F_1$ output 1 kHz

F₂ input

1.001 MHz BFO CW signal

F₂ output

FIGURE 6-15 Heterodyne detector frequencies.

output is the sum and difference signals. If the two signals are identical, the output frequency is the same as the input. This is generally called *zero beating*. There is no developed beat frequency when this occurs.

When the two RF input signals are slightly different, beat frequencies occur. At one instant the signals move in the same direction. The resulting output is the sum of the two frequencies. When one signal is rising and one is falling at a different rate, there will be times when the signals cancel each other. The addition and difference of the two signals are therefore frequency dependent. B_1 and B_2 are combined in a single RF wave. The amplitude of this wave varies according to frequency difference in the two signals. In a sense, the mixing process causes the amplitude of the RF wave to vary at an audio rate.

The diode of a heterodyne dectector is also responsible for rectification of the applied signal frequencies. The rectified output of the added beat frequency is of particular importance. It contains both RF and AF. Figure 6-16 shows how the diode responds. Initially, the RF signals are rectified. The output of the diode is a series of RF pulses varying in amplitude at an AF rate. C_5 added to the output of the diode responds as a C input filter. Each RF

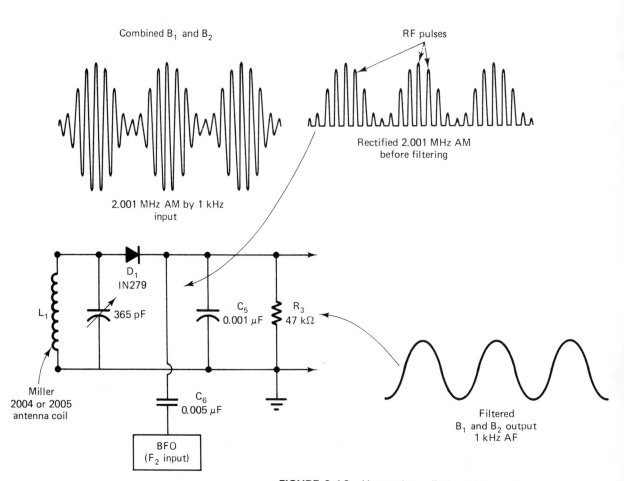

FIGURE 6-16 Heterodyne diode output.

pulse charges C_5 to its peak value. During the off period C_5 discharges through R_3. The output is a low-frequency AF signal of 1000 Hz. This signal occurs only when the incoming CW signal has been keyed. The AF signal is representative of the keyed information imposed on the RF signal at the transmitter.

Beat Frequency Oscillator.   Nearly any basic oscillator circuit could be used as the *BFO* of a CW receiver. Figure 6-17 shows a *Hartley oscillator.* Feedback is provided by a tapped coil in the base circuit. C_1 and T_1 are the frequency-determining components. Note that this particular oscillator has variable-frequency capabilities. C_1 is usually connected to the front panel of the receiver. Adjustment of C_1 is used to alter the tone of the beat frequency signal. When the BFO signal is equal to the coded CW signal, no sound output occurs. This is where zero beating takes place. When the frequency of the BFO is slightly above or below the incoming signal frequency, a low AF tone is produced. Increasing the BFO frequency causes the pitch of the AF tone to increase. Adjustment of the AF output is a matter of personal preference. In practice, an AF tone of 400 to 1 kHz is common.

Output of the BFO is coupled to the diode through capacitor C_6. In a communications receiver, BFO output is controlled by a switch. With the switch on, the receiver produces an output for CW signals. With the switch off, the receiver responds to AM and possibly FM signals. The BFO is needed only to receive coded CW signals.

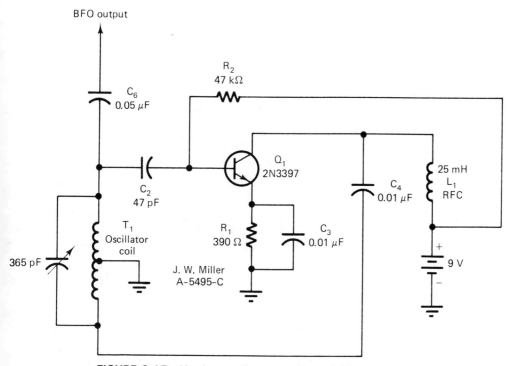

FIGURE 6-17   Hartley oscillator used as a BFO.

AF Amplification. The *AF amplifier* of a CW receiver is responsible for increasing the level of the developed sound signal. The type and amount of signal amplification varies a great deal among different receivers. Typically, a small signal amplifier and a power amplifier are used. The small-signal amplifier responds as a voltage amplifier. This amplifier is designed to increase the signal voltage to a level that will drive the speaker. The power output of a communications receiver rarely ever exceeds 5 W. A number of the AF amplifier circuits discussed in Unit 3 could be used in a CW receiver.

AMPLITUDE MODULATION COMMUNICATION

Amplitude modulation, or AM, is an extremely important form of communication. It is achieved by changing the physical size or amplitude of the RF wave by the intelligence signal. Voice, music, data, and picture intelligence can be transmitted by this method. The intelligence signal must first be changed into electrical energy. Transducers such as microphones, phonograph cartridges, tape heads, and photoelectric devices are designed to achieve this function. The developed signal is called the *modulating component*. In an AM communication system, the RF transmitted signal is much higher in frequency than the modulating component. The RF component is an uninterrupted CW wave. In practice, this part of the radiated signal is called the *carrier wave*.

An example of the signal components of an AM system is shown in Fig. 6-18. The unmodulated RF carrier is a CW signal. This signal is generated by an oscillator. In this example, the carrier is 1000 kHz, or 1.0 MHz. A signal of this frequency is in the *standard AM broadcast band* of 535 to 1620 kHz. When listening to this station, a receiver would be tuned to the carrier frequency. Assume now that the RF signal is modulated by the indicated 1000-Hz tone. The RF component changes 1000 cycles for each AF sine wave. The amplitude of the RF signal varies according to the frequency of the modulating signal. The resulting wave is called an *amplitude-modulated RF carrier*.

In order to achieve amplitude modulation, the RF and AF components are applied to a nonlinear device. A solid-state device operating in its nonlinear region is used to produce modulation. In a sense, the two signals are mixed, or heterodyned, together. This operation causes beat frequencies to be developed. For the signals of Fig. 6-18, there will be two *beat frequencies*. Beat frequency 1 is the sum of the two signal frequencies. B_1 is 1,000,000 Hz plus 1000 Hz, or 1,001,000 Hz. B_2 is 1,000,000 Hz minus 1000 Hz, or 999,000 Hz. The resulting AM signal therefore contains three RF signals. These are the 1.0-MHz carrier, 0.999 MHz, and 1.001 MHz.

When the modulating component of an AM signal is music,

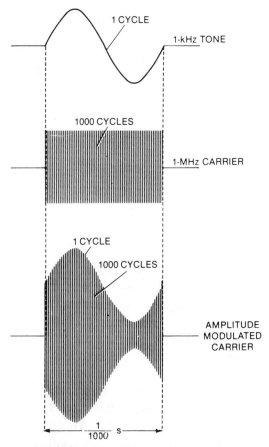

1 CYCLE
1-kHz TONE

1000 CYCLES
1-MHz CARRIER

1 CYCLE
1000 CYCLES
AMPLITUDE MODULATED CARRIER

$\frac{1}{1000}$ s

FIGURE 6-18 AM signal components.

the resulting beat frequencies become quite complex. As a rule, music involves a range or a band of frequencies. In AM systems these are called *sidebands*. B_1 is the upper sideband and B_2 is the lower sideband.

The space that an AM signal occupies with its frequency is called a *channel*. The *bandwidth* of an AM channel is twice the highest modulating frequency. For our 1-kHz modulating component, a 2-kHz bandwidth is needed. This is 1 kHz above and below the carrier frequency of 1 MHz. In commercial AM broadcasting a station is assigned a 10-kHz channel. This limits the AM modulation component to a frequency of 5 kHz. Figure 6-19 shows the sideband produced by a standard AM station.

It is interesting to note that the carrier wave of an AM signal contains no modulation. All the modulation appears in the sidebands. If the modulating component is removed, the sidebands disappear. Only the carrier is transmitted. The sidebands are directly related to the carrier and the modulating component. The carrier has a constant frequency and amplitude. The sidebands vary in frequency and amplitude according to the modulation component. In AM radio, the receiver is tuned to the carrier wave.

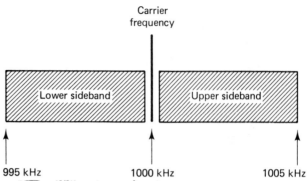

FIGURE 6-19 Sideband frequencies of an AM signal.

Percentage of Modulation

In AM radio the percentage of modulation is not permitted by law to exceed 100%. Essentially, this means that the modulating component cannot cause the RF component to vary over 100% of its unmodulated value. Figure 6-20 shows an AM signal with three different *levels of modulation*.

When the peak amplitude of modulating signal is less than the peak amplitude of the carrier, modulation is less than 100%. If the modulation component and carrier amplitudes are equal, 100% modulation is achieved. A modulating component greater than the carrier causes overmodulation. An overmodulated wave has an interrupted spot in the carrier wave. *Overmodulation* causes increased signal bandwidth and additional sidebands to be generated. This causes interference with adjacent channels.

Modulation percentage can be calculated or observed on an indicator. When operating voltage values are known, the *percentage of modulation* can be calculated. The formula is

$$\% \text{ modulation} = \frac{V_{\max} - V_{\min}}{2\text{-}V \text{ carrier}}$$

Using the values of Fig. 6-20(a), compute the percentage of modulation. Compute the percentage of modulation for part (b). If the $V_{\min}$ value of the overmodulated signal is considered to be 0 V, compute the modulation percentage.

AM Communication System

A block diagram of AM communication system is shown in Fig. 6-21. The transmitter and receiver respond as independent systems. In a one-way communication system, there are one transmitter and an infinite number of receivers. Commercial AM radio is an example of *one-way communication. Two-way communication* systems have a transmitter and receiver at each location. CB radio systems are of the two-way type. The operating principles are basically the same for each system.

The transmitter of an AM system is responsible for signal

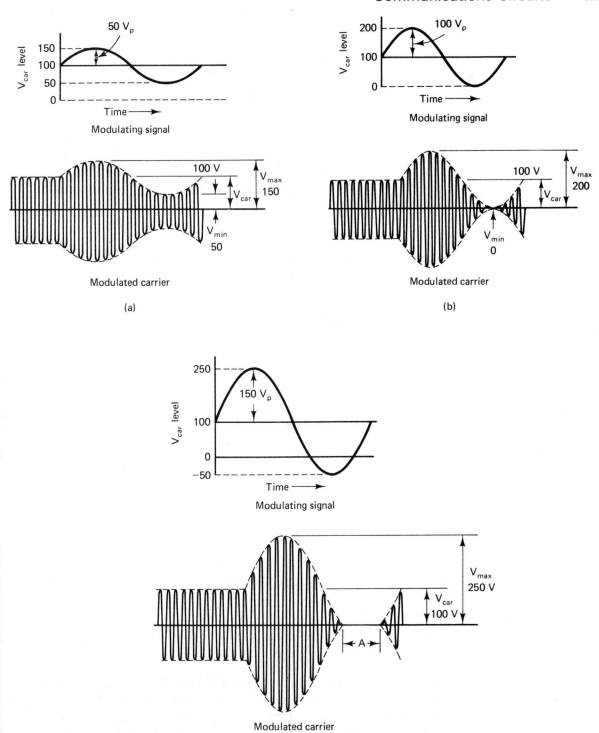

FIGURE 6-20 AM modulation levels: (a) undermodulation; (b) 11%
modulation; (c) overmodulation.

generation. The RF section of the transmitter is primarily the same
as that of a CW system. The modulating signal component is, however, a unique part of the transmitter. Essentially, this function is
achieved by an AF amplifier. A variety of amplifier circuits can

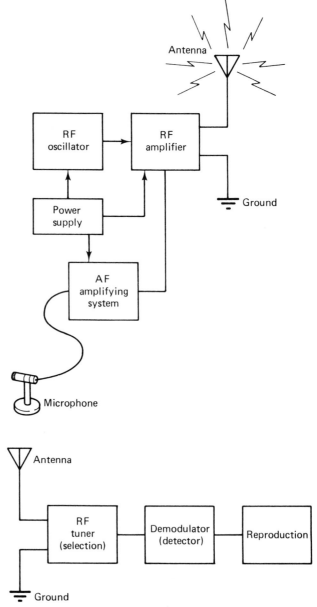

FIGURE 6-21 AM communication system.

be used. Typically, one or two small-signal amplifiers and a power amplifier are suitable for low-power transmitters. The developed modulation component power must equal the power level of the RF output. Modulation of the two signals can be achieved at a number of places. High-level modulation occurs in the RF power amplifier. Low-level modulation is achieved after the RF oscillator. All amplification following the point of modulation must be linear. Only high-level modulation is discussed in this unit.

The receiver of an AM system is responsible for signal interception, selection, demodulation, and reproduction. Most AM receivers employ the heterodyne principle in their operation. This type of receiver is known as a *superheterodyne* circuit. It is somewhat different from the heterodyne detector of the CW receiver. A large part of the circuit is the same as the CW receiver. We discuss only the new functions of the AM receiver.

AM Transmitters. A wide range of AM transmitters is available today. Toy "walkie-talkie" units with an output of less than 100 mW are very popular. *Citizen's band (CB) transmitters* are designed to operate at 27 MHz with a 5-W output. AM amateur radio transmitters are available with a power output of up to 1 kW of power. Commercial AM transmitters are assigned power-output levels from 250 W to 50 kW. These stations operate between 535 and 1620 kHz. In addition to this, there are mobile communication systems, military transmitters, and public radio systems that all use AM. As a general rule, frequency allocations and power levels are assigned by the Federal Communications Commission.

An AM transmitter has a number of fundamental parts regardless of its operational frequency or power rating. It must have an oscillator, an audio-signal component, an antenna, and a power supply. The oscillator is responsible for the RF carrier signal. The audio component is responsible for the intelligence being transmitted. The modulation function can be achieved in a variety of ways. The signal is radiated from the antenna.

A simplified *AM transmitter* with a minimum of components is shown in Fig. 6-22. Transistor Q_1 is an AF signal amplifier. The

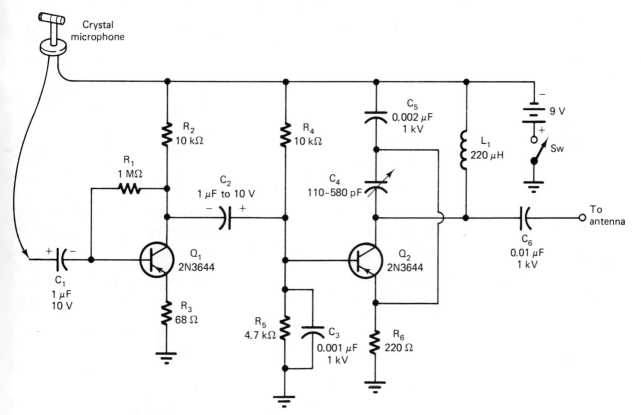

FIGURE 6-22 Simplified AM transmitter.

sound signal being amplified is developed by a *crystal microphone.* It is then applied to transistor Q_2. This transistor is a *Colpitts oscillator.* L_1, C_4, and C_5 determine the RF frequency of the oscillator. The amplitude of the oscillator varies according to the AF component. Resistor R_3 is used to adjust the amplitude level of the AF signal. The developed AM output signal is applied to the transmitting antenna. With proper design of the transmitter, it can be tuned to the standard AM broadcast band. The frequency of the system is adjusted by capacitor C_4. The operating range of this unit is several hundred feet.

A schematic of an improved low-power AM transmitter is shown in Fig. 6-23. Transistors Q_1, Q_2, and Q_3 of this circuit are responsible for the RF signal component. The AF component is developed by Q_4, Q_5, and Q_6. High-level modulation is achieved by this transmitter. The AF signal modulates the RF power amplifier Q_3.

Transistor Q_1 is the active device of a Hartley oscillator. This particular oscillator has a variable-frequency output of 1 to 3.5 MHz. The frequency-determining components are C_1 and L_{T1}. The output of the oscillator is coupled to the base of Q_2 by capacitor C_3. The emitter-follower output of the oscillator has a very low impedance.

Transistor Q_2 is an RF signal amplifier. This transistor is primarily responsible for increasing the signal level of the oscillator. It is also used to isolate the RF load from the oscillator. This is needed to improve oscillator frequency stability. When Q_2 is used in this regard, it is called a *buffer amplifier.*

The signal output of Q_2 must be capable of driving the power amplifier. In high-power transmitters, several RF signal amplifiers may be found between the oscillator and the power amplifier. Each stage is responsible for increasing the signal level to a suitable level. RF amplifiers of this type are often called *drivers.* Q_3 is an RF power amplifier. It is designed to increase the power level of the RF signal applied to its input. The output is used to drive the transmitting antenna. The load of Q_3 is a tuned circuit composed of C_{10}, L_5, and C_{11}. This is a pi-section filter. A filter of this type is used to remove signals other than those of the resonant frequency. C_{11} is the output capacitor of the filter. It is adjusted to match the impedance of the antenna. C_{10} is the input capacitor. It is used to resonate the filter to the applied carrier frequency. Resonance of the tuning circuit occurs when the collector current meter dips to its lowest value. In the broadcasting field an adjustment of this type is called *dipping the final.* Q_3 is operated as a class C power amplifier.

The modulating component of the transmitter is developed by an AF amplifier. Q_4 and Q_5 are push-pull AF power amplifiers. The developed AF signal is applied to the modulation transformer T_2. This signal causes the collector voltage of Q_3 to vary at an AF rate instead of being dc. This causes the RF output to vary in amplitude according to the AF component. The output signal has

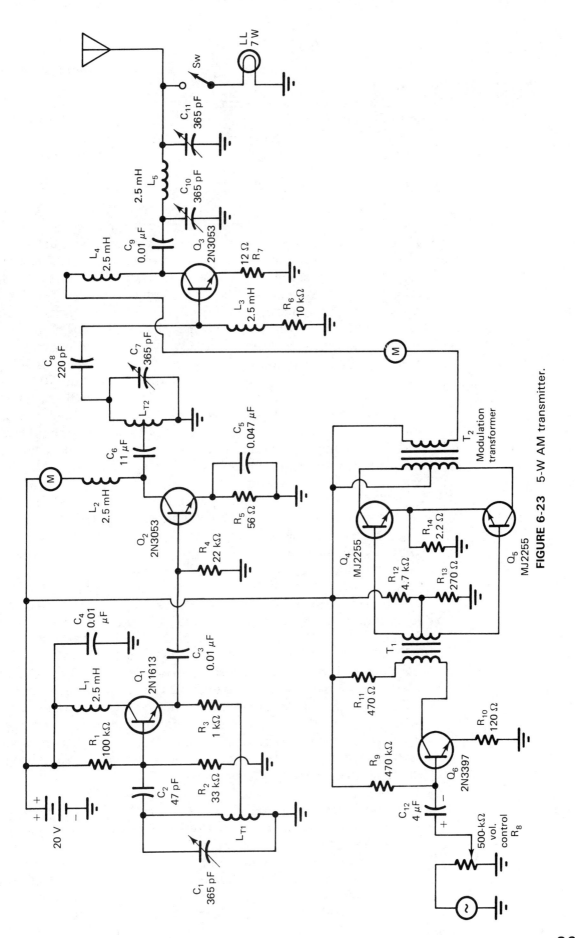

FIGURE 6-23 5-W AM transmitter.

a carrier and two sidebands. The power output is approximately 5 W.

Do not connect the output of this transmitter to an outside antenna unless you hold a valid radio-telephone operator's license. A load lamp is used for operational testing. The intensity of the load lamp is a good indication of the RF power developed by the transmitter.

A Simple AM Receiver. An AM receiver has four primary functions that must be achieved in order for it to be operational. No matter how complex or involved the receiver is, it must accomplish the following functions: *signal interception, selection, detection,* and *reproduction.* Figure 6-24 shows a diagram of an AM receiver that accomplishes these functions. Note that the circuit does not employ an amplifying device. No electrical power source is needed to make this receiver operational. The signal source is intercepted by the antenna-ground network. The receiver is energized by the intercepted RF signal energy. A receiver of this type is generally called a *crystal radio.* The detector is a crystal diode.

The functional operation of our crystal diode radio receiver is very similar to the CW receiver. This particular circuit does not employ an RF amplifier, a beat frequency oscillator, or an AF amplifier. A strong AM signal is needed to make this receiver operational. Signal interception and selection are achieved in the same way as in the CW receiver.

The detection, or *demodulation,* function of an AM receiver is responsible for removing the AF component from the RF signal. The detector is essentially a half-wave rectifier for the RF signal. Germanium diodes are commonly used as detectors. They are more sensitive than a silicon diode to RF signals.

The waveforms of Fig. 6-24 show how the crystal diode responds. The selected AM signal is applied to the diode. *Detection* is accomplished by rectification and filtering. The detected wave is a half-wave rectification version of the input. C_2 responds as an RF filter. It charges to the peak value of each RF pulse. It then discharges through the resistance of the earphone when the diode is reverse biased. The average value of the RF component appears across C_2. This is representative of the AF signal component. It energizes a small coil in the earphone. A thin metal disk in the earphone fluctuates according to coil energy. The earphone therefore changes electrical energy into sound waves. AF signal reproduction is achieved by the earphone.

Superheterodyne Receivers. Practically all AM radio receivers in operation today are of the superheterodyne type. This type of system accomplishes all the basic receiver functions. It has a number of circuit modifications that provide improved reception capabilities. It has excellent *selectivity* and is very sensitive to long-distance-signal reception. Figure 6-25(a) shows a block diagram of an AM superheterodyne receiver. This particular receiver

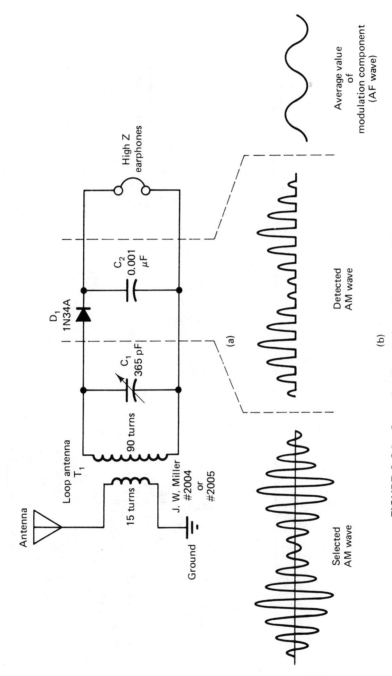

FIGURE 6-24 Crystal radio receiver: (a) circuit; (b) waveforms.

369

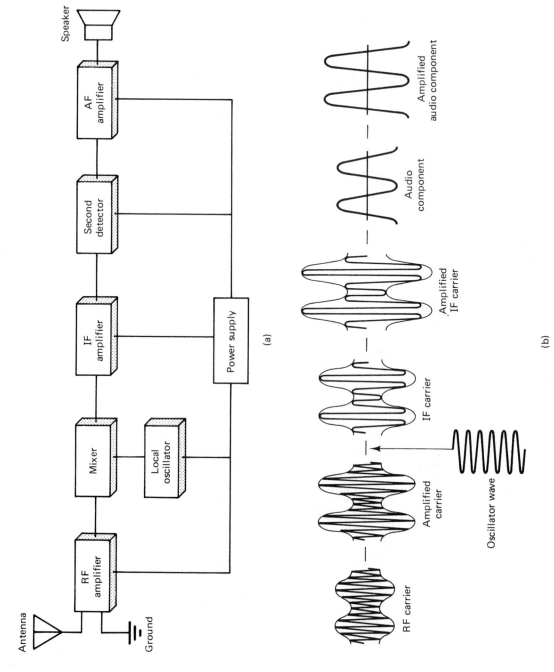

(a)

(b)

Amplified audio component

Audio component

Amplified IF carrier

IF carrier

Amplified carrier

Oscillator wave

RF carrier

Speaker

AF amplifier

Second detector

IF amplifier

Power supply

Mixer

Local oscillator

RF amplifier

Antenna

Ground

FIGURE 6-25 Superheterodyne receiver: (a) block diagrams; (b) waveforms.

is designed for signal reception between 550 and 1605 kHz, which is the *AM band.*

Superheterodyne operation is somewhat unusual compared with other methods of radio reception. Special RF amplifiers that are tuned to a fixed frequency are used in this circuit. These amplifiers are called *IF amplifiers.* Each incoming station frequency is changed into the IF. The IF amplifier responds in the same manner to all incoming signals. Each receiver has a tunable CW oscillator. This local oscillator generates an unmodulated RF signal of constant amplitude. This signal and the selected station are then mixed together. Mixing, or *heterodyning,* these two signals produces the IF signal. The IF contains all the modulation characteristics of the incoming RF signal. The IF signal is the difference beat frequency. After suitable amplification, the IF signal is applied to the detector. The resulting AF output signal is then amplified and applied to the speaker for sound reproduction.

Figure 6-26 shows a schematic diagram of a representative AM superheterodyne receiver. Transistor Q_1 is a tuned RF amplifier. Adjustment of capacitor C_1 selects the desired RF signal frequency. Q_2 is the mixer stage. The selected RF signal and the local oscillator are mixed together in Q_2. Q_3 is the local oscillator. The oscillator signal is fed to the base Q_2 through C_8. Transistor Q_4 is the IF amplifier. The output of Q_2 is coupled to Q_4 by transformer T_4. This transformer is tuned to pass only an IF of *455 kHz.* T_5 is the output IF transformer. The IF signal is coupled to the diode detector D_1 through this transformer. The detector rectifies the IF signal and develops the AF modulating component. C_{15} is the RF bypass filter capacitor and R_{12} is the volume control. Q_5 is the first AF signal amplifier. Push-pull AF power amplification is achieved by transistors Q_6 and Q_7. The AF output signal is changed into sound by the speaker. The entire circuit is energized by a 9-V battery. Switch SW-1 is connected to the volume control and turns the circuit on and off.

Practically all the superheterodyne receiver circuits have been discussed in conjunction with other system functions. The local oscillator, for example, is a variable-frequency Hartley oscillator. Mixing of the oscillator signal and selected RF signal has been described as heterodyning. Its output contains two input frequencies and two beat frequencies. The detector is essentially a crystal diode. An AF amplifying system is used to develop the audio signal component. The speaker was previously described as a transducer. All these operations remain the same when used in the superheterodyne receiver. Operational frequencies and circuit performance are generally somewhat different than previously described. We therefore direct our attention to specific circuit performance.

Using Figs. 6-25 and 6-26, we can now see how the superheterodyne responds when tuned to a specific frequency. Assume now that a number of AM signals are intercepted by the antenna-ground network. The desired station to be received occupies a fre-

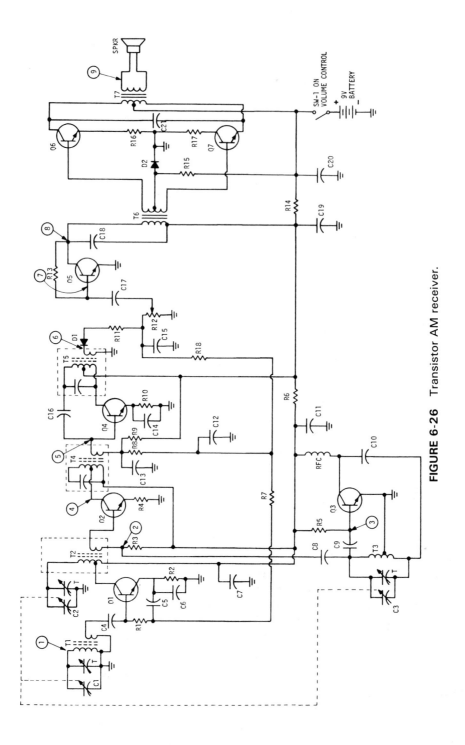

FIGURE 6-26 Transistor AM receiver.

quency of 1000 kHz. The tuner dial is adjusted 1000 kHz. The AF signal component then produces sound from the speaker. The volume control is adjusted to a desired signal level. The receiver is operational and performing its intended function.

In order for our receiver to develop a sound signal, a number of operational steps must be performed. Adjustment of the tuning dial changes the LC circuit of the RF amplifier. Capacitors C_1, C_2, and C_3 are all *ganged* together. One tuner adjustment alters the three tuned circuits at the same time. C_1 and C_2 tune the input and output of the RF amplifier to 1000 kHz. C_2 adjusts the frequency of the oscillator to 1455 kHz. The LC components of the oscillator are designed to generate a frequency that will be 455 kHz higher than the selected RF signal.

The 1000-kHz signal now passes through the input tuner and is applied to the base of Q_1. This common-emitter amplifier increases the voltage level of the applied signal. C_2 and T_2 are also tuned to pass the signal to the base of Q_2. The oscillator signal is coupled to the base of Q_2 through capacitor C_8. The incoming signal is AM and the oscillator signal is CW. By heterodyning action the collector of Q_2 has four signals. The *fundamental* frequencies are 1000 kHz AM and 1455 kHz CW. The *beat* frequencies are 2455 kHz AM and 455 kHz AM.

The IF amplifier has fixed tuning in its input and output circuits. T_4 is the input IF transformer. It is tuned to be resonant at 455 kHz. This frequency passes into the base of Q_4 with a minimum of opposition. The other three signals encounter a very high impedance. 455 kHz will be amplified by Q_4. The signal is then coupled to the detector through transformer T_5. The detector recovers the AF component and applies it to the AF amplifier system for sound reproduction.

Assume now that the receiver is tuned to select a station at 1340 kHz. C_1, C_2, and C_3 all change to the new signal frequency. The oscillator develops a CW signal of 1795 kHz. The mixer has 1340 kHz and 1795 kHz applied to its input. The collector of Q_2 has these two fundamental frequencies plus 3135 kHz and 455 kHz. The IF amplifier section processes the 455-kHz signal and applies it to the detector. The AF component is then recovered and processed for reproduction.

Suppose now that the receiver is tuned to a station located at 600 kHz. This frequency is amplified and applied to the mixer. The oscillator now sends a CW signal of 1055 kHz to the mixer. The collector of Q_2 has signals of 600, 1055, 1655, and 455 kHz. The IF amplifer again passes only the 455-kHz signal for reproduction.

The basic operation of an AM superheterodyne receiver should be obvious by this time. The oscillator is designed so that it will always be higher in frequency than the incoming station frequency. In effect, its frequency is the incoming station frequency plus the IF. The standard IF for AM receivers is 455 kHz. Through this type of circuit design the IF can be adjusted to a

FIGURE 6-27 AM receiver IF transformers. (Courtesy of J.W. Miller Div./Bell Industries.)

specific frequency. Figure 6-27 shows two IF transformers that are used in transistor AM receivers. The metal can surrounding the transformer shields it from stray electromagnetic fields. This type of transformer is designed for printed-circuit-board installations.

FREQUENCY MODULATION COMMUNICATION

Frequency modulation (FM) has a number of unique advantages over other communication systems. It uses very low audio signal levels for modulation, has excellent frequency reproduction capabilities, and is nearly immune to noise and interference. FM is commonly used in the VHF range of 30 to 300 MHz and extends into the UHF range of 300 to 3000 MHz. The commercial FM band is 88 to 108 MHz. FM is the dominant form of communication for private two-way mobile communication services.

In an FM communication system, intelligence is superimposed on the carrier by variations in frequency. An FM signal does not effectively change in amplitude. The modulating component does, however, cause the carrier to shift above and below its *center frequency.* Each FM station has an assigned center frequency. Receivers are tuned to this frequency for signal reception. When the carrier is unmodulated, it rests at the center frequency. Frequency allocations for FM are made by the FCC.

Examine the FM signal of Fig. 6-28. Note that the modulating component is low-frequency AF and the carrier is RF. In commercial FM, the modulating component could be an audio signal of 20 to 15 kHz. The carrier would be of some RF value between 88 and 108 MHz. As the modulating component changes from 0 to 90°, it causes an increase in the carrier frequency. The carrier rises above the center frequency during this time. Between 90 and 180° of the audio component, the carrier decreases in frequency. At 180° the carrier returns to the center frequency. As the audio signal changes from 180° to 270°, it causes a decrease in carrier frequency. The carrier drops below center frequency during this period. Between 270 and 360° the carrier rises again to the center

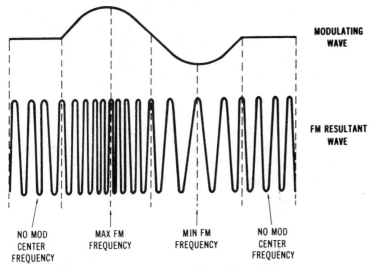

FIGURE 6-28 Frequency modulation waves.

frequency. This shows us that without modulation applied, the carrier rests at the center frequency. With modulation, the carrier shifts above and below the center frequency.

The amplitude of the modulating component is also extremely important in the shifting operation. The larger the modulating signal, the greater the frequency shift. For example, if the center frequency of an FM signal is 100 MHz, a weak audio signal may cause a 1-kHz change in frequency. The carrier would change from 100.001 to 99.999 MHz. A strong audio signal could cause the carrier to change as much as 75 kHz. This action would cause a shift of 100.075 to 9.925 MHz. It is important to note that the amplitude of the carrier does not change. Figure 6-29 shows how the amplitude of the audio component changes the frequency of the carrier.

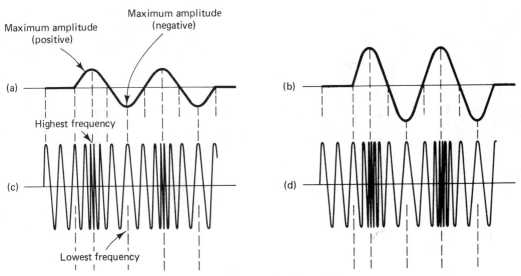

FIGURE 6-29 Influence of amplitude on FM.

The frequency of the modulation component is another important consideration in FM. Modulating frequency determines the rate at which the frequency change takes place. With a 1-kHz modulating component the carrier will swing back and forth 1000 times per second. A modulating component of 10 kHz will cause the carrier to shift 10,000 times per second.

Figure 6-30 shows how the FM carrier responds to two different frequencies. Here t_1 has two cycles of modulation in the same time that t_2 has three cycles; t_2 is of slightly higher frequency than t_1. Note how the carrier changes during the same period of time for each modulating component.

FM Communication System

A block diagram of an FM communication system is shown in Fig. 6-31. The transmitter and receiver respond as independent systems. In commercial FM each station has a transmitter. There can be an infinite number of FM receivers. In two-way mobile communication systems, there is a transmitter and receiver at each location. FM mobile communications systems are classified as *narrow-band FM*. This type of system transmits only voice signals. The carrier of a narrowband FM system deviates by only ± 5 kHz. Commercial FM is considered to be wideband FM. The carrier of a commercial FM system deviates ± 75 kHz at 100% modulation. The operational principles of narrow-band and wideband FM are primarily the same for each system.

The transmitter of an FM system is primarily responsible for signal generation (see Fig. 6-31(a)). Note that it employs an oscillator, an RF signal amplifier, and a power amplifier. The modulating component of an FM system is applied directly to the oscillator. AF changes in the applied signal cause the oscillator to

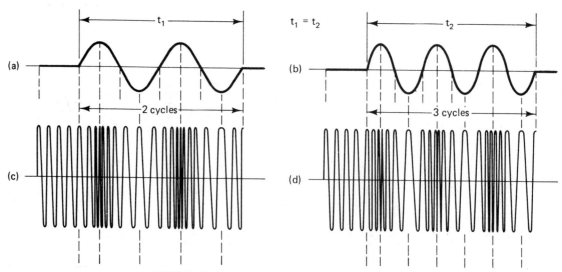

FIGURE 6-30 FM due to frequency changes in modulating component.

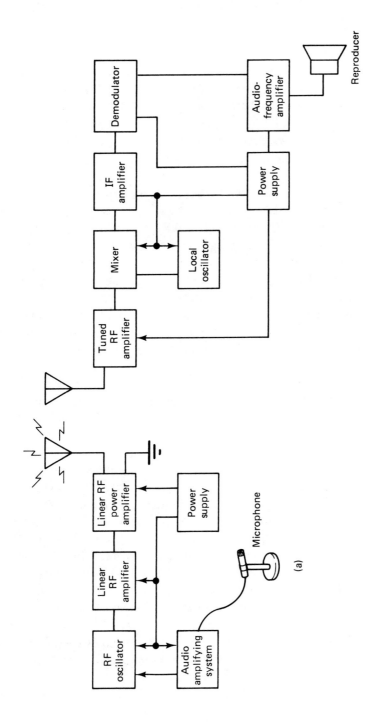

FIGURE 6-31 FM communication system: (a) direct FM transmitter; (b) superheterodyne FM receiver.

377

deviate. Amplifiers following the oscillator are classified as linear devices.

The receiver of an FM system is responsible for signal interception, selection, demodulation, and reproduction (see Fig. 6-31(b)). Most FM receivers are of the superheterodyne type. These receivers are similar in operation to the AM superheterodyne circuit. The selection frequency is much higher (88 to 108 MHz) for commercial FM. A standard IF of 10.7 MHz is used in an FM receiver today. The demodulator of an FM receiver is uniquely different from an AM detector. The FM demodulator is designed to respond to changes in the carrier frequency. The modulating component is recovered from these frequency changes. The remainder of the circuit is the same as the AM superheterodyne.

FM Transmitters.   There is a rather extensive list of different FM transmitters available today. Wireless microphones have an output of approximately 100 mW. Maritime mobile FM communication systems operate with a few watts of power. Two-way land mobile communication employs low-power transmitters. Educational FM transmitters can operate with a power of several hundred thousand watts. Commercial FM transmitters have a similar power rating. In addition to these types are public service radio and military FM. The frequency allocation and power capabilities of an FM transmitter are allocated and regulated by the FCC.

An FM transmitter regardless of its power rating or operational frequency has a number of fundamental parts. The oscillator is responsible for the development of the RF carrier. A large number of FM transmitters employ direct oscillator modulation. This means that the frequency of the oscillator is shifted by direct application of the modulation component. The RF signal is then processed by a number of amplifiers. The power amplifier ultimately feeds the FM carrier into the antenna for radiation.

A simplified direct FM transmitter is shown in Fig. 6-32. A basic Hartley oscillator is used as the RF signal source. With power applied, the oscillator generates the RF carrier. The frequency of the oscillator is determined by the values of C_4 and T_1. Without modulation, the oscillator would generate a constant frequency. This is representative of the unmodulated carrier wave.

Modulation of the oscillator is achieved by voltage changes across a *varicap diode*. The capacitance of this diode changes with the value of the reverse-bias voltage. Bias voltage is used to establish a dc operating level for this diode. A change in audio signal voltage causes the capacitance of the diode to change at an audio rate. Note that the series network of C_1, D_1, and C_2 is connected in parallel with capacitor C_4 of the oscillator tank circuit. A change in audio signal level causes a corresponding change in tank circuit capacitance. This in turn causes the oscillator frequency to change according to the modulating component. The modulating compo-

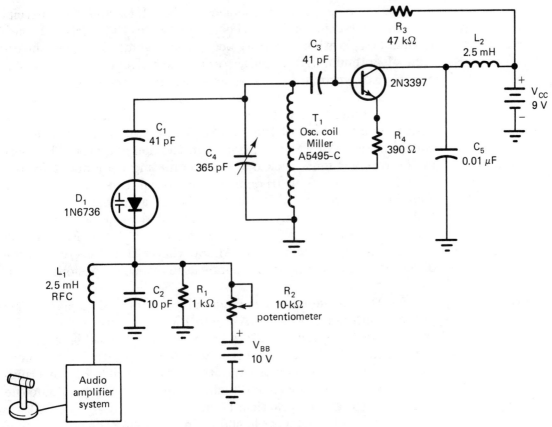

FIGURE 6-32   Direct FM transmitter.

nent is applied directly to the oscillator's frequency-determining components.

Direct FM can be achieved in a number of ways. The varicap diode method is commonly found in new transmitter equipment. This application is only one of a large number of circuit variations of the varicap diode. Circuits of this type are particularly well suited for voice communication in narrow-band FM.

The power output of an FM transmitter is dependent primarily on its application. In some systems the oscillator output can be applied directly to the antenna for radiation. In other systems the power level must be raised to a higher level. The output of the oscillator is applied to linear amplifiers for processing. The amplification function is primarily the same as that achieved by CW and AM systems. We do not repeat the discussion of RF power amplification for FM transmitters. The primary difference in FM power amplification is the resonant frequency of the tuned circuits. FM usually operates in the VHF band. This generally calls for smaller capacitance and inductance values in the resonant circuits.

FM Receivers.   FM signal reception is achieved in basically the same way as AM and CW. An FM superheterodyne circuit is designed to respond to frequencies in the VHF band. Commercial FM signal reception is in the range 88 to 108 MHz. Higher-frequency operation generally necessitates some change in the design of antenna, RF amplifier, and mixer circuits. These differences are due primarily to the increased frequency rather than the FM signal. The RF and IF sections of an FM receiver are somewhat different. They must be capable of passing a 200-kHz-bandwidth signal instead of the 10-kHz AM signal. The most significant difference in FM reception is in the demodulator. This part of the receiver must pick out the modulating component from a signal that changes in frequency. In general, this circuit is more complicated than the AM detector. The AF amplifier section of an FM receiver is generally better than an AM receiver. It must be capable of amplifying frequencies of 30 to 15 kHz. Figure 6-33 shows a block diagram of an FM superheterodyne receiver. Some differences in FM reception are indicated by each functional block.

FM Demodulation.   The demodulator of an FM receiver is the primary difference between AM and FM superheterodyne receivers. A number of demodulators have been used to achieve this receiver function. Today the *ratio detector* seems to be the most widely used. This circuit is an outgrowth of earlier FM discriminators. Solid-state diodes and transformers design have made the ratio detector a very practical circuit.

A basic ratio detector is shown in Fig. 6-34. The operation of this detector is based on the phase relationship of an FM signal that is applied to the input. This signal comes from the IF amplifier section of the receiver (see Fig. 6-33). It is 10.7 MHz and deviates in frequency. Design of the circuit causes two IF signal components to appear in the secondary winding of the transformer. One part of this component is coupled by capacitor C_1 to the center connection of the transformer. This resulting voltage is considered to be V_1. Note the location of V_1 in the circuit and voltage line diagrams.

The second IF signal component is developed inductively by

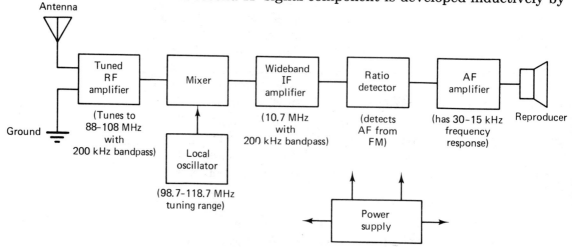

FIGURE 6-33 Block diagram of an FM receiver.

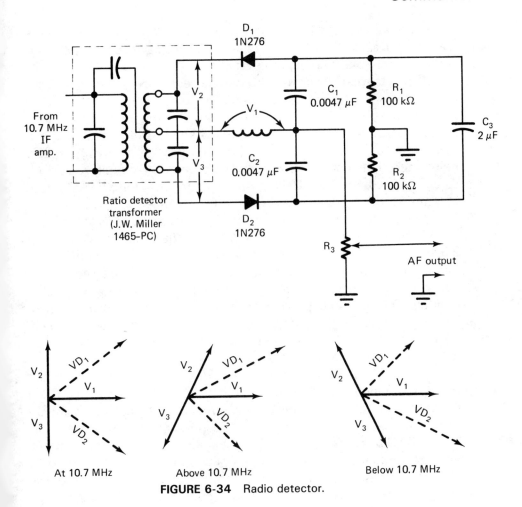

FIGURE 6-34 Radio detector.

transformer action. Design of the secondary winding causes two voltage values to be developed. These voltages are labeled V_2 and V_3 in the circuit and voltage line diagrams. They are of equal amplitude and 180° out of phase with each other. The center-tap connection of the secondary winding is used as the common reference point for these voltage values.

The resulting secondary voltage is based on the combined component voltage values of V_1, V_2, and V_3. V_{D1} and V_{D2} are the resulting secondary voltages. These two voltage values appear across diodes D_1 and D_2. The developed voltage for each diode is based on the phase relationship of the two IF components. Note the location of V_{D1} and V_{D2} in the circuit and the voltage line diagrams.

Operation of the ratio detector is based on voltage developed by the transformer for diodes D_1 and D_2. With no modulation applied to the FM carrier, the transformer has a 10.7-MHz IF applied to the transformer. The developed voltage values are shown by the center-frequency voltage line diagram. This method of voltage display is generally called a *phasor diagram*. A phasor shows the relationship of voltage values (line length) and their phase relationship (direction). At the center frequency note that the resulting diode voltage values (V_{D1} and V_{D2}) are of the same length. This means that each diode receives the same voltage value. D_1 and D_2 conduct an equal amount of current.

When the carrier swings above the center frequency, it causes the IF to swing above its resonant frequency. Note the resulting phasor for this change in frequency. V_{D1} is longer than V_{D2}. This means that D_1 conducts more current than D_2. In the same regard, note how the phasor changes when the frequency swings below resonance. This condition causes D_1 to be less conductive and D_2 to be more conductive. Essentially, this means that the IF signal is translated into different diode voltage values.

Let us now see how the input RF voltage is used to develop an AF signal. The two diodes of the ratio detector are connected in series with the secondary winding and capacitors C_1 and C_2. For one alternation of the input signal the two diodes are reverse biased. No conduction occurs during this alternation (see Fig. 6-35(a)). For the next alternation both diodes are forward biased. The input signal voltage is then rectified (see Fig. 6-35(b)). Essentially, this means that the incoming signal is changed into a pulsating waveform for one alternation.

Figure 6-36 shows how the *ratio detector* responds when the input signal deviates above and below the center frequency. Keep in mind that conduction occurs only for one alternation of the input. Part (a) shows how the circuit responds when the input is at its 10.7-MHz *center frequency*. Each diode has the same input voltage value for this condition of operation. Capacitors C_1 and C_2 charge to equal voltages, as indicated. With respect to ground, the output voltages are -2 V. Note this point on the output voltage waveform in part (d).

Assume now that the input IF signal swings to 10.8 MHz. This condition causes D_1 to receive more voltage than D_2. C_1 charges to -3 V, whereas C_2 charges to -1 V. With respect to ground, the output voltage rises to -1 V. Note this point on the output voltage waveform.

The input IF signal then swings to 10.6 MHz. This condition causes D_2 to receive more voltage than D_1. C_2 charges to -3 V, while C_1 charges to -1 V. With respect to ground, the output voltage drops to -3 V. See this point on the output voltage waveform.

The output of the ratio detector is an AF signal of 2 V p-p. This signal corresponds to the frequency changes placed on the carrier at the transmitter. In effect, we have recovered the AF component from the FM carrier signal. This signal can then be amplified by the AF section for reproduction.

TELEVISION COMMUNICATION

Nearly everyone has had an opportunity to view a television communication system in operation. This communication process plays a very important role in our lives. Very few people spend a

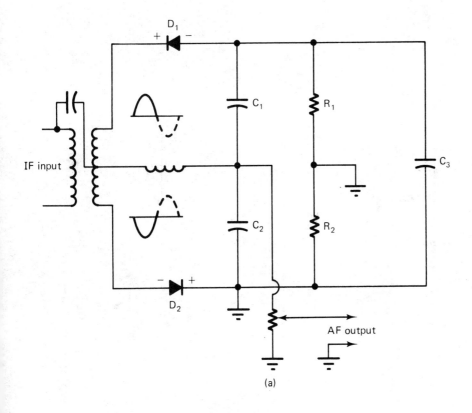

(a)

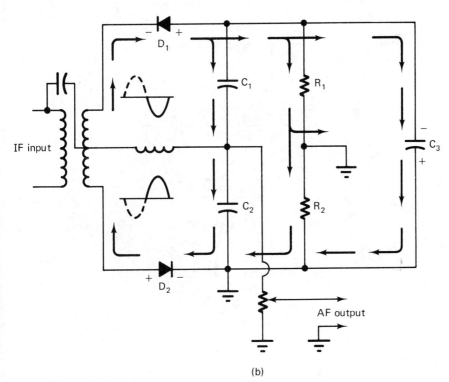

(b)

FIGURE 6-35 Diode conduction of a ratio detector: (a) nonconduction; (b) conduction.

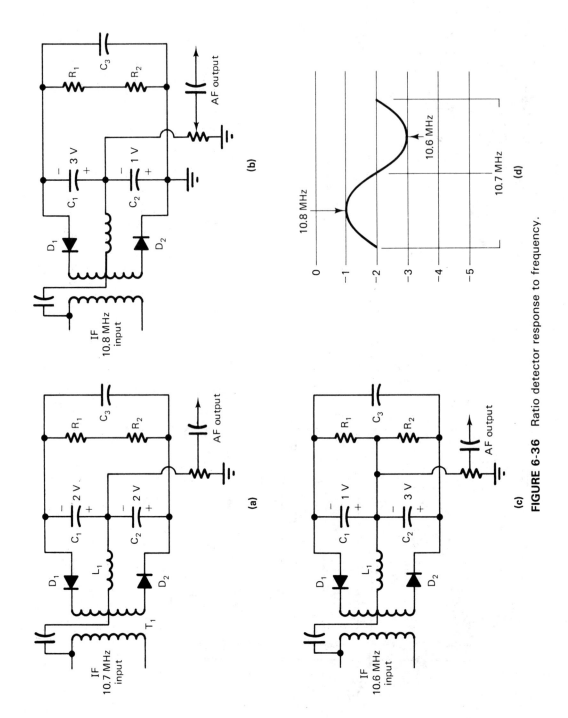

FIGURE 6-36 Ratio detector response to frequency.

384

day without watching television. It is probably the most significant application of electronics today.

The signal of a television system is quite complex. It is made up of a number of unique parts, or components. Basically, the transmitted signal has a *picture carrier* and a *sound carrier*. These two signals are transmitted at the same time from a single antenna. The picture carrier is *amplitude modulated*. The sound carrier is *frequency modulated*. A television receiver intercepts these two signals from the air. They are tuned, amplified, demodulated, and ultimately reproduced. Sound is reproduced by a speaker. Color and picture information are reproduced by a picture tube. Television is primarily a one-way communication system. There are a central transmitting station and an infinite number of receivers. A simplification of the television communication system is shown in Fig. 6-37.

The television camera of our system is basically a transducer. It changes the light energy of a televised scene into electrical signal energy. Light energy falling on a highly sensitive surface varies the conduction of current through a resistive material. The resulting current flow is proportional to the brightness of the scene. An electron beam scans horizontally and vertically across the light-sensitive surface. Vidicon tubes and plumbicons are employed in TV cameras today.

Figure 6-38 show a simplification of the television camera tube. Note that a complete circuit exists between the cathode and power supply. Electrons are emitted from the heated cathode. These are formed into a very thin beam and directed toward the back side of the photoconductive layer. Conduction through the layer is based on the intensity of light from the scene being televised. A bright or intense light area becomes low resistant. Dark areas have a higher resistance. As the electron beam scans across the back of the photoconductive layer it sees different resistance values. Conduction through the layer is based on these resistances. A discrete area with low resistance causes a large current through the layer. Dark areas cause less current flow. Current flow is directly related to the light intensity of the televised scene. Output current flow appears across the load resistor (R_L). Voltage developed across R_L is amplified and ultimately used to modulate the picture carrier. In practice, the developed camera tube voltage is called a *video signal*. *Vide* means something to see. A camera tube sees things electronically.

The scene being televised by a camera tube must be broken into very small parts called *picture elements*. For this to take place, it is necessary to scan the light-sensitive surface of a camera tube with a stream of electrons. This process is very similar to reading a printed page. Letters, words, and sentences are placed on the page by printing. We do not determine what is on a printed page at one instant. Our eyes must scan the page one line at a time, starting at the upper left-hand corner. They move left to right, drop down one line, quickly return to the left, and then scan right

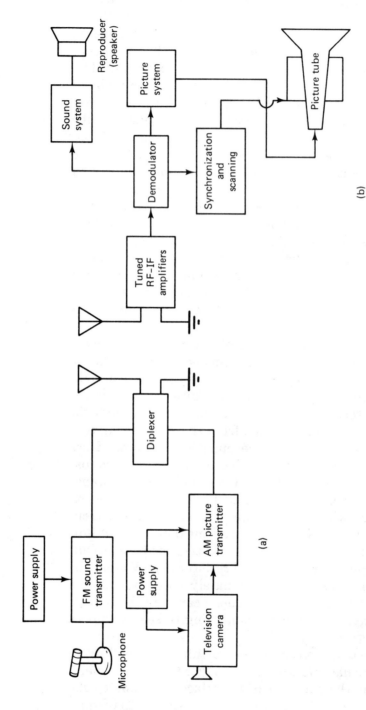

FIGURE 6-37 TV communication system: (a) transmitter; (b) receiver.

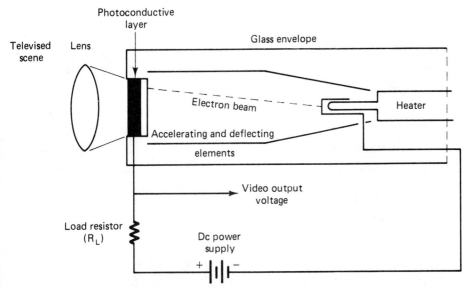

FIGURE 6-38 Simplification of a TV camera tube.

again for the next line. The process continues until all lines are scanned.

In a similar way the electron beam of a camera tube scans the back surface of the photoconductive layer. The electron beam is deflected horizontally and vertically by an electromagnetic field. A coil fixture known as a *deflection yoke* is placed around the neck of the camera tube. This coil deflects the electron beam. Current flow in the deflection yoke is varied so that the field rises to a peak value and then drops to its starting value. Figure 6-39 shows the deflection yoke current and the resulting electron beam scanning action. Notice that each line has a *trace* and *retrace* time. During the trace period, the line is scanned from left to right. This takes a rather large portion of the complete sawtooth wave. Retrace occurs when the beam returns from right to left. Notice that this takes only a small portion of the total waveform. The same condition applies to the vertical sweep waveform.

Figure 6-39 shows another rather unique difference in the scanning lines. During the trace time, the scanning line is solid. This indicates that the electron beam is conducting during this time. It also shows a broken line during the retrace period. This indicates that the electron beam is nonconductive during this period. In effect, conduction occurs during the trace time and no conduction occurs during retrace. This same condition applied to the vertical trace and retrace time.

The *scanning* operation of a camera tube requires two complete sets of deflection coils, one set for horizontal deflection and one for vertical. The current needed to produce deflection comes from two sawtooth oscillators. A vertical blocking oscillator and a horizontal multivibrator could be used for this operation. In U.S. television, the horizontal sweep frequency is 15,750 Hz and the vertical frequency is 60 Hz.

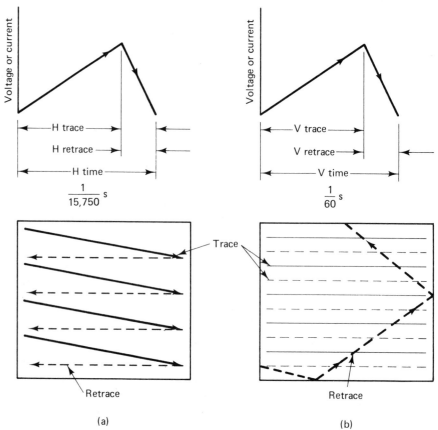

FIGURE 6-39   Scanning and sweep signals; (a) horizontal; (b) vertical.

To produce a moving television scene, there must be a least 30 complete pictures produced per second. In a TV system a complete picture is called a *frame*. The U.S. television system has 525 horizontal lines in a frame. These lines are not scanned consecutively. Rather, they are divided into two *fields*. One field contains 262.5 odd-numbered lines, lines 1, 3, 5, 7, 9, 11, ... The other field has 262.5 even-numbered lines, lines 2, 4, 6, 8, 10, ... A complete picture or frame has one *odd-lined field* and *one even-lined field*.

Picture production in a TV system employs *interlace scanning*. To produce one frame, the odd-numbered-line field is scanned first. After scanning all the odd-numbered lines, the electron beam is deflected from the bottom position to the top. The even-numbered-line field is then scanned. The odd-line field starts with a complete line and ends with a half line. The even-line field starts with a half line and ends with a complete line.

The picture repetition rate of U.S. television is 30 *frames per second*. Since two fields are needed to produce one frame, the vertical frequency is 2 × 30 Hz, or 60 Hz. This particular frequency was chosen to coincide with the ac power line frequency. In some foreign countries the vertical frequency is 50 Hz.

In television signal production, the vertical and horizontal sweep circuits must be properly synchronized in order to produce a picture. The signal sent out by the transmitter must contain *syn-chronization*, or sync, information. This signal is used to keep the oscillator of the receiver in step with the correct signal frequency. Separate generators are used to develop the sync signal, which is added to the video signal developed by the camera.

Picture Signal

The picture signal of a TV transmitter contains a number of important parts. Each part of the signal plays a specific role in the operation of the system. The *video signal,* for example, is developed by the camera tube. It represents instantaneous variations in scene brightness. Figure 6-40(a) shows the video signal for one horizontal line.

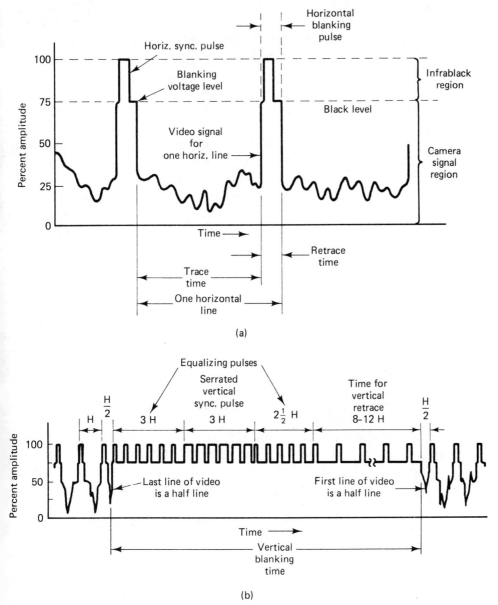

FIGURE 6-40   Composite TV picture signal.

The video signal of a television system has *negative picture phase*. This means that the part of the signal with highest amplitude corresponds to the darkest picture area. Bright picture areas have the lowest amplitude level. Signal levels that are 75% of the total amplitude range are considered to be in the black region. All light disappears in this region. Signal-level amplitude percentages are shown in Fig. 6-40(b).

In addition to the video information, the picture signal must also provide some way of cutting off the electron beam at certain times. When scanning occurs, the electron beam is driven to cut off during the retrace period. Horizontal retrace occurs at the end of one line and vertical retrace occurs at the end of each field. A rectangular pulse of sufficient amplitude is needed to reach *cutoff*. This condition permits the electron beam to retrace without producing unwanted lines. This part of the signal is called *blanking*. A composite signal has both horizontal and vertical blanking pulses. Figure 6-40(a) shows the location of a *horizontal blanking* pulse. *Vertical blanking* occurs after 262.5 horizontal lines. Figure 6-40(b) shows the vertical blanking time.

A *composite TV picture signal* also has vertical and horizontal synchronization pulses. These pulses ride on the top of the blanking pulses. *Horizontal sync pulses* are shown in both parts of Fig. 6-40. The serrated pulses of part (b) provide continuous horizontal sync during the vertical retrace time. The *vertical sync pulse* is made up of six rectangular pulses near the center of the vertical blanking time. The width of a vertical sync pulse is much greater than that of the horizontal sync pulse. All these pulses, plus blanking and the video signal, are described as a composite picture signal.

Television Transmitter

A television transmitter is divided into two separate sections, or divisions, that feed outputs into a common antenna. The *video section* is responsible for the picture part of the signal. A crystal oscillator is used for carrier wave generation. As a general rule, the frequency is multiplied to bring it up to an allocated channel in the VHF or UHF band. An intermediate power amplifier and a final power amplifier follow the last multiplier. The modulating component is a composite picture signal. It contains video, blanking pulses, sync, and equalizing pulses. The composite signal is amplified and ultimately applied to the final power amplifier. The final is amplitude modulated by the composite picture signal. This section of the transmitter is essentially the same as that of a commercial AM station. There is an obvious difference in frequency and power output. Figure 6-41 shows a block diagram of a black-and-white TV transmitter.

A unique difference in TV and commercial AM transmitter circuitry appears after the final power amplifier. The TV output signal is applied to a *vestigial sideband* filter. This filter is de-

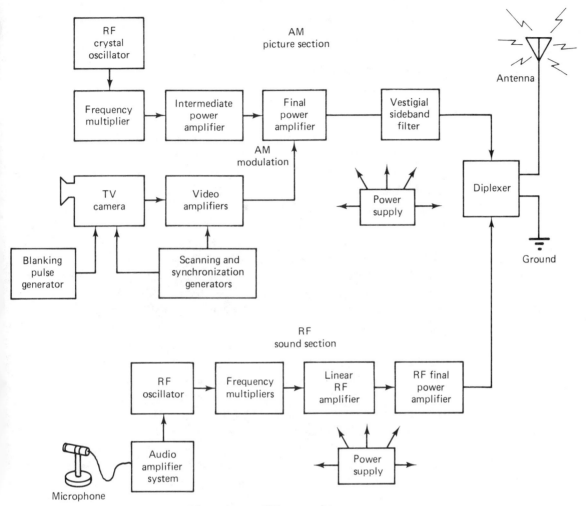

FIGURE 6-41 Monochrome TV transmitter.

signed to remove all sideband frequencies 1.25 MHz below the carrier frequency. The entire upper sideband of the signal is transmitted. A large portion of the lower sideband is suppressed. This is done purposely to reduce the frequency occupied by a channel. With the lower sideband suppressed, the bandwidth of a TV channel is 6 MHz.

The vestigial sideband signal is then applied to a *diplexer*. The diplexer is a filter circuit that isolates the picture and sound carriers. Essentially, the sound carrier will not pass into the picture section and the picture carrier will not pass into the sound section. This prevents undesirable interaction between the two carrier signals.

The *sound section* of a TV system is primarily an FM transmitter. It is very similar to a commercial FM transmitter. The center frequency of the FM carrier is always 4.5 MHz above the picture carrier. Carrier deviation is ±25 kHz in a TV system. Modulation is normally applied to the oscillator of the FM sound system. The remainder of the sound section is similar to that of a commercial FM transmitter. See the FM sound section of the transmitter in Fig. 6-41.

Television Receiver

A television receiver is designed to intercept the electromagnetic waves sent out by the transmitter and use them to develop sound and a picture. The received signals are in the VHF or UHF band. The FCC has allocated a 6-MHz bandwidth for each TV channel. Channels 2 through 13 are the *VHF band.* These frequencies are from 54 to 216 MHz. Channels 14 to 83 are in the *UHF band.* This ranges from 470 to 890 MHz. All channels in the immediate area induce a signal into the antenna. The desired station is selected by altering a tuning circuit. This LC circuit passes only the selected channel and rejects the others.

A functional block diagram of a black-and-white television receiver is shown in Fig. 6-42. Practically all this circuitry has been used in other communication systems. The front end of a TV receiver for example, is a superheterodyne circuit. It has a tuned RF amplifier, a mixer, and an oscillator. This section of the receiver is called the *tuner.* It is housed in a shielded metal container to reduce interference. The output of the tuner is an IF signal. A standard IF for TV is 41.25 MHz for the sound and 45.75 MHz for the picture. The IF must pass both sound and picture carriers. This necessitates a 6-MHz bandpass for the IF amplifiers.

The *demodulation* function of a TV receiver is achieved by a diode. This circuit is primarily an AM detector. The output of the detector has all the picture information placed on the carrier at the transmitter. After demodulation, the picture carrier is discarded. Three different kinds of signal information appear at the output of the detector. It recovers the video signal and sync signals for immediate use. The sound carrier passes into the sound IF section.

The *video signal* recovered by the detector is processed by the video amplifier. The output of this section is then used to control the *brightness* of the electron beam of the picture tube. A dark spot detected by the TV camera will cause a corresponding reduction in picture tube brightness. A bright spot or an intense picture element causes the picture tube to conduct very heavily. This causes a corresponding bright spot to appear on the picture tube face. The video signal developed by the camera of the transmitter is accurately reproduced on the picture tube screen.

The *sync signal* of the video detector is used to synchronize the vertical and horizontal sweep oscillators. These oscillators develop the sawtooth waves that are used to deflect the electron beam. The electron beam must be in step with the transmitted signal in order for the picture to be usable. The transmitted sync signal is recovered and used to trigger the two receiver sweep oscillators.

An *AM video detector* does not respond effectively to fre-

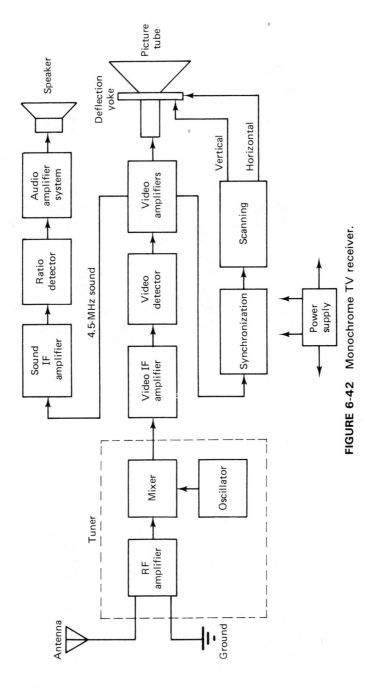

FIGURE 6-42 Monochrome TV receiver.

393

quency changes in the IF sound carrier. It does, however, heterodyne the two signals together. The difference between the 45.75- and 41.25-MHz IF signals is 4.5 MHz. This signal takes on the FM modulation characteristic of the sound carrier. 4.5 MHz is called the sound IF. The sound IF deviates ± 25 kHz above and below 4.5 MHz.

A ratio detector can be used to demodulate the sound IF signal. The recovered audio component is then processed by an AF amplifier. It is ultimately used to drive a loudspeaker for sound reproduction. The FM sound section of a TV receiver is very similar to that of a standard FM broadcast receiver.

The picture tube of a TV receiver is responsible for changing an electrical signal into light energy. A picture tube is also called a *cathode-ray tube* (CRT). The intensity of an electron beam changes as it scans across the face of the tube. The inside face of the tube is coated with phosphor. When the electron beam strikes tiny grains of phosphor, it produces light. A combination of different light and dark phosphor grains causes a picture to appear on the inside of the face area. The resulting picture can be observed by viewing the front of the face area.

Figure 6-43 shows a simplification of the CRT of a TV receiver. The tube is divided into three parts. The *gun area* is responsible for electron-beam production. The *coil fixture* attached to the neck of the tube deflects the electron beam. The *viewing area* changes electrical energy into light energy.

A CRT is a *thermionic emission* device. Heat applied to the cathode causes it to emit electrons. These electrons form into a

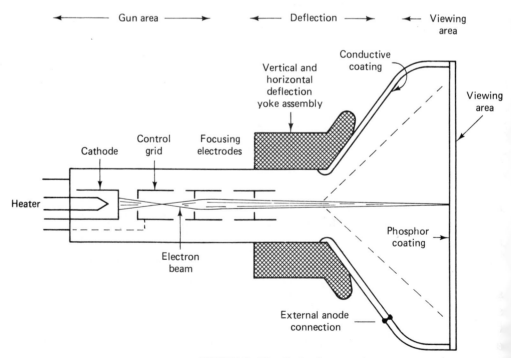

FIGURE 6-43 Cathode-ray tube.

fine beam and move toward the face of the tube. A high positive charge is placed on a conductive coating on the inside of the glass housing. This coating serves as the anode. The electron beam is attracted to this area of the tube because of its positive charge. Voltage is supplied to the coating by an external anode connection terminal. The anode voltage of a black and white CRT is normally 15 kV.

After leaving the cathode, electrons pass through the control grid. Voltage applied to the grid controls the flow of electrons. A strong negative voltage causes the electron beam to be cut off. Varying values of grid voltage cause the screen to be illuminated. The video signal developed by the receiver can be used to control the intensity level of the electron beam. Focusing of the electron beam is achieved by the next two electrodes. They shape the electron beam into a very fine trace. These electrodes are generally called *focusing anodes*.

Electromagnetic deflection of the electron beam is achieved by the yoke assembly. This coil fixture produces an electromagnetic field, and the electron beam causes deflection of the beam. Both vertical and horizontal deflection are needed to produce scanning.

Color Television

The transmission of a color television picture complicates the system to some extent. The principles involved in transmission are primarily the same as those of *monochrome,* or black-and-white television. One of the problems of color TV is that it must be *compatible* with monochrome. This means that programs designed for color reception must also be received on a monochrome receiver. The transmitted signal must therefore contain color information as well as the monochrome signal. All of this must fit into the 6-MHz channel allocation.

The picture portion of the color signal is basically the same as the monochrome signal. The video signal is, however, made up of three separate color signals. Each color signal is produced by an independent camera tube. One camera tube is used to develop a signal voltage that corresponds to the *red* content of the scene being televised. *Blue* and *green* camera tubes are used in a similar arrangement to produce the other two color signals. Red, green, and blue are considered to be the primary colors of the video signal. White is a mixture of all three colors. Black is the absence of all three primary colors.

Color Cameras.   Figure 6-44 shows a simplified color television camera. The scene being televised is focused by a lens onto a special mirror. This mirror reflects one-third of the light and passes two-thirds of the light. The reflected image goes through a filter and is applied to the *blue camera* tube. Two-thirds of the image passing through the mirror is applied to a *divider mirror.* One-third

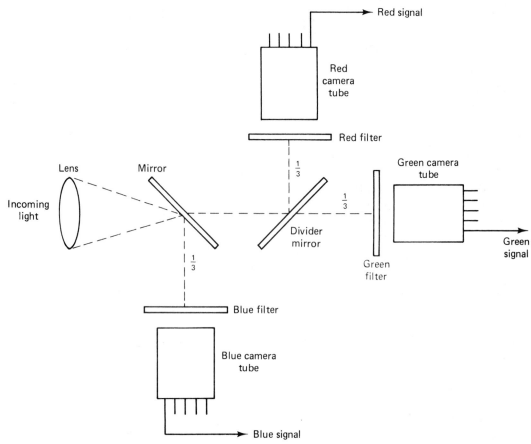

FIGURE 6-44 Simplified color TV camera.

of it is applied to the *red camera* and one-third to the *green camera*. Each camera tube provides an output signal that is proportional to the light level of the primary color. The brightness or *luminance* signal is a mixture of the three primary colors. These proportions of the color signal are 59% green, 30% red, and 11% blue. The luminance signal is generally called a *Y signal*. A monochrome receiver responds only to the Y signal and produces a standard black-and-white picture.

The human eye needs two stimuli to perceive color. One of these is called *hue*. Hue refers to a specific color. Green leaves have a green hue and a red cap has a red hue. The other consideration is called *saturation*. This refers to the amount, or level, of color present. A vivid color is highly saturated. Light or weak colors are diluted with white light. A scene being televised by a color camera has hue, saturation, and brightness.

Color Transmitters. Figure 6-45 shows a block diagram of a color transmitter. The color section adds significantly to the transmitter. The remainder of the diagram has been reduced to a few blocks. This part of the transmitter is essentially the same for either a color or a black-and-white system.

In a color transmitter three separate color signals are devel-

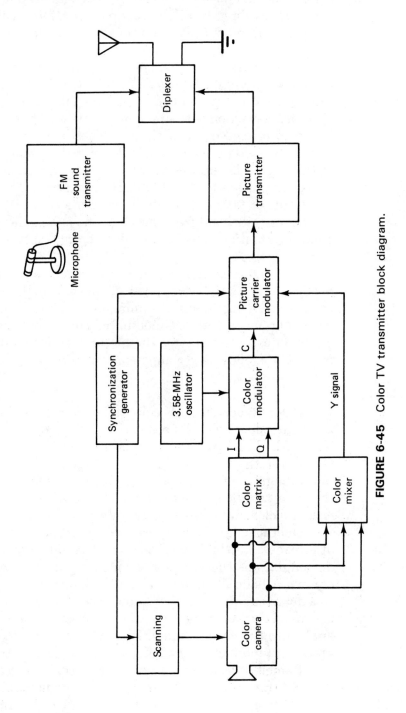

FIGURE 6-45 Color TV transmitter block diagram.

397

oped by the TV camera. These colors are applied to a *matrix* and a *mixer* circuit. The mixer develops the luminance signal. Green, red, and blue are mixed in correct proportions. Luminance is the equivalent of a black-and-white signal. The output of the mixer (Y signal) goes directly to the carrier modulator. This signal is the same as the modulating component of a black-and-white transmitter.

The matrix is a rather specialized mixing circuit. It combines the three color signals into an I and a Q signal. These two signals can be used to represent any color developed by the system. The *Q signal* corresponds to green or purple information. The *I signal* refers to orange or cyan signals. These two signals are used to amplitude modulate a 3.58-MHz subcarrier signal. The carrier part of this signal is suppressed to prevent interference. The I signal is kept in phase with the 3.58-MHz subcarrier. The Q signal is one quadrant, or 90°, out of phase with the *subcarrier.* The I and Q signals are sideband frequencies of the subcarrier. Only this sideband information is used to modulate the transmitter picture carrier.

A synchronization signal is also generated by the color transmitter. This signal is applied to both the modulator and the camera-scanning circuit. Scanning of the three camera tubes must be synchronized with the modulated carrier output. The sync signal is an essential part of the picture carrier modulation component. The modulating component contains I, Q, Y, and sync signals.

Color Receivers. A color television receiver picks up the transmitted signal and uses it to develop sound and a picture. It has all the basic parts of monochrome receiver plus those needed to recover the original three color signals. Figure 6-46 shows a block diagram of the color TV receiver. The shaded blocks denote the color section of the receiver. This is where the primary difference in color and monochrome TV receivers occurs.

Operation of the color receiver is primarily the same as that of a black-and-white receiver up to the video demodulator. After demodulation, the composite video signal is then divided into *chroma (C)* and *Y signals.* The Y, or luminance, signal is coupled to a delay circuit. This slows down the Y signal. It thus arrives at the matrix at the same time as the chroma signal. The C signal is applied to the chroma amplifier. This signal must pass through a great deal more circuitry before reaching the matrix.

Remember that the chroma signal contains I and Q color and a 3.58-MHz suppressed carrier. To demodulate this signal, the 3.58-MHz signal must be reinserted. Notice that the 3.58-MHz oscillator is connected to the I and Q demodulators. This is where the carrier reinsertion function takes place.

The *I demodulator* receives a 3.58-MHz signal directly from the oscillator. This is where the in-phase signal is derived. The 3.58-MHz signal fed into the Q demodulator is shifted 90°. This is where

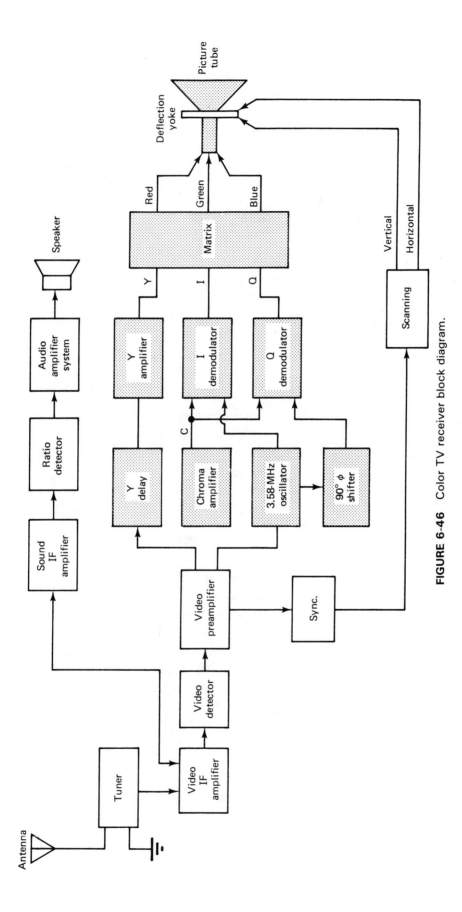

FIGURE 6-46 Color TV receiver block diagram.

399

the quadrature signal is derived. I and Q color signals appear at the output of the demodulator.

The matrix of the receiver has three signals applied to its input. The Y signal contains the luminance information. The I and Q signals are the demodulated color signals. The matrix combines these three signals in proper proportions. Its output comprises the original red, green, and blue color signals. These signals are applied to the R, G, and B guns of the picture tube. The electron beam of each gun strikes closely spaced red, green, and blue phosphor spots on the face of the tube. These spots glow in differing amounts according to the signal level. Any color can be reproduced on the face of the tube by the three primary colors.

SELF-EXAMINATION

1. Which radiates best from an antenna, RF or AF? _____

2. Three common communications systems are _____, _____, and _____.

3. _____ waves are reflected by the ionosphere.

4. _____ code is used for CW transmission.

5. The _____ of a communications system intercepts the RF signal and recovers the information.

6. The ability of a receiver to pick out a desired RF signal is called _____ _____.

7. The process of mixing two AC signals in a receiver is called _____.

8. A heterodyne detector circuit is called a _____.

9. The mixture of AF and an RF carrier wave is called _____.

10. The AM broadcast band is _____ to _____.

11. Three levels of modulation are _____, _____, and _____.

12. The _____ of a communications system is responsible for signal generation.

13. AM transmitter parts include the _____, _____, _____, and _____.

14. AM receiver parts include the _____, _____, _____, and _____.

15. The detector of a crystal AM radio is a _____.

16. Most AM receivers today are the _____ type.

17. Superheterodyne receivers use special tuned RF amplifiers called _____ amplifiers.

18. The FM band is _____ to _____.

19. With modulation, an FM carrier shifts above and below its _____ _____.

20. A standard IF of _____ MHz is used in FM receivers.

21. A _____ detector may be used in FM receivers.

22. The U.S. television system has _____ horizontal lines in a frame.

23. Television signals are in _____ and _____ bands.

24. A TV picture tube is called a _____ tube.

25. Color TVs have _____, _____, and _____ color signals.

ANSWERS

1. RF
2. AM, FM, CW
3. Sky
4. Morse
5. Receiver
6. Signal selection
7. Heterodyning
8. Mixer
9. Modulation
10. 535 kHz; 1620 kHz
11. Under; over; 100%
12. Transmitter
13. Antenna; oscillator; AF component; power supply
14. Antenna; RF tuner, demodulator; power supply
15. Diode
16. Superheterodyne
17. IF
18. 88 MHz; 108 MHz
19. Center frequency
20. 10.7
21. Ratio
22. 525
23. VHF; UHF
24. Cathode-ray
25. Red; green; blue

EXPERIMENT 6-1
CW TRANSMITTERS

The simplest type of radio frequency (RF) transmitter in operation today is a single-stage unit designed to radiate a continuous wave (CW) signal. The essential parts include an RF oscillator, a code key to turn it on and off, an antenna, and a power supply. By opening and closing the key, international Morse code signals can be produced and radiated from the antenna. CW transmitters are commonly used today in radio telegraph communication systems and amateur radio.

OBJECTIVES

1. To construct a CW transmitter with an RF oscillator circuit.
2. To key information into the transmitter circuit.
3. To observe the output of the transmitter on an oscilloscope.

EQUIPMENT

Variable power supply, 0–15 V
Oscilloscope
Variable capacitor: 365/125 pF, two-ganged
Capacitors: 0.01 μF, 47 pF, 100 pF
NPN transistor

PROCEDURE

1. Connect the RF oscillator circuit of Fig. 6-1A.
2. Turn on the variable dc power supply and adjust it to 10V before turning on the keying switch.
3. Prepare the oscilloscope for operation with the horizontal sweep selector at 1-μ s/cm or 1-MHz sweep range. Set the trigger or sync selector for internal sync.
4. Connect the vertical probe to test point A. Connect the common lead to ground. If the circuit is operating properly, an RF signal of approximately 600 kHz should be displayed on the oscilloscope.
5. Turn off the keying switch while observing the oscilloscope. Turn it on and off several times while observing

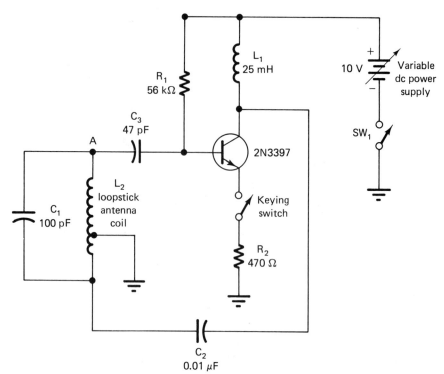

FIGURE 6-1A   RF oscillator circuit.

the displayed signal. How does keying alter the oscillator? _____

_____

6. Using the international Morse code, send several letters of code with the switch while observing the signal on the oscilloscope.

7. With the switch open (off), remove the 100-pF capacitor from the circuit. In its place, connect the large plated side (365-pF section) of the two-ganged variable capacitor.

8. Close the switch again and adjust the variable capacitor to produce the highest oscillator frequency. This occurs when the rotary plates are fully extended.

9. Test the output of the circuit again while observing the oscilloscope. How does the oscilloscope show this output to be of a higher frequency than the original circuit?

_____

_____

10. If a communications receiver with CW reception capabilities is available, place it near the oscillator. Remove the oscilloscope from the output terminal and connect a 24-in. length of wire to the output terminal to serve as an antenna.

11. Key a coded signal into the oscillator and tune in the

signal on the receiver. The receiver must be in CW position to receive this signal.

12. If an AM radio receiver is available, turn it on and place it near the CW transmitter circuit. Tune the receiver to a strong local station.

13. Turn on the oscillator and adjust the variable capacitor until a whistling sound is heard on the receiver.

14. Try keying a CW signal into the oscillator by opening and closing the switch. Reception of the CW signal in this case is produced by heterodyning or beating an AM station carrier with a CW carrier.

ANALYSIS

1. How is coded information transmitted by a CW transmitter? _____

   _____

2. What are the fundamental stages or parts of a CW transmitter? _____

   _____

3. What type of signal is radiated from a CW transmitter?

   _____

EXPERIMENT 6-2
AM PRINCIPLES

Amplitude modulation (AM) occurs when the RF carrier of a transmitter changes in amplitude with an audio signal. Through this modulation process, it is possible to transmit voice and music signals through space. The sound signal or AF information is commonly called the modulation component of the transmitted signal. When modulation takes place, the RF signal changes in amplitude with the modulating signal.

OBJECTIVES

1. To construct a diode modulator circuit and an AM transmitter circuit.
2. To observe both AF and RF signals and the resulting modulated waves on an oscilloscope.
3. To alter the level of modulation of an AM signal.

EQUIPMENT

Variable power supply, 0–15 V
Oscilloscope
Signal generator
NPN transistors (2)
Oscillator coil
Inductor: 25 mH
Variable capacitor: 365/125 pF, two-ganged
Capacitors: 47 pF, 100 pF, 0.01 μF, 0.68 μF, 33 μF
Diode
Crystal microphone
Potentiometer: 100 kΩ
Resistors: 470Ω, 1 kΩ, 56 kΩ, 100 kΩ
SPST switch

PROCEDURE

Part A

1. Construct the amplitude modulation circuit of Fig. 6-2A. The RF source of the circuit is a Hartley oscillator. Construct the *RF oscillator* part of the circuit first.

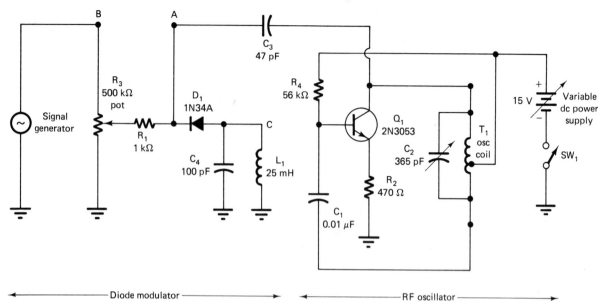

FIGURE 6-2A AM circuit.

2. Turn on the power supply and adjust it to 15 V dc before turning on the circuit switch.

3. Prepare the oscilloscope for operation. Connect the ground lead to the negative and the vertical probe to the collector of Q_1.

4. Set the sweep selector of 1 s/cm or a frequency of 1 MHz. A sine wave should be observed on the oscilloscope if the oscillator is functioning properly.

5. Adjust the variable capacitor C_2 through its range. The highest frequency occurs when the capacitor is

_____.

6. Adjust C_2 to the center of its adjustment range. The peak-to-peak amplitude of the RF signal observed by the oscilloscope at the 47-pF capacitor is _____ V p-p. Turn off the circuit switch and connect the *diode modulator* part of the circuit.

7. With the circuit switch still in the *off* position, turn on the signal generator. Connect the vertical probe of the oscilloscope to test point B. Adjust the signal generator to an audio frequency 1 kHz. Adjust its output control to a level that is equal to that of the RF oscillator of step 6.

8. Change the oscilloscope horizontal sweep frequency to 500 Hz, or 0.5 ms/cm. Connect the vertical probe to test point A. The amplitude of the AF signal is variable at this test point according to the setting of R_3.

9. Turn on the circuit switch so that both RF and AF signals are applied to the diode.

100 percent
modulated _____
signal

Unmodulated _____
signal

50 percent
modulated _____
signal

FIGURE 6-2B   Waveform diagrams.

10. Move the vertical probe of the oscilloscope to test point *C*.

11. Adjust R_3 while observing the modulated signal. The modulation level is adjusted by R_3. At one extreme position, no modulation occurs. At the other extreme, the signal is overmodulated. This control also changes the amplitude of modulated output signal to some extent.

12. Make a sketch of a 100% modulated, unmodulated, and a 50% modulated signal appearing at test point *C* in the spaces provided in Fig. 6-2B.

13. Turn off circuit switch, power supply, oscilloscope, and signal generator. The oscillator of this part of the experiment is used in Part B.

ANALYSIS

1. What signal of the modulator circuit represents the carrier wave? _____

2. What signal represents the modulating wave? _____

3. If a 50-V peak-to-peak RF carrier changes in amplitude from 75 V p-p to 25 V p-p when modulated, what is the percent of modulation? _____

4. What causes an AM signal to be overmodulated?
_____
_____

Part B

1. Connect the AM transmitter of Fig. 6-2C.

2. Turn on the power supply and adjust it to 15 V dc before turning on SW_1.

3. Connect a 24-in. length of wire to the antenna terminal of the transmitter.

4. Prepare the oscilloscope for operation. Connect the ground lead to the negative and the vertical probe to the antenna of the AM circuit. Make a sketch of the ob-

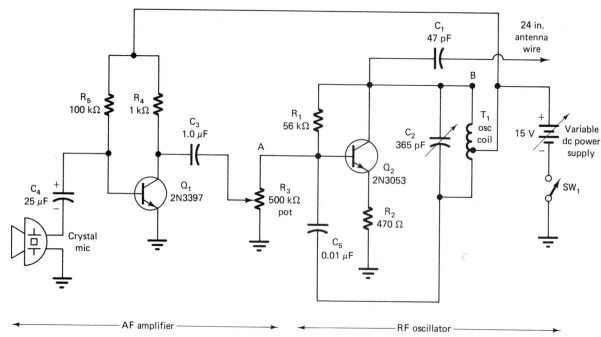

FIGURE 6-2C AM transmitter circuit.

served waveform of the unmodulated output in the space of Fig. 6-2D.

5. Connect the vertical probe of the oscilloscope to test point *A*. Adjust the sensitivity of the microphone to a position that will produce a signal when you are speaking into it. This represents a modulated input signal.

6. Make a sketch of a modulated input signal in Fig. 6-2D. As you will note, this signal is very irregular.

7. Move the vertical probe of the oscilloscope to test point *B*. Speak into the microphone while observing the output. How does modulation influence the output of the transmitter? _____

8. In Fig. 6-2D, make a sketch of the waveform appearing at the output when modulation occurs. Note that this output is very irregular.

9. If an AM radio receiver is available, place it near the antenna wire. Speak into the microphone and adjust the

Unmodulated output _____   Modulated input signal _____

Modulated output signal _____

FIGURE 6-2D Waveform diagrams.

receiver to pick up the transmitted signal. If a local station interferes with the signal, adjust the variable capacitor of the circuit to a different frequency.

10. The receiver frequency setting is approximately _____ kHz.

11. Move the receiver to a location about 10 ft from the transmitter and test its operation. Try orienting the antenna wire vertically, then horizontally. Reception of the signal is best with which antenna polarity?

_____

ANALYSIS

1. If the modulated signal is stronger than the RF signal of the transmitter, what occurs? _____

_____

2. How does a crystal microphone produce electrical energy from sound waves? _____

_____

3. Make a sketch in the space provided in Fig. 6-2E, showing an unmodulated RF wave, 50% modulated wave, 100% modulated wave, and an overmodulated RF wave.

Unmodulated _____   50 percent modulated _____
RF wave   RF wave

100 percent modulated _____   Overmodulated _____
RF wave   RF wave

FIGURE 6-2E Waveform diagrams.

EXPERIMENT 6-3
AM RECEIVERS

In an AM communication system, information is applied to the input of the transmitter and radiated into space. The RF waves carry the information to the receiver through the air. The receiver part of the system then picks up the wave, selects the correct frequency, rectifies the RF signal and reproduces the AF signal through the earphones. Either voice or music signals can be transmitted through this type of system.

OBJECTIVES

1. To construct an AM receiver that demonstrates the principles of signal pickup, tuning, demodulation, and reproduction.
2. To study the primary functions of an AM receiver and observe its operation with an oscilloscope.
3. To tune in an AM signal and see how it responds to different modulation levels.

EQUIPMENT

Signal generator
Transistor loopstick antenna coil
Variable capacitor (365/125 pF)
Diode
Resistor: 22 kΩ
Earphone
Capacitor: 0.1 μF

PROCEDURE

Part A: AM Receiver Function

1. Connect the diode AM receiver circuit of Fig. 6-3A.
2. Turn on the signal generator and adjust it to produce an AM signal of 100 kHz. Connect the generator output of T_1 as indicated.
3. Prepare the oscilloscope for operation with the horizontal sweep frequency set to 100 Hz or 10 ms/cm. Connect

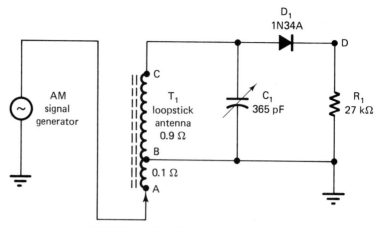

FIGURE 6-3A AM receiver circuit.

the ground lead to negative and the vertical probe to point A.

4. Adjust the modulation control of the signal generator to produce 100% or its highest modulation level.

5. Make a sketch of the observed input waveform in the space of Fig. 6-3B.

6. Move the vertical probe of the oscilloscope to test point C with the ground probe remaining at point B. Adjust capacitor C_1 to produce a maximum signal indication on the oscilloscope. What does this indicate about the tuning circuit? _____

_____

7. Move the vertical probe of oscilloscope to test point D with the ground probe at point B. Make a sketch of observed waveform in the space of Fig. 6-3B. How does this wave compare with observed wave in step 5?

_____

8. Connect a 0.22-μF capacitor (C_2) across test points D–B of the demodulator circuit. Make a sketch of the waveform displayed on the oscilloscope in the space of Fig. 6-3B. How does this display compare with step 7?

_____

Modulated
input
waveform at _____
points A–B

Demodulated
waveform at
points D–B _____
without C_2

Demodulated
waveform at
points D–B _____
with C_2

FIGURE 6-3B Waveform diagrams.

9. Connect a set of earphones across test point *D-B* with the capacitor remaining in the circuit. What do you hear?

   _____

10. Reduce the modulation level of the RF signal generator. What effect does this have on the sound output?

   _____

11. Turn off the modulation so that only RF is supplied to the demodulator circuit. What effect does this have on sound output? _____

   _____

12. Disconnect the signal generator from the circuit. Connect point *A* to an outside antenna and point *B* to a good ground connection. Adjust the tuner through its range to see if it will pick up a local AM station.

Part B: AM Receiver Block Diagram

1. Refer to the block diagram of an AM superheterodyne receiver in Fig. 6-3C. Note that each block is labeled to designate the specific function to be performed.

2. In the space provided, briefly explain the function that each block of an AM receiver performs.
   a. RF amplifier _____

   _____

   _____

   b. Local oscillator _____

   _____

   _____

   c. Mixer _____

   _____

   _____

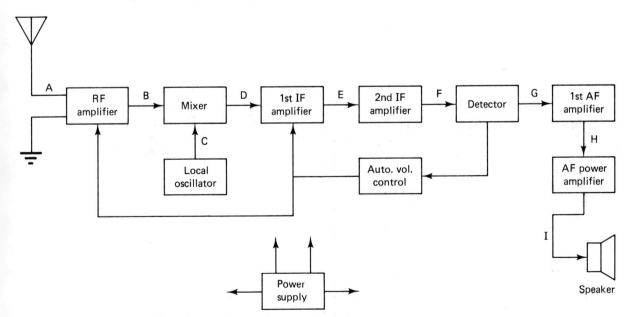

FIGURE 6-3C   AM superheterodyne receiver circuit.

d. IF amplifiers _____

_____

e. Detector _____

_____

f. First AF amplifier _____

_____

g. AF output amplifier _____

_____

h. Speaker _____

_____

3.   Refer again to the block diagram of the AM receiver of Fig. 6-3C. Assume now that a 400-Hz audio tone is used to modulate an AM transmitter operating at 900 kHz. If your receiver is tuned to 900 kHz and brought into the circuit at test point *A*, what frequencies will appear at points *B, D, E, F, G, H,* and *I*? _____

_____

We must assume that the local oscillator is generating a frequency of 1355 kHz when the receiver is tuned to the designated frequency. This represents the incoming station frequency (900 kHz) plus the intermediate frequency (455 kHz).

4.   If the receiver is tuned to 1200 kHz, what frequency would the local oscillator generate? _____ kHz.

5.   If a representative AM receiver diagram is available, study it and note where the block diagram functions are achieved by the circuit.

Part C: AM Receiver Kit Construction

1.   A number of AM receiver kits are available. If you are interested and elect to build an AM receiver from one of these kits, you will have the opportunity to learn to adjust tuner circuits, trace the signal path, and test receiver operation.

ANALYSIS

1.   Explain how the diode circuit of this experiment achieves demodulation of an applied AM signal. _____

_____

_____

2. What is the main function of C_2 in the demodulator circuit of this experiment? ———————————

————————————————————————————

————————————————————————————

3. How is the IF of an AM receiver produced? ————————

————————————————————————————

————————————————————————————

EXPERIMENT 6-4
FM PRINCIPLES

In frequency modulation (FM), a modulating signal causes only the RF carrier to shift in frequency. The amplitude of the RF signal of this type of transmitter therefore remains at a constant level. When modulation is applied, its amplitude causes the RF carrier to shift above or below its center resting frequency. The frequency of the modulating signal determines the rate at which the RF signal shift occurs. Since FM is not influenced by amplitude changes, it is practically immune to noise caused by weather and electrical interference.

OBJECTIVES

1. To construct an FM signal demonstration circuit and observe the signal in operation.
2. To become familiar with FM principles.

EQUIPMENT

Variable power supply, 0–15 V
Signal generator
Oscilloscope
NPN transistor
Varicap diode
Capacitors: 47 pF, 100 pF, 0.001 μF, 0.01 μF, 0.1 μF
Potentiometer: 100 kΩ
Resistors: 470Ω, 56kΩ, 470 kΩ
Oscillator coil
Inductor: 25 mH

PROCEDURE

1. Connect the FM demonstrator circuit of Fig. 6-4A. Connect the RF oscillator part of the circuit first; then test its operation before proceeding.
2. Turn on the variable dc power supply and adjust it to 10 V before turning on the circuit switch.
3. Prepare the oscilloscope for operation with the horizontal sweep selector switch set at 1-μs/cm or 1-MHz frequency.

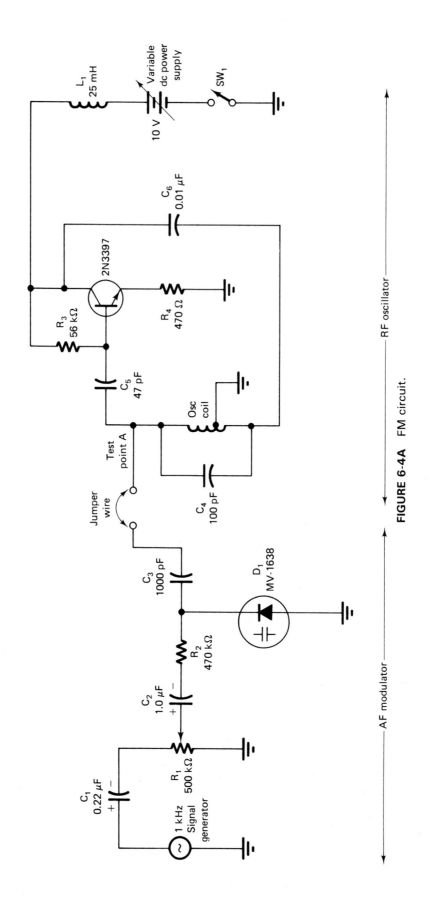

FIGURE 6-4A FM circuit.

420

Set the sync selector switch to the internal (INT) position. Connect the ground probe to the circuit ground and the vertical probe to test point A.

4. If the oscillator is functioning, an RF sine wave should appear on the oscilloscope. Adjust the trace of the oscilloscope to display a few sine waves in order to observe the FM effect of this circuit.

5. Connect the AF modulator section of the circuit and attach a jumper wire between test point A and C_3.

6. Turn on the AF signal source and adjust it to produce a frequency of 10 to 30 Hz.

7. Adjust the potentiometer until FM occurs. Describe the appearance of the RF waveform with FM compared with the RF wave of step 4.

8. If possible, reduce the frequency to a lower value while making the next observation on the oscilloscope. Adjust the AF amplitude control (R_1) while observing the output. Explain the effect that amplitude has upon the waveform. _____

_____

9. With the amplitude control at its maximum output, change the frequency to 60 Hz while observing the oscilloscope display. What influence does frequency of the AF source have upon display? _____

_____

ANALYSIS

1. If the amplitude of the modulating signal increases, what influence does this have upon the output signal? _____

_____

2. What influence does the frequency of the modulating signal have upon the output of an FM signal? _____

_____

_____

3. Briefly compare AM and FM transmitters with respect to the influence that modulating signal amplitude and frequency has upon the output signal. _____

_____

_____

EXPERIMENT 6-5
FM RECEIVERS

FM receivers employ the heterodyne principle of operation that is commonly used in AM receivers. This includes such things as an RF amplifier, local oscillator, mixer, IF amplifiers, detector, AF amplification, speaker, and a common power supply. The tuned-station frequency, IF, and method of detection are the three primary differences between AM and FM receiver types.

OBJECTIVES

1. To study the block diagram of an FM receiver.
2. To explain the primary function of each part of the block diagram.
3. To calculate specific frequencies that will appear at selected points in an FM receiver.

PROCEDURE

1. Refer to block diagram of an FM receiver in Fig. 6-5A. Note that each block is labeled to designate the function performed by the block.

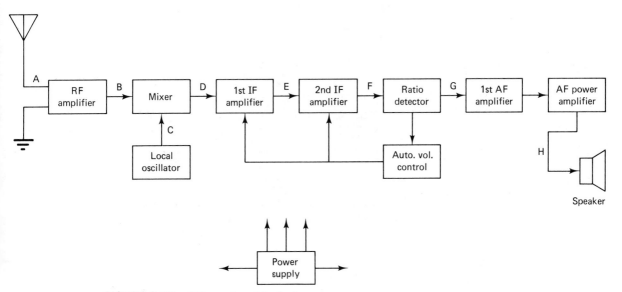

FIGURE 6-5A   FM receiver block diagram.

2. The function of each specific block of the FM receiver is nearly the same as those of the AM receiver with exception of the ratio detector. In each space, briefly explain what each block of FM receiver does.

   a. The RF amplifier _____

   _____

   _____

   b. Mixer _____

   _____

   _____

   c. Local oscillator _____

   _____

   d. First and second IF amplifiers _____

   _____

   _____

   e. Ratio detector _____

   _____

   _____

   f. Automatic volume control _____

   _____

   _____

   g. First AF amplifier _____

   _____

   _____

   h. AF power output amplifier _____

   _____

   _____

   i. Speaker _____

   _____

   _____

   j. Power supply _____

   _____

   _____

3. If a 1-kHz audio tone used to modulate an FM station operating at 101 MHz is picked up at point A, what frequencies will appear at points *B, D, E, F, G,* and H? Assume that the local oscillator is producing a frequency of 90.3 MHz and operates at the received station frequency minus the IF. _____

   _____

   _____

4. If an FM station were tuned in at a frequency of 98 MHz, what frequency would local oscillator produce? What four frequencies would appear at point *D*? _____

   _____

5. A number of FM receiver kits are available to construct. You could build an FM receiver, align its tuned circuits, and test the receiver operation.

ANALYSIS

1. What are the primary differences in AM and FM superheterodyne receivers? _____

   _____

   _____

2. If an FM receiver tunes in a signal of 90 MHz and a similar AM receiver tunes in a 1000-kHz signal, what is the
   primary difference in the components used for the tuned
   circuits? _____

   _____

   _____

EXPERIMENT 6-6
SOUND TRANSDUCERS

Sound transducers are devices designed to change electrical energy into sound energy or sound into electrical energy. Speakers and headphones (earphones) are two important sound transducers used in communication equipment today. Headphones provide private listening. Speakers provide sound for many people to hear over a large area.

OBJECTIVES

1. To test headphones and a speaker to see how they operate.
2. To use a signal generator and listen to the signal with earphones and a speaker.

EQUIPMENT

Variable dc power supply
6.3-V, 60-Hz ac source
Signal generator
Headphones
Speaker
Multimeter

PROCEDURE

1. Install a headset on your head and position it over your ear.
2. Prepare a multimeter to measure resistance. Connect the ohmmeter test probes to the headset leads. What do you hear? _____
3. The dc resistance of the headset coil measures ____ Ω.
4. Remove the headset and carefully unscrew the top cover cap from one earphone. Slide the metal diaphragm to one side and remove it.

5. Make a sketch of the inside structure of the earphone in the space below. Label the parts.

6. Return the diaphragm to the top of the earphone. What holds it in place? _____

_____

7. Turn on the variable dc power supply and adjust it to produce 10 V dc.

8. Connect one lead of the earphone to the negative terminal and momentarily touch the positive lead to the other lead several times. Do you see the diaphragm physically move? _____ Describe the action. _____

_____

9. Disconnect the dc power supply and plug the filament transformer into the power line.

10. Connect 6.3 V ac to the earphone. Gently place your finger on the diaphragm. Describe the action of the diaphragm. _____

_____

11. How do you account for this action? _____

_____

12. Disconnect the earphone from the ac source. Return the cap to the earphone.

13. Again, place the headset over your ears and connect the leads to an AF signal generator. Increase the signal amplitude to a level that can be heard. Run a frequency-response test to find the reproduction range of earphone: Sound range of the earphone is _____ Hz to _____ Hz.

14. Disconnect the headset from signal generator and set it aside.

15. Obtain a speaker and position it at a location on top of a table.

16. With a multimeter measure the resistance of the speaker voice coil: voice coil resistance = _____ Ω. Gently place your finger on the cone of the speaker, and then touch the ohmmeter to it. Do you feel the cone

move? _____ What do you hear?
_____

17. Adjust the power supply to 3 V dc. Connect the negative lead to one speaker terminal. Gently place your finger on the cone of the speaker. Then, touch the positive lead of the power supply to the other speaker terminal. Describe the action. _____

18. Reverse the polarity of the dc power source and repeat step 17. Describe the action. _____

19. Disconnect the dc source from the speaker. Connect the speaker to 6.3 V ac with your finger touching the cone. Describe the action. _____

20. Disconnect the ac source from the speaker and prepare the oscilloscope for operation. Set the horizontal sweep selector to the 20-ms range or 10- to 100-Hz sweep frequency. Adjust the vertical input to its most sensitive range (lowest volts/division).

21. Connect the oscilloscope probes to the speaker terminals. Gently press the cone of the speaker down with your finger, and then release it. Repeat this test several times. Gently rub your finger over the speaker cone a few times. There should be a noticeable oscilloscope action. What causes generation of a signal voltage? _____

ANALYSIS

1. What is meant by the term *transducer?* _____
_____

2. Explain how sound is produced by a speaker. _____
_____

3. Why is an earphone more sensitive than a speaker?
_____

EXPERIMENT 6-7
MONOCHROME TV RECEIVERS

A monochrome TV receiver employs many basic electronic communication principles in a single unit. The *RF signal input* (front end) of a TV receiver, for example, is a basic superheterodyne circuit. Both AM picture information and FM sound signals are processed through this part of the receiver at the same time. *RF amplifier, local oscillator,* and *mixer* circuits are all placed in the channel-selecting chassis unit (tuner). The output of the tuner is applied to the *IF amplifier* section. Both FM and AM detectors are connected to the output of the last IF amplifier.

OBJECTIVES

1. To become familiar with the basic controls on a monochrome television receiver.
2. To become familiar with the basic operation of a monochrome receiver.

EQUIPMENT

Monochrome TV receiver (a color TV can be used, but it is more complex)

PROCEDURE

1. Remove the back cover so that chassis of the TV receiver is exposed. If possible, remove the receiver from its cabinet so that internal circuits can be observed. Place the knobs on each control to improve receiver tuning.
2. *DO NOT* apply electrical power to the receiver. *DO NOT* touch any part of the set!
3. Locate the picture tube (CRT) of receiver. Note the front viewing area, back of the tube, neck, and connection socket. Where are the electron gun and tube elements located in the tube? _____
   _____
4. Locate the deflection yoke on the CRT. How would you describe this part? _____
   _____
   _____

5. The tuner of a TV receiver is mounted on a separate chassis unit. There should be one VHF tuner and one UHF tuner. Locate these units. What function does the tuner perform? _____

6. How is the tuner manipulated to achieve its function? _____

7. List the receiver controls other than the tuner that are accessible from the front of the cabinet. As a rule, these are called operator controls. Briefly explain the function of each control. _____

8. Refer to the rear panel on the back of the receiver. List all the controls found at this location. These controls are commonly called service controls. _____

9. Locate the position of the speaker. What is the approximate size of the cone area of the speaker? _____

10. How is ac supplied to the TV receiver? _____

11. Why is an interlock plug used on the ac power line cord? _____

12. How would this receiver be classified: solid state or vacuum tube? _____

13. Make a sketch of the chassis of the TV receiver used in this experiment, pointing out the location or position of the operator controls, service controls, tuners, and transformers. Use the space below.

14. Briefly describe the general construction of the receiver circuit, pointing out such things as printed circuit boards, transformer locations, and the like. _____

ANALYSIS

1. Explain how a TV picture is produced by the *scanning* process. Point out such things as fields, frames, number of lines, retrace or flyback, and trace time. _____

---

2. How is the horizontal frequency of a TV receiver derived? _____

---

3. What is the function of the deflection yoke? _____

EXPERIMENT 6-8
MONOCHROME TV CONTROLS

Monochrome TV receiver controls have three major classifications: operator controls, service controls, and alignment controls. Operator controls are readily accessible from the front of the receiver. As a rule, these controls provide the operator some personal preference in selecting sound level, brightness, channel selection, contrast, and power on or off. Service controls generally are less accessible to the operator. Adjustment of these controls usually requires some degree of technical understanding of receiver functions. Alignment controls are more complex and require specialized circuit information and test equipment. Avoid adjustment of these controls unless components are being replaced.

OBJECTIVES

1. To work with a TV monochrome receiver and see how various controls affect its operation.
2. To compare the schematic (if available) of the TV with the actual layout of the set.

EQUIPMENT

Monochrome TV receiver (a color TV can be used, but it is more complex)
Receiver service literature and schematic (if available)

PROCEDURE

1. Turn on the TV receiver and tune in a station. Keep the sound level (volume control) very low.
2. Adjust each listed operator control and briefly explain the influence this control has on the receiver.
   a. Channel selector (VHF or UHF) _____

   _____

   b. Fine tuning _____

   _____

   c. Brightness _____

_____

_____

d. Sound or volume _____

_____

e. Vertical hold _____

_____

f. Contrast _____

_____

g. On/off switch _____

_____

h. Additional controls (list them) _____

_____

3. Adjust the following service controls over a range that will permit you to notice the specific effects they produce. Note the original control position of each with a mark or sketch and return it to this position on completion of the adjustment. Describe the effect of each.
   a. Horizontal hold _____

   _____

   b. AGC _____

   _____

   c. Vertical height _____

   _____

   d. Vertical linearity _____

   _____

   e. Horizontal size or width _____

   _____

   f. Focus _____

   _____

4. List other service controls that may be located on rear panel of the receiver but not indicated in step 3. Do not attempt to adjust these controls. _____

   _____

   _____

   _____

   _____

5. Refer to the schematic diagram of the receiver (if available) and locate the operator controls of step 2. Indicate which circuit of the receiver is effected by adjustment of the control.

   a. Channel selector _____

   _____

   b. Fine tuning _____

   _____

   c. Brightness _____

   _____

   d. Sound or volume _____

   _____

   e. Vertical hold _____

   _____

   f. Contrast _____

   _____

   g. On/off switch _____

   _____

   h. Additional controls _____

   _____

6. Locate the service controls of step 3 and indicate which circuit of the receiver is affected by adjustment of the control.

   a. Horizontal hold _____

   _____

   b. AGC _____

   _____

   c. Vertical height _____

   _____

   d. Vertical linearity _____

   _____

   e. Horizontal size or width _____

   _____

   f. Focus _____

   _____

7. Refer to the schematic diagram of the receiver and locate some alignment controls. Where are these controls located in the receiver? _____

   _____

   _____

   _____

   _____

ANALYSIS

1. What is the primary difference between operator, service, and alignment controls? _____

_____

_____

2. What receiver controls are used to change sweep frequencies? _____

_____

_____

3. What is the difference between brightness and contrast?

_____

_____

UNIT 6 EXAMINATION
COMMUNICATIONS CIRCUITS

Instructions: For each of the following, circle the answer that most correctly completes the statement.

1. The percentage of AM modulation represented by the signal shown in Fig. E-6-1 is:

    a. 0%   b. 50%

    c. 100%   d. 150%

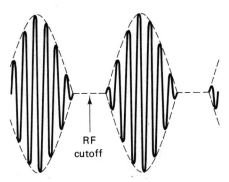

RF cutoff

FIGURE E-6-1

2. The signal shown in Fig. E-6-2 that is an unmodulated carrier is:

    a. A   b. B

    c. C   d. D

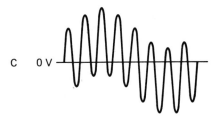

FIGURE E-6-2

3. The assigned carrier frequency band of FM broadcast stations in the United States is:
   a. 20–20 kHz   b. 535–1605 kHz
   c. 88–108 MHz   d. 535–1605 MHz

4. The simplest type of FM detector is the:
   a. Ratio detector   b. Slope detector
   c. Double-tuned detector   d. Foster-seeley discriminator

5. The standard intermediate frequency of an FM receiver is:
   a. 10 kHz   b. 455 kHz
   c. 1605 kHz   d. 10.7 MHz

6. The frequency considered to be a fixed value in a properly aligned superheterodyne receiver is:
   a. RF   b. IF
   c. Audio frequency   d. Local oscillator frequency

7. The first step when aligning a superheterodyne receiver is to:
   a. Adjust the IF transformers.
   b. Tune the local oscillator.
   c. Tune the RF amplifier.
   d. Tune the AF amplifier.

8. On the diagram shown in Fig. E-6-8, the block that represents the RF amplifier is:
   a. A   b. B   c. C   d. D   e. E

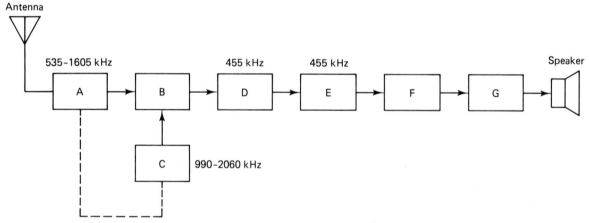

FIGURE E-6-8

9. On the diagram shown in Fig. E-6-8, the block that represents the mixer is:
   a. A   b. B   c. C   d. D   e. E

10. On the diagram shown in Fig. E-6-8, the block that represents the local oscillator is:
   a. A   b. B   c. C   d. D   e. E

11. The type of receiver depicted by the block diagram shown in Fig. E-6-8 is:
    a. Tuned radio frequency
    b. Super regenerative
    c. Super heterodyne
    d. Reflex

12. The following is a characteristic of a IF amplifier:
    a. Has RC coupling
    b. Has high power output
    c. Has a voltage gain of 1
    d. Has a transformer between stages

13. The bandwidth of a superheterodyne receiver is determined by the:
    a. RF amplifier   b. IF amplifier
    c. Local oscillator   d. Audio amplifier

14. The standard IF of an AM receiver is:
    a. 10 kHz   b. 455 kHz
    c. 1605 kHz   d. 10.7 MHz

15. The FM detector that is not sensitive to amplitude changes of the IF input signal is the:
    a. Ratio detector   b. Slope detector
    c. Double-tuned detector   d. Foster-Seeley discriminator

16. The type of detector shown in Fig. E-6-16 is the:
    a. Ratio detector   b. Slope detector
    c. Double-tuned detector   d. Foster-Seeley detector

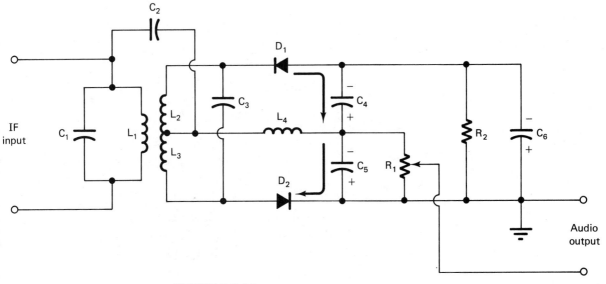

FIGURE E-6-16

17. The frequency of the upper sideband of a 1-MHz carrier modulated by a 10-kHz tone is:
    a. 10 kHz   b. 1 MHz
    c. 990 kHz   d. 1010 kHz

18. The circuit shown in Fig. E-6-18 is a/an:
    a. Base modulator   b. Diode modulator
    c. Emitter modulator   d. Collector modulator

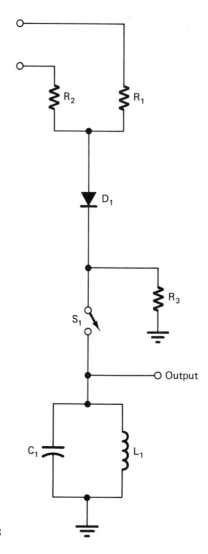

19. The assigned carrier frequency band of AM broadcast stations in the United States is:
    a. 20 Hz–20 kHz   b. 535–1620 kHz
    c. 88–108 MHz   d. 535–1620 MHz

20. An AM waveform has a maximum amplitude of 100 V p-p and a minimum amplitude of zero. The percent of modulation is:
    a. $33\frac{1}{3}\%$ b. 50%
    c. 75%   d. 100%

21. The frequencies that result when 1000 Hz and 550 kHz
    are mixed are:
    a. 500 kHz, 599.5 Hz, 550.5 Hz, and 501 kHz
    b. 1 Hz, 549.5 kHz, 550.5 kHz, and 501 kHz
    c. 1 kHz, 549 kHz, 550 kHz, and 551 kHz
    d. 551 kHz, 1 kHz, 251 kHz, and 249 kHz
22. The result when two frequencies are mixed is:
    a. The two frequencies only
    b. The sum of the frequencies only
    c. The difference of the frequencies only
    d. The two frequencies and the sum and difference
23. The two types of modulation used in the broadcast in-
    dustry are:
    a. AM and CW   b. CW and FM
    c. FM and SSB   d. FM and AM
24. The process of adding information to a radio frequency
    carrier is known as:
    a. Modulation   b. Demodulation
    c. Detection   d. Sidebands
25. The frequency on which a broadcast station transmits
    information is called:
    a. Audio   b. Carrier
    c. Modulation   d. Demodulation

APPENDIX A

Electronic Symbols

Symbols for Electricity

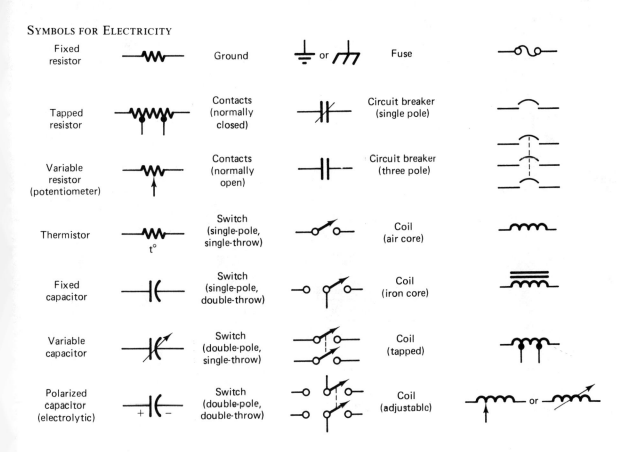

Fixed resistor	Ground	Fuse
Tapped resistor	Contacts (normally closed)	Circuit breaker (single pole)
Variable resistor (potentiometer)	Contacts (normally open)	Circuit breaker (three pole)
Thermistor	Switch (single-pole, single-throw)	Coil (air core)
Fixed capacitor	Switch (single-pole, double-throw)	Coil (iron core)
Variable capacitor	Switch (double-pole, single-throw)	Coil (tapped)
Polarized capacitor (electrolytic)	Switch (double-pole, double-throw)	Coil (adjustable)

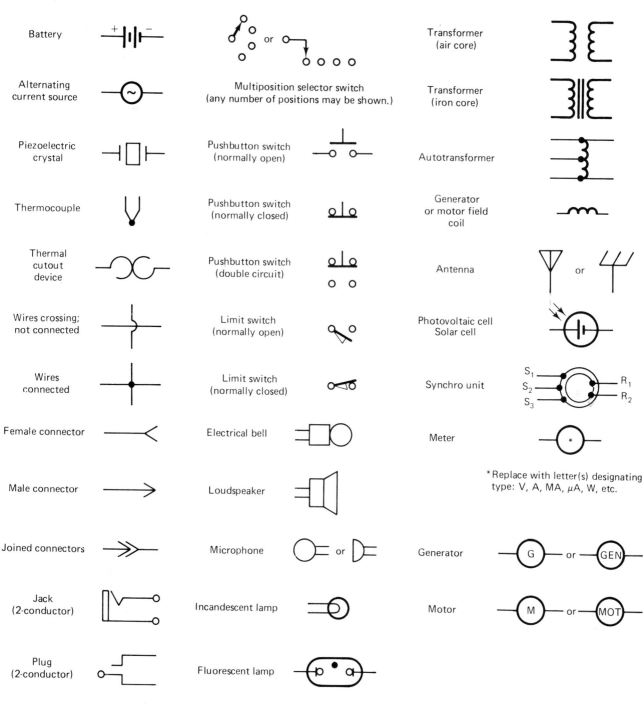

Battery

Alternating current source

Piezoelectric crystal

Thermocouple

Thermal cutout device

Wires crossing; not connected

Wires connected

Female connector

Male connector

Joined connectors

Jack (2-conductor)

Plug (2-conductor)

Multiposition selector switch (any number of positions may be shown.)

Pushbutton switch (normally open)

Pushbutton switch (normally closed)

Pushbutton switch (double circuit)

Limit switch (normally open)

Limit switch (normally closed)

Electrical bell

Loudspeaker

Microphone

Incandescent lamp

Fluorescent lamp

Transformer (air core)

Transformer (iron core)

Autotransformer

Generator or motor field coil

Antenna

Photovoltaic cell Solar cell

Synchro unit

Meter

*Replace with letter(s) designating type: V, A, MA, μA, W, etc.

Generator

Motor

Semiconductor symbols:

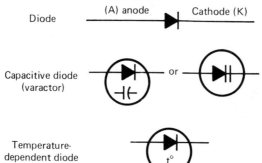

Diode

Capacitive diode (varactor)

Temperature-dependent diode

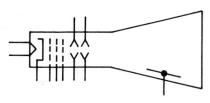

Cathode-ray tube
(electrostatic deflection)

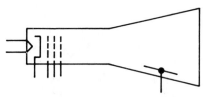

Cathode-ray tube
(electromagnetic deflection)

Thyristor,
bidirectional-
diode type

Thyristor,
bidirectional-
triode type (triac)

Bipolar
transistor

Photodiode

Light emitting
diode (LED)

Zener
diode

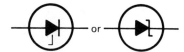

Thyrector
diode

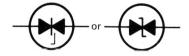

Tunnel
diode

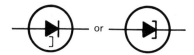

Trigger diac.
unidirectional

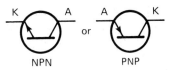

Thyristor,
reverse-
blocking-
diode type

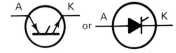

Thyristor,
reverse-blocking-
triode type
(solid-state thyratron, or SCR)

P-channel MOSFET,
enhancement type

Vacuum-type
diode

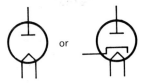

Gas-filled
diode

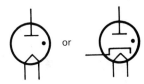

Phototransistor

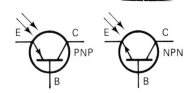

Darlington
transistor

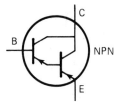

Unijunction
transistor

N-type base   P-type base

N-channel
JFET

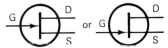

P-channel
JFET

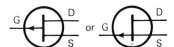

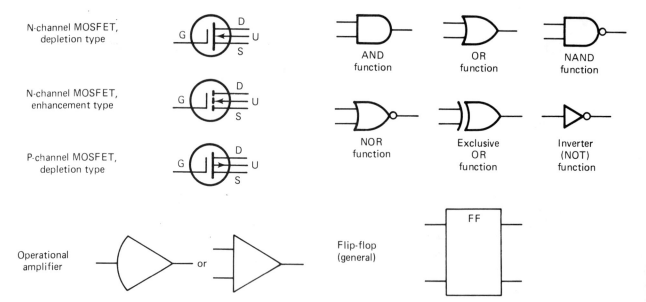

N-channel MOSFET, depletion type

N-channel MOSFET, enhancement type

P-channel MOSFET, depletion type

AND function

OR function

NAND function

NOR function

Exclusive OR function

Inverter (NOT) function

Operational amplifier — or

Flip-flop (general) — FF

APPENDIX B

Electrical Safety

Electrical safety is very important. Many dangers are not easy to see. For this reason, safety should be based on understanding basic electrical principles. Common sense is also important. The physical arrangement of equipment in the electrical lab or work area should be done in a safe manner. Well-designed electrical equipment should always be used. Electrical equipment is often improvised for the sake of economy. It is important that all equipment be made as safe as possible. This is especially true for equipment and circuits that are designed and built in the lab or shop.

Work surfaces in the shop or lab should be covered with a material that is nonconducting, and the floor of the lab or shop should also be nonconducting. Concrete floors should be covered with rubber tile or linoleum. A fire extinguisher that has a nonconducting agent should be placed in a convenient location. Extinguishers should be used with caution. Their use should be explained by the teacher.

Electrical circuits and equipment in the lab or shop should be plainly marked. Voltages at outlets require special plugs for each voltage. Several voltage values are ordinarily used with electrical lab work. Storage facilities for electrical supplies and equipment should be neatly kept. Neatness encourages safety and helps keep equipment in good condition. Tools and small equipment should be maintained in good condition and stored in a tool panel or marked storage area. Tools that have insulated handles should be used. Tools and equipment plugged into convenience outlets should be wired with three-wire cords and plugs. The purpose of the third wire is to prevent electrical shocks by grounding all metal parts connected to the outlet.

Soldering irons are often used in the electrical shop or lab. They can be a fire hazard. They should have a metal storage rack. Irons should be unplugged while not in use. Soldering irons can also cause burns if not used properly. Rosin-core solder should always be used in the electrical lab or shop.

Adequate laboratory space is needed to reduce the possibility of accidents. Proper ventilation, heat, and light also provide a safe working environment. Wiring in the electrical lab or shop should conform to specifications of the National Electrical Code (NEC®). The NEC® governs all electrical wiring in buildings.

LAB OR SHOP PRACTICES

All activities should be done with low voltages whenever possible. Instructions should be written, with clear directions, for performing lab activities. All lab or shop work should emphasize safety. Experimental circuits should always be checked before they are plugged into a power source. Electrical lab projects should be constructed to provide maximum safety when used.

Disconnect electrical equipment from the source of power before working on it. When testing electronic equipment, such as TV sets or other 120-V devices, an isolation transformer should be used. This isolates the chassis ground from the ground of the equipment and eliminates the shock hazard when working with 120-V equipment.

ELECTRICAL HAZARDS

A good first-aid kit should be in every electrical shop or lab. The phone number of an ambulance service or other available medical services should be in the lab or work area in case of emergency. Any accident should be reported immediately to the proper school officials. Teachers should be proficient in the treatment of minor cuts and bruises. They should also be able to apply artificial respiration. In case of electrical shock, when breathing stops, artificial respiration must be immediately started. Extreme care should be used in moving a shock victim from the circuit that caused the shock. An insulated material should be used so that someone else does not come in contact with the same voltage. It is not likely that a high-voltage shock will occur. However, students should know what to do in case of emergency.

Normally, the human body is not a good conductor of electricity. When wet skin comes in contact with an electrical conductor, the body is a better conductor. A slight shock from an electrical circuit should be a warning that something is wrong. Equipment that causes a shock should be immediately checked and repaired or replaced. Proper grounding is important in preventing shock.

Safety devices called ground-fault circuit interrupters (GFIs) are now used for bathroom and outdoor power receptacles. They have the potential of saving many lives by preventing shock. GFIs immediately cut off power if a shock occurs. The National Electric Code specifies where GFIs should be used.

Electricity causes many fires each year. Electrical wiring with too many appliances connected to a circuit overheats wires. Overheating may set fire to nearby combustible materials. Defective and worn equipment can allow electrical conductors to touch one another and cause a short circuit, which causes a blown fuse. It could also cause a spark or arc which might ignite insulation or other combustible materials or burn electrical wires.

FUSES AND CIRCUIT BREAKERS

Fuses and circuit breakers are important safety devices. When a fuse blows, it means that something is wrong in the circuit. Causes of blown fuses could be:

1. A short circuit caused by two wires touching
2. Too much equipment on the same circuit
3. Worn insulation allowing bare wires to touch grounded metal objects such as heat radiators or water pipes

After correcting the problem a new fuse of proper size should be installed. Power should be turned off to replace a fuse. Never use a makeshift device in place of a new fuse of the correct size. This destroys the purpose of the fuse. Fuses are used to cut off the power and prevent overheated wires.

Circuit breakers are now very common. Circuit breakers operate on spring tension. They can be turned on or off like wall switches. If a circuit breaker opens, something is wrong in the circuit. Locate and correct the cause and then reset the breaker.

Always remember to use common sense whenever working with electrical equipment or circuits. Safe practices should be followed in the electrical lab or shop as well as in the home. Detailed safety information is available from the National Safety Council and other organizations. It is always wise to be safe.

APPENDIX C

Electronic Equipment and Parts Sales

Listed below are several of the names and addresses of several companies that sell electronic equipment and parts. You can write to these companies and obtain catalogs and price lists for purchasing the equipment you need.

All Electronics Corp.
905 S. Vermont Ave.
P.O. Box 20406
Los Angeles, CA 90006
800-826-5432

Allied Electronics
1355 N. McLean Blvd.
Elgin, IL 60120
800-433-5700

Brodhead-Garrett
4560 E. 71st St.
Cleveland, OH 44105
216-341-0248

Cal West Supply Inc.
31320 Via Colinas, Suite 105
Westlake Village, CA 91362
800-892-8000

Circuit Specialist Co.
P.O. Box 3047
Scottsdale, AZ 85257

Digi-Key Corp.
P.O. Box 677
Thief River Falls, MN 56701

Edlie Electronics
2700 Hempstead Twp
Levittown, L.I., NY 11756-1443
800-645-4722

ETCO Electronics
North Country Shopping Ctr.
Plattsburgh, NY 12901

F.W. Bell
6120G Hanging Moss Rd.
Orlando, FL 32807

Heath Co.
Benton Harbor, MI 49022
800-253-0570

Hewlett-Packard
1501G Page Mill Rd.
Palo Alto, CA 94305

Hickok Teaching Systems
2 Wheeling Ave.
Woburn, Mass. 01801
617-935-5850

Hughes-Peters
4865 Duck Creek Rd.
P.O. Box 27119
Cincinnati, OH 45227
800-543-4483

Jameco Electronics
1355 Shoreway Rd.
Belmont, CA 94002

Kelvin Electronics, Inc.
P.O. Box #8
1900 New Hwy.
Farmingdale, NY 11735
800-645-9212

Lab Volt
Buck Engineering Co.
Farmingdale, N.J. 07727

MCM Electronics
858 E. Congress Park Dr.
Centerville, OH 45459

Merlin P. Jones & Assoc.
P.O. Box 12685
Lake Park, FL 33403-0685
305-848-8236

Mouser Electronics
2401 Hwy. 287 N
Mansfield, TX 76063

Omnitron Electronics
770 Amsterdam Ave.
New York, NY 10025
800-223-0826

Priority One Electronics
21622 Plumer St.
Chatsworth, CA 91311
800-423-5922

RNJ Electronics
805 Albany Ave.
Lindenhurst, NY 11757
800-645-5833

Satco
924 S. 19th Ave.
Minneapolis, MN 55404
800-328-4644

Techni-Tool
P.O. Box 368
Plymouth Meeting, PA 1946
215-825-4990

Tektronix, Inc.
P.O. Box 1700
Beaverton, OR 97075

Triplett Corp.
1G Triplett Dr.
Bluffon, OH 45817

URI Electronics
P.I. Burks Co.
842 S. 7th St.
Louisville, KY 40203

APPENDIX D

Soldering Techniques

Soldering is an important skill for electrical technicians. Good soldering is important for the proper operation of equipment.

Solder is an alloy of tin and lead. The solder that is most used is 60/40 solder. This means that it is made from 60% tin and 40% lead. Solder melts at a temperature of about 400°F.

For solder to adhere to a joint, the parts must be hot enough to melt the solder. The parts must be kept clean to allow the solder to flow evenly. Rosin flux is contained inside the solder. It is called rosin-core solder.

A good mechanical joint must be made when soldering. Heat is then applied until the materials are hot. When they are hot, solder is applied to the joint. The heat of the metal parts (not the soldering tool) is used to melt the solder. Only a small amount of heat should be used. Solder should be used sparingly. The joint should appear smooth and shiny. If it does not, it could be a "cold" solder joint. Be careful not to move the parts when the joint is cooling. This could cause a "cold" joint.

When parts that could be damaged by heat are soldered, be very careful not to overheat them. Semiconductor components, such as diodes and transistors, are very heat sensitive. One way to prevent heat damage is to use a *heat sink,* such as a pair of pliers. A heat sink is clamped to a wire between the joint and the device being soldered. A heat sink absorbs heat and protects delicate devices. Printed circuit boards, such as the one shown in Fig.

D-1, are also very sensitive to heat. Care should be taken not to damage PC boards when soldering parts onto them.

Several types of soldering irons and soldering guns are shown in Figs. D-2 to D-7. Small, low-wattage irons should be used with PC boards and semiconductor devices.

Below are some rules for good soldering:

1. Be sure that the tip of the soldering iron is clean and tinned.
2. Be sure that all the parts to be soldered are heated. Place the tip of the soldering iron so that the wires and the soldering terminal are heated evenly.
3. Be sure not to overheat the parts.
4. Do not melt the solder onto the joint. Let the solder flow onto the joint.
5. Use the right kind and size of solder and soldering tools.
6. Use the right amount of solder to do the job but not enough to leave a "blob."
7. Be sure not to allow parts to move before the solder joint cools.

Figure D-1 Printed circuit (PC) board. [Courtesy of TRW/UTC Transformers.]

Figure D-2 Low-wattage soldering iron kit with several soldering tips. [Courtesy of The Cooper Group.]

Figure D-4 Cordless soldering iron kit. [Courtesy of The Cooper Group.]

Figure D-3 Cordless soldering iron with charger. [Courtesy of The Cooper Group.]

Figure D-5 High-wattage soldering gun kit with 60/40 solder and spare tips. [Courtesy of The Cooper Group.]

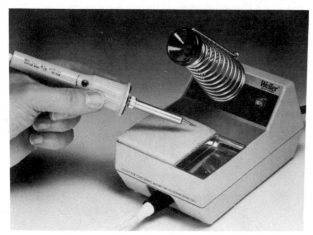

Figure D-6   Soldering iron with temperature control station and cleaning pad. [Courtesy of The Cooper Group.]

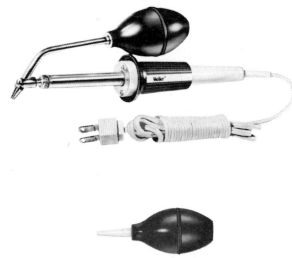

(b)

Figure D-7   (a) Desoldering tool; (b) desoldering bulb. [Courtesy of The Cooper Group.]

APPENDIX E

Electronic Tools

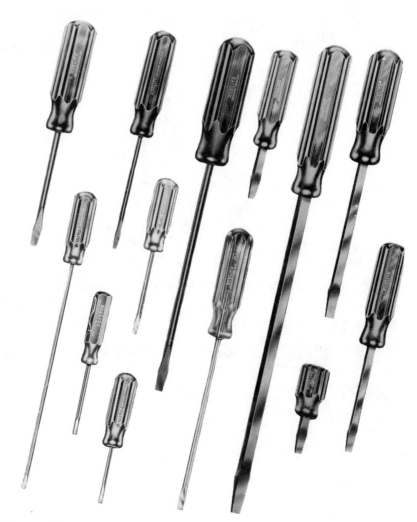

Figure E-1 Various sizes of standard screwdrivers.
[Courtesy of The Cooper Group.]

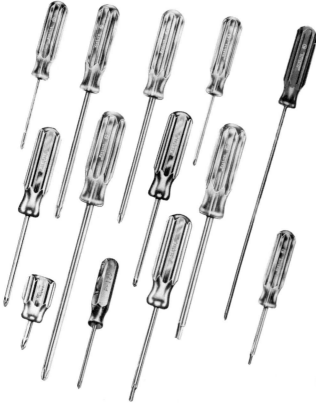

Figure E-2 Various sizes of Phillips screwdrivers. [Courtesy of The Cooper Group.]

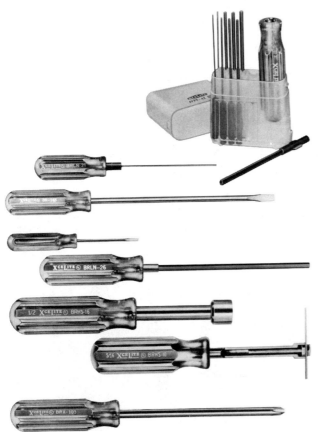

Figure E-3 (a) Allen setscrew drivers; (b) standard screwdrivers; (c) nut drivers; (d) Phillips screwdriver. [Courtesy of The Cooper Group.]

Figure G-4. Driver combination kits. [Courtesy of The Cooper Group.]

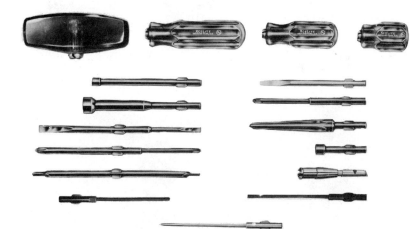

Figure G-5. Screwdriver kits. [Courtesy of The Cooper Group

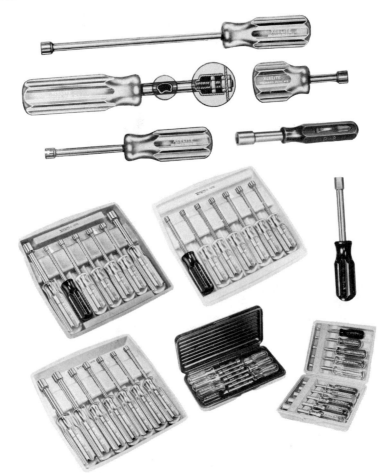

Figure E-6 Nut-driver kits. [Courtesy of The Cooper Group.]

Figure E-7 (a) Round-nose diagonal wire cutters; (b) wire strippers. [Courtesy of Hunter Tools, Division of K-D Manufacturing Co.]

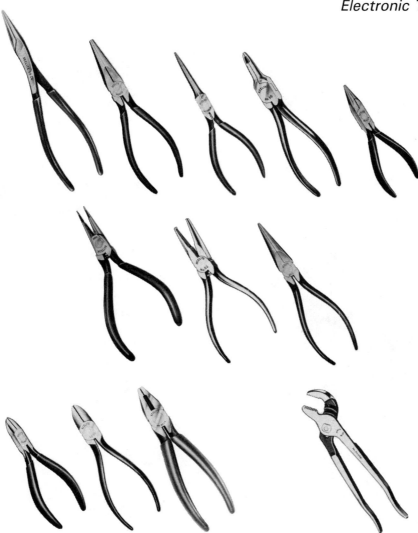

Figure E-8 (a) Long-nose or needle-nose pliers; (b) diagonal cutting pliers; (c) channel lock pliers. [Courtesy of The Cooper Group.]

Figure E-9 Wire crimping tool with stripping and bolt cutting capability. [Courtesy of Vaco Products Co.]

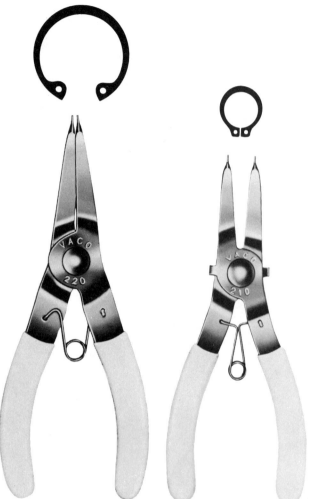

Figure E-10.   C-rings and pliers. [Courtesy of Vaco Products Co.]

Figure E-11.   Cheater cord for electronic equipment repair. [Courtesy of GC Electronics.]

Figure E-12.   Roll of tools for servicing. [Courtesy of The Cooper Group.]

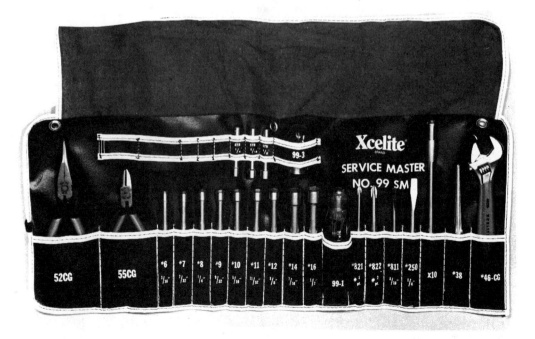

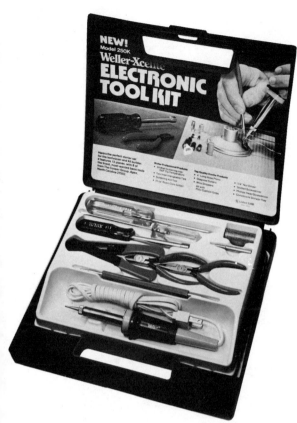

Figure E-13 Electronic tool kit. [Courtesy of The Cooper Group.]

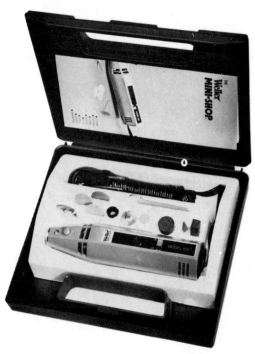

Figure E-14 Portable multipurpose tool. [Courtesy of The Cooper Group.]

APPENDIX F

Troubleshooting

Troubleshooting is a method of finding out why something doesn't work properly. If you follow logical steps, you will be able to locate most difficulties that occur in electronic equipment. Sometimes the trouble is so complex that it requires many hours of concentration and work. Other problems are easy to solve and require only a brief time.

There are some resources that most people find helpful in troubleshooting. These include:

1. Using a "common sense" approach;
2. Knowing how electronic systems work;
3. Knowing how to use test equipment;
4. Knowing how to use schematic diagrams effectively; and
5. Being able to find the trouble through a logical sequence.

To begin any kind of troubleshooting, you should first determine possible courses of action. Without some system, the procedure of troubleshooting becomes a guessing game. You should be aware that no one system of troubleshooting is perfect. In the process of troubleshooting, keep in mind that most problems are usually component failures. If you know what each component is supposed to do, then you will be aware of the troubles they can cause.

During the process of troubleshooting, it is important that you use proper tests. Much of your time is used in locating the trouble. You must have some suitable approach in order to save time. As you become more familiar with troubleshooting, it becomes less time consuming.

As you continue your troubleshooting effort, you must constantly keep aware of circuit or system operations you have already tested. Make a list or remember the probable troubles that have been tested.

Troubleshooting is a method of finding out why something doesn't work properly. If you follow logical steps, you will be able to locate most difficulties that occur in electronic equipment. Sometimes the trouble is so complex that it requires many hours of concentration and work. Other problems are easy to solve and require only a brief time. An important part of troubleshooting skills is initial inspection.

Initial inspection involves looking for the obvious in troubleshooting. There are several things that should be done before getting into actual circuit or system testing. In the initial inspection of any equipment, first open the equipment to look at it. There are several things you should observe. These include:

1. Burned resistors. They are often obvious, may be charred, blistered or bulged, have discolored color bands, and even holes.
2. Broken parts. These may come in the form of cracks, wires pulled out of parts, or completely destroyed parts.
3. Broken wires and poor connections.
4. Smoke or heat damage. Parts may smoke when equipment is turned on; this identifies defective parts (but not the cause).
5. Oil leaks and water leaks.
6. Loose, damaged, or worn parts. These are determined by visual and tactile examination.
7. Noisy parts. Uncommon noises indicate defective parts.

When initial inspection is used properly, many troubles can be located without having to go through unnecessary steps. Initial inspection involves all of the senses of sight, touch, smell, and hearing. It is important to organize your thoughts to solve the problem. If you suspect a part to be the source of the problem, take a closer look. If you suspect a specific part, turn the equipment off and smell it, touch it, examine it closely. For example, a transformer that is good doesn't have an odor, but a burned transformer does. Initial inspection can help to locate the trouble in any defective electronic equipment or circuits.

Final solution of the problem involves application of your *knowledge of electronic circuit operation* and *understanding of proper use of test equipment*. Remember that troubleshooting is a *systematic* procedure.

APPENDIX G

Power Supply Construction Project

The following is a step-by-step procedure for constructing a 9-V power supply using printed-circuit (PC) board. If PC board is not available, the parts may be soldered together using any type of circuit board. Materials for circuit board construction are available through many electronic parts dealers.

Power-Supply Project

1. To construct a 9-V regulated power supply, using the circuit of Fig. G-1.

2. A representative PC-board layout is shown in Fig. G-2. The circuit board may require some modification according to your part selection. Make a prototype layout of the PC board on a piece of card stock. Use the actual dimensions of the parts being used.

3. Lay out the PC board using tape and PC resist dots.

4. Etch the board.

5. Drill holes at the designated locations.

6. Solder components in place.

7. Wire the external circuitry.

8. Assemble the external circuitry and PC board.

9. Test the operation of the circuit with a voltmeter and/or oscilloscope. The output should be 9 V dc.

10. Assemble the entire unit in a cabinet and test it again for operation.

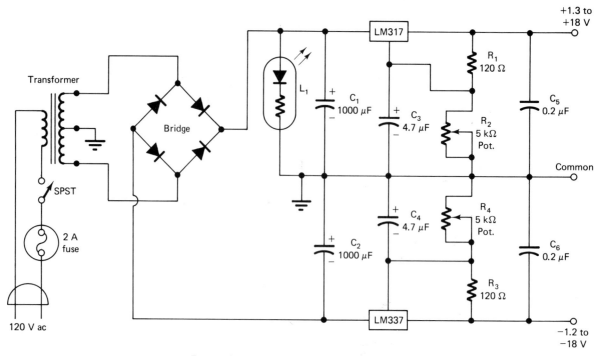

Figure G-1 9-V regulated power supply.

Parts

C_1 and C_2	1000 μF, 35 V
C_3 and C_4	4.7 μF, 35 V
C_5 and C_6	0.2 μF, 50 V
R_1 and R_3	120 Ω, 1/4 W
R_2 and R_4	5 kΩ Pot.
L_1	Indicator lamp LED and resistor (12 V)

1 LM317 (+ Reg)

1 LM337 (− Reg)

1 Bridge rectifier

1 Transformer 25 V, 2 A, CT secondary

1 Fuse/holder (2 A)

1 SPST switch

1 Line cord/plug

3 Binding posts

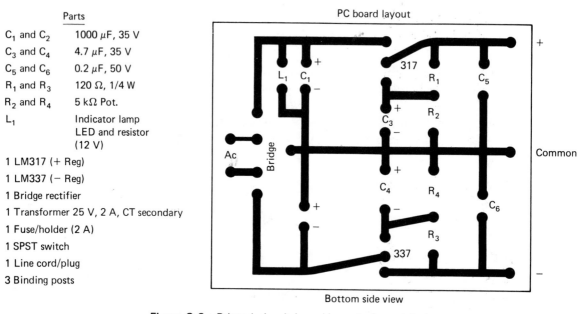

PC board layout

Bottom side view

Figure G-2 Printed-circuit-board layout of regulated power supply: (a) bottom view; (b) top view.

APPENDIX H

Table of Common Logarithms

N	0	1	2	3	4	5	6	7	8	9
10	0000	0043	0086	0128	0170	0212	0253	0294	0334	0374
11	0414	0453	0492	0531	0569	0607	0645	0682	0719	0755
12	0792	0828	0864	0899	0934	0969	1004	1038	1072	1106
13	1139	1173	1206	1239	1271	1303	1335	1367	1399	1430
14	1461	1492	1523	1553	1584	1614	1644	1673	1703	1732
15	1761	1790	1818	1847	1875	1903	1931	1959	1987	2014
16	2041	2068	2095	2122	2148	2175	2201	2227	2253	2279
17	2304	2330	2355	2380	2405	2430	2455	2480	2504	2529
18	2553	2577	2601	2625	2648	2672	2695	2718	2742	2765
19	2788	2810	2833	2856	2878	2900	2923	2945	2967	2989
20	3010	3032	3054	3075	3096	3118	3139	3160	3181	3201
21	3222	3243	3263	3284	3304	3324	3345	3365	3385	3404
22	3424	3444	3464	3483	3502	3522	3541	3560	3579	3598
23	3617	3636	3655	3674	3692	3711	3729	3747	3766	3784
24	3802	3820	3838	3856	3874	3892	3909	3927	3945	3962
25	3979	3997	4014	4031	4048	4065	4082	4099	4116	4133
26	4150	4166	4383	4200	4216	4232	4249	4265	4281	4298
27	4314	4330	4346	4362	4378	4393	4409	4425	4440	4456
28	4472	4487	4502	4518	4533	4548	4564	4579	4594	4609
29	4624	4639	4654	4669	4683	4698	4713	4728	4742	4757
30	4771	4786	4800	4814	4829	4843	4857	4871	4866	4900
31	4914	4928	4942	4955	4969	4983	4997	5011	5024	5038
32	5051	5065	5079	5092	5105	5119	5132	5145	5159	5172
33	5185	5198	5211	5224	5237	5250	5263	5276	5289	5302

(continued)

N	0	1	2	3	4	5	6	7	8	9
34	5315	5328	5340	5353	5366	5378	5391	5403	5416	5428
35	5441	5453	5465	5478	5490	5502	5514	5527	5539	5551
36	5563	5575	5587	5599	5611	5623	5635	5647	5658	5670
37	5682	5694	5705	5717	5729	5740	5752	5763	5775	5786
38	5798	5809	5821	5832	5843	5855	5866	5877	5888	5899
39	5911	5922	5933	5944	5955	5966	5977	5988	5999	6010
40	6021	6031	6042	6053	6064	6075	6085	6096	6107	6117
41	6128	6138	6149	6160	6170	6180	6191	6201	6212	6222
42	6232	6243	6253	6263	6274	6284	6294	6304	6314	6325
43	6335	6345	6355	6365	6375	6385	6395	6405	6415	6425
44	6435	6444	6454	6464	6474	6484	6493	6503	6513	6522
45	6532	6542	6551	6561	6571	6580	6590	6599	6609	6618
46	6628	6637	6646	6656	6665	6675	6684	6693	6702	6712
47	6721	6730	6739	6749	6758	6767	6776	6785	6794	6803
48	6812	6821	6830	6839	6848	6857	6886	6875	6884	6893
49	6902	6911	6920	6928	6937	6946	6955	6964	6972	6981
50	6990	6998	7007	7016	7024	7033	7042	7050	7059	7067
51	7076	7034	7093	7101	7110	7118	7126	7135	7143	7152
52	7160	7168	7177	7185	7193	7202	7210	7218	7226	7235
53	7243	7251	7259	7267	7275	7284	7292	7300	7308	7316
54	7324	7332	7340	7348	7356	7364	7372	7380	7388	7396
55	7404	7412	7419	7427	7435	7443	7451	7459	7466	7474
56	7482	7490	7497	7505	7513	7520	7528	7536	7543	7551
57	7559	7566	7574	7582	7589	7597	7604	7612	7619	7627
58	7634	7642	7649	7657	7664	7672	7679	7686	7694	7701
59	7709	7716	7723	7731	7738	7745	7752	7760	7767	7774
60	7782	7789	7796	7803	7810	7818	7825	7832	7839	7846
61	7853	7860	7868	7875	7882	7889	7896	7903	7910	7917
62	7924	7931	7938	7945	7952	7959	7966	7973	7980	7987
63	7993	8000	8007	8014	8021	8028	8035	8041	8048	8055
64	8062	8069	8075	8082	8089	8096	8102	8109	8116	8122
65	8129	8136	8142	8149	8156	8162	8169	8176	8182	8189
66	8195	8202	8209	8215	8222	8228	8235	8241	8248	8254
67	8261	8267	8274	8280	8287	8293	8299	8306	8312	8319
68	8325	8331	8338	8344	8351	8357	8363	8370	8376	8382
69	8388	8395	8401	8407	8414	8420	8426	8432	8439	8445
70	8451	8457	8463	8470	8476	8482	8488	8494	8500	8506
71	8513	8519	8525	8531	8537	8543	8549	8555	8561	8567
72	8573	8579	8585	8591	8597	8603	8609	8615	8621	8627
73	8633	8639	8645	8651	8657	8663	8669	8675	8681	8686
74	8692	8698	8704	8710	8716	8722	8727	8733	8739	8745
75	8751	8756	8762	8768	8774	8779	8785	8791	8797	8802
76	8808	8814	8820	8825	8831	8837	8842	8848	8854	8859
77	8865	8871	8876	8882	8887	8893	8899	8904	8910	8915
78	8921	8927	8932	8938	8943	8949	8954	8960	8965	8971
79	8976	8982	8987	8993	8998	9004	9009	9015	9020	9025
80	9031	9036	9042	9047	9053	9058	9063	9069	9074	9079
81	9085	9090	9096	9101	9106	9112	9117	9122	9128	9133
82	9138	9143	9149	9154	9159	9165	9170	9175	9180	9186
83	9191	9196	9201	9206	9212	9217	9222	9227	9232	9238
84	9243	9248	9253	9258	9263	9269	9274	9279	9284	9289
85	9294	9299	9304	9309	9315	9320	9325	9330	9335	9340
86	9345	9350	9355	9360	9365	9370	9375	9380	9385	9390

N	0	1	2	3	4	5	6	7	8	9
87	9395	9400	9405	9410	9415	9420	9425	9430	9435	9440
88	9445	9450	9455	9460	9465	9469	9474	9479	9484	9489
89	9494	9499	9504	9509	9513	9518	9523	9528	9533	9538
90	9542	9547	9552	9557	9562	9566	9571	9576	9581	9586
91	9590	9595	9600	9605	9609	9614	9619	9624	9628	9633
92	9638	9643	9647	9652	9657	9661	9666	9671	9675	9680
93	9685	9689	9694	9699	9703	9708	9713	9717	9722	9727
94	9731	9736	9741	9745	9750	9754	9759	9763	9768	9773
95	9777	9782	9786	9791	9795	9800	9805	9809	9814	9818
96	9823	9827	9832	9836	9841	9845	9850	9854	9859	9863
97	9868	9872	9877	9881	9886	9890	9894	9899	9903	9908
98	9912	9917	9921	9926	9930	9934	9939	9943	9948	9952
99	9956	9961	9965	9969	9974	9978	9983	9987	9991	9996

Index